The Chemistry
and Biochemistry of
Plant Proteins

Annual Proceedings of the Phytochemical Society

ANNUAL PROCEEDINGS OF THE PHYTOCHEMICAL SOCIETY NUMBER 11

The Chemistry and Biochemistry of Plant Proteins

PROCEEDINGS OF THE
PHYTOCHEMICAL SOCIETY SYMPOSIUM
UNIVERSITY OF GHENT, BELGIUM
SEPTEMBER, 1973

Edited by

J. B. HARBORNE

Department of Botany,
University of Reading, England

and

C. F. VAN SUMERE

Laboratorium voor Plantenbiochemie,
University of Ghent, Belgium

1975

ACADEMIC PRESS

LONDON   NEW YORK   SAN FRANCISCO

A Subsidiary of Harcourt Brace Jovanovich, Publishers

ACADEMIC PRESS INC. (LONDON) LTD.
24/28 Oval Road,
London NW1

United States Edition published by
ACADEMIC PRESS INC.
111 Fifth Avenue
New York, New York 10003

Library of Congress Catalog Card Number: 74-17992
ISBN: 0-12-324668-7

PRINTED IN GREAT BRITAIN BY
WILLIAM CLOWES & SONS LIMITED
LONDON, BECCLES AND COLCHESTER

Contributors

J. ALBRECHT, *Laboratorium voor Plantenbiochemie, Rijksuniversiteit, Ghent, Belgium* (p. 211).

D. BOULTER, *Department of Botany, University of Durham, Durham, England* (p. 1).

ORIO CIFERRI, *Institute of Plant Physiology, University of Pavia, Italy* (p. 113).

JEAN DAUSSANT, *Physiologie des Organes Végétaux, C.N.R.S., 92-Bellevue, France* (p. 31).

ANDRÉE DEDONDER, *Laboratorium voor Plantenbiochemie, Rijksuniversiteit, Ghent, Belgium* (p. 211).

H. DE POOTER, *Laboratorium voor Plantenbiochemie, Rijksuniversiteit, Ghent, Belgium* (p. 211).

G. E. INGLETT, *Northern Regional Research Laboratory, Agricultural Research Service, U.S. Department of Agriculture, Peoria, Illinois, U.S.A.* (p. 265).

R. KRAUSPE, *Institute of Plant Biochemistry, Research Centre for Molecular Biology and Medicine, Academy of Sciences, 401 Halle/E. German Democratic Republic* (p. 167).

C. J. LEAVER, *Department of Botany, University of Edinburgh, Scotland* (p. 137).

R. LONTIE, *Laboratorium voor Biochemie, Katholieke Universiteit te Leuven, Louvain, Belgium* (p. 89).

D. MUNSCHE, *Institute of Plant Biochemistry, Research Centre for Molecular Biology and Medicine, Academy of Sciences, 401 Halle/E. German Democratic Republic* (p. 167).

B. PARTHIER, *Institute of Plant Biochemistry, Research Centre for Molecular Biology and Medicine, Academy of Sciences, 401 Halle/E. German Democratic Republic* (p. 167).

IRMA PÉ, *Laboratorium voor Plantenbiochemie, Rijksuniversiteit, Ghent, Belgium* (p. 211).

GISÈLE PRÉAUX, *Laboratorium voor Biochemie, Katholieke Universiteit te Leuven, Louvain, Belgium* (p. 89).

J. A. M. RAMSHAW, *Department of Botany, University of Durham, Durham, England* (p. 1).

HERMANN STEGEMANN, *Biochemical Institute, Messeweg 11, D33 Brunswick, West Germany* (p. 71).

C. F. VAN SUMERE, *Laboratorium voor Plantenbiochemie, Rijksuniversiteit, Ghent, Belgium* (p. 211).

J. G. VAUGHAN, *Biology Department, Queen Elizabeth College, Campden Hill Road, London, England* (p. 281).

R. WOLLGIEHN, *Institute of Plant Biochemistry, Research Centre for Molecular Biology and Medicine, Academy of Sciences, 401 Halle/E. German Democratic Republic* (p. 167).

Preface

With the continuing expansion of the world population and the recognition that a balanced and sufficient protein intake, mainly from plant, not animal, sources, is the key to feeding mankind in the future, a new book on plant proteins really requires no apology. Much has been written in recent years about proteins in general; yet there seems to be no modern account dealing specifically with the proteins of higher plants, a gap which the present book should at least help to fill. This is an account of recent progress in studying their chemistry and biochemistry, which in the short space of 300 pages can only touch on some of the highlights in our present endeavours in the field.

One could argue that "proteins are proteins" so that there is no need for a separate treatment of those in plants as distinct from those in other living organisms. However, there are some important differences and a perusal of the various chapters in this volume should make this clear. Right at the beginning when one is considering their extraction and isolation, one meets certain difficulties not encountered with other proteins. G. Préaux and R. Lontie make this clear in their chapter on barley proteins. In spite of the many advances in separation techniques, the solubilization and fractionation of cereal proteins still pose considerable technical difficulties. The storage proteins in seeds, besides being of value to us as a source of food, play an important physiological role in plants and an account of the variations they undergo with a changing environment is given by H. Stegemann, with reference to those in cereal grains and in the potato tuber. Immunochemistry is a relatively new technique which can be applied to the separation and study of storage proteins, and this is the subject of the second chapter in the volume, by J. Daussant.

Another major problem research workers face when isolating proteins from plant tissues is their frequent contamination during extraction with other plant constituents, especially phenolic substances. A full discussion of the many interactions proteins and phenolics undergo is covered in a later chapter by C. Van Sumere and his colleagues. These were the first workers to show that one particular phenolic, ferulic acid, is an integral part of the structure, as an end group, of some barley proteins. Because of the difficulties in isolating plant proteins in quantity, sequence studies have seriously lagged behind similar studies of proteins from other sources. Indeed, there is only one laboratory in the world where it can be said that the sequencing of plant proteins has become a matter of routine. This is that of D. Boulter at Durham and it is thus very appropriate that the first chapter of the book, covering amino acid sequence analysis, should be written by D. Boulter and his colleague J. A. M. Ramshaw. Most of the sequences at present known, for eight plant proteins variously from about fifty sources, were determined at Durham.

Turning to the study of protein biosynthesis in higher plants, one finds again that considerable problems have arisen because of difficulties in cellular fractionation and in removal of contaminants, which were not encountered in bacterial or mammalian systems. Another facet of protein synthesis in plants, not at first envisaged by molecular biologists, is that there are different synthesizing systems in the cytoplasm, in the chloroplast and in the mitochondrion. To cover this topic here, a general account of protein synthesis in higher plants (O. Cifferi) is followed by specific accounts of the macromolecular machinery of plant mitochondria (C. J. Leaver) and of chloroplasts (B. Parthier *et al.*).

When one considers the various biological properties ascribed to plant proteins, one again finds a number of ways in which plant and animal proteins differ. Perhaps the most surprising property for a plant protein is that of having a sweet taste. According to G. E. Inglett, proteins from two African plants, called Serendipity berry and Miracle fruit, are now being actively studied for potential use as natural sweeteners. An account of recent discoveries in this field is contained in Chapter 9. The final chapter by J. G. Vaughan could be said to cover another practical application of plant proteins, since it describes the use that has been made of protein patterns in seeds for solving problems in plant taxonomy.

The present volume is based on a series of review lectures presented at an International Symposium organized by one of us (C.F.V.S.) on behalf of the Phytochemical Society at the University of Ghent, Belgium, in September 1973. We would like to record here our debt to a number of bodies who provided essential financial support for this meeting, namely the Belgian Government, the University of Ghent, the Belgian foundation for the advancement of the Sciences, Artois Breweries, and Janssen Pharmaceutica–Aldrich Europe. As editors, we would thank our contributors, who have responded generously and promptly to our various demands, and we are grateful to the staff of Academic Press for their expert assistance in preparing this volume for publication.

September, 1974

J. B. HARBORNE
C. F. VAN SUMERE

Contents

CHAPTER 1

Amino Acid Sequence Analysis of Proteins

D. Boulter and J. A. M. Ramshaw

CHAPTER 2

Immunochemical Investigations of Plant Proteins

Jean Daussant

CHAPTER 3

Properties of and Physiological Changes in Storage Proteins

Hermann Stegemann

CHAPTER 4

The Proteins of Barley

Gisèle Préaux and R. Lontie

CHAPTER 5

Mechanism of Protein Synthesis in Higher Plants

Orio Ciferri

CHAPTER 6

The Biogenesis of Plant Mitochondria

C. J. Leaver

CHAPTER 7

The Biogenesis of Chloroplasts

B. Parthier, R. Krauspe, D. Munsche and R. Wollgiehn

CHAPTER 8

Plant Proteins and Phenolics

C. F. Van Sumere, J. Albrecht, Andrée Dedonder, H. De Pooter
and Irma Pé

CHAPTER 9

Protein Sweeteners

G. E. Inglett

CHAPTER 10

Proteins and Taxonomy

J. G. Vaughan

CHAPTER 1

Amino Acid Sequence Analysis of Proteins

D. BOULTER AND J. A. M. RAMSHAW*

Department of Botany, University of Durham, Durham, England

I. INTRODUCTION

Proteins are involved in many biological functions including catalysis, regulation, transport, structure, disease resistance, "recognition" and nutrition. These functions are usually specific and this specificity is due in each case to the possession of a precise three-dimensional structure which results from the folding of a polypeptide chain(s) into its (their) most stable form; this form is dictated by a linear sequence of amino acid residues which, in turn,

* Present address:
Dept. of Inorganic Chemistry, Univ. of Sydney, New South Wales 2006, Australia.

are the result of the translation of a particular cistron (gene). The biological function normally involves the binding of the protein to a small molecule, for example a substrate, regulator, antigen, ligand, or to another protein, and it is the surface topology of the protein that determines the selectivity of these interactions. A major task of biochemistry, therefore, is to explain the many functions of proteins in simple and understandable chemical terms.

The first step in this understanding is often to know the amino acid sequence, the "primary" structure of the protein. Amino acid sequence determinations are no longer confined to a few major laboratories, and many different kinds of research programmes now require the determination of the sequence of the major protein involved. This paper describes briefly some of the modern technology of sequence determination. The methods which are discussed are those shown to be generally applicable from our experience with plant cytochromes c and plastocyanins; inevitably, there is some bias in the choice of methods. The overall strategy of an investigation is stressed and estimates of the logistics of different procedures (amounts of material, time taken, etc.) will be given. Mention will be made of some of the major uses to which amino acid sequence data of plant proteins have been put.

II. Purification of Proteins

Before attempting primary structure elucidation, it is essential to have homogeneous starting material. A variety of procedures can be used to isolate and purify proteins and several of these can also be used to establish the homogeneity or otherwise of the protein preparation (see, for example, Van Holde, 1970). None of these methods, either separately or in combination, definitively proves the presence of only a single protein, but before proceeding with a sequence programme, one should at least be sure that the protein pre- paration is a single component on electrophoresis at different pHs on acryl- amide gels, on sodium dodecyl sulphate (SDS) acrylamide gels, and on iso- electric focusing gels. In addition, the numbers of different N- and C-terminal amino acids should be consistent with the number of subunits (polypeptide chains) in the protein. Even so, the protein preparation may not be pure and one of the uses of automatic protein and peptide sequencers is to act as a final check on purity. More and more proteins which hitherto were assumed to have been homogeneous substances, by using a combination of physical methods, are being shown by the most recent procedures to consist of more than one similar subunit. When proteins are made up of different subunits these must normally be separated and sequenced separately.

III. AMINO AND CARBOXY TERMINAL ANALYSIS

A. AMINO-TERMINUS

The general principle for amino-terminal determination is the introduction of a coloured or fluorescent marker group specifically onto the amino group of the N-terminal residue, followed by isolation and identification of the amino acid derivatized. The original method of Sanger (1945) used 1-fluoro-2,4-dinitrobenzene (FDNB) which gives yellow dinitrophenyl (DNP) amino acid derivatives. This method has been superseded by the "dansyl" method (Gray and Hartley, 1963a, b) for both peptides and proteins. If a quantitative analysis is required, dansyl derivatives are not entirely suitable and an alternative procedure such as the cyanate method (see Stark, 1972) is preferred. Of the many other alternative methods for N-terminal analysis which have been described (see for example, Narita, 1970), none has any significant advantages over these methods.

1. *The Dansyl Chloride Method* (Gray and Hartley, 1963a)

Dansyl chloride (1-dimethylaminonaphthalene-5-sulphonyl chloride) gives a strong yellow fluorescence upon sulphonamide formation with the amino group of amino acids and peptides. This is by far the most useful N-group method; its main advantages are a 100-fold increase in sensitivity compared with the DNP amino acids, so that less than 1 nmol of sample can be used, the dansyl amino acids are relatively stable to acid hydrolysis, and the separation and identification of the dansyl amino acids is fairly easy.

Dansyl chloride not only reacts with amino groups, but also with imino, phenolic, thiol and imidazole groups (see Gray, 1967). The reaction mixture for optimum labelling of amino groups in peptides is dansyl chloride in acetone (5 mg/ml)–0·2 M sodium bicarbonate (1:1, v/v). The labelling is usually complete after 1 h at 37 °C (Gray, 1972). The conditions used with proteins are somewhat different; the labelling is carried out in the presence of a denaturing solvent (Gros and Labouesse, 1969) or with addition of SDS (Weiner *et al.*, 1972). After drying, the reaction products are hydrolysed in constant boiling HCl at 105 °C for 14–18 h. These hydrolysis conditions are not ideal for all derivatives; dansyl dipeptides with isoleucine, leucine or valine as N-terminal residue are particularly resistant to hydrolysis and require much longer hydrolysis times. On the other hand, dansyl-proline is less stable to hydrolysis and the time must be reduced to 6 h for good recovery. Dansyl-tryptophan is completely degraded and dansyl-asparagine and dansyl-glutamine are hydrolysed to the dansyl parent acid. Since N-terminal asparagine and glutamine cannot be detected using this method they are identified by thin-layer chromatography of the phenylthiohydantoin (PTH) amino acids, derived after subjecting the peptide to an Edman degradation step (see below). In addition, it is possible to deduce their presence in a peptide from the mobility of the peptide at pH 6·5 (see below) and from the change in mobility of the peptide at pH 6·5 on successive steps of Edman degradation.

2. Identification of Dansyl Amino Acids

Normally the derivatized amino acid can be analysed directly without extraction of the sulphonic acid (DNS-OH), produced by hydrolysis of excess dansyl chloride from the reaction mixture. Separation and identification of dansyl amino acids is accomplished by polyamide sheet chromatography (Woods and Wang, 1967). The dansyl amino acids are spotted by using aq. 95% (v/v) ethanol followed by 1 M ammonia. Development in the first dimension is with aq. 90% (v/v) formic acid–water (3:200, v/v) for 35 min and in the second dimension with toluene-acetic acid (9:1, v/v) for 40 min (Fig. 1A). This is followed by butyl acetate–methanol–glacial acetic acid (30:20:1 by vol.) for 40 min in the second dimension to resolve dansyl-serine from dansyl-threonine and dansyl-aspartic acid from dansyl-glutamic acid (Ramshaw et al., 1970) (Fig. 1B). More recently it has been found possible to reduce development times considerably by using smaller sheets (i.e. 5 cm × 5 cm instead of 15 cm × 15 cm). Several other useful solvent systems, in addition to those given, have also been described (see Gray, 1972). The identification of the mono dansyl derivatives of basic amino acids can be ambiguous on polyamide sheets. These derivatives are easily identified by electrophoresis at pH 1·9 (Gray, 1967). If, after electrophoresis, α-dansyl-lysine, ε-dansyl-lysine or α-dansyl-histidine are present, these derivatives may then be eluted and after labelling with dansyl chloride identified as the bis-dansyl derivatives by polyamide sheet chromatography.

Several other systems have been described for the identification of dansyl amino acids. These include paper chromatography (Boulton and Bush, 1964), thin-layer chromatography on silica or alumina using various solvents (see Seiler, 1970) and paper electrophoresis using different buffers (Gray, 1967). None of these however, is as useful as the polyamide chromatography system for dealing rapidly with the large numbers of samples that occur during a sequence analysis. The extreme sensitivity of the dansyl method has been demonstrated by Bruton and Hartley (1970), who have described a micro-dansyl method which requires only about 10 pmol of starting material.

The dansyl method is semi-quantitative and attempts to quantitate it completely have often met with difficulty; for example, direct fluorescence scanning of chromatograms may be affected by traces of moisture which cause fluorescence quenching (Hartley, 1970). Alternative methods based on the elution of dansyl spots from chromatograms have been proposed (Seiler and Wiechmann, 1966; Gros and Labouesse, 1969).

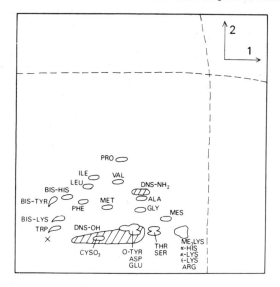

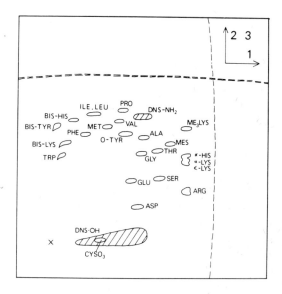

FIG. 1. Separation of dansyl amino acids by polyamide sheet chromatography. (Fig. 1A. Development (1) in aq. 90% (v/v) formic acid–water (3:200, by vol.) and (2) in toluene–acetic acid (9:1, v/v). Fig. 1B. Additional development (3) in butyl acetate–methanol–acetic acid (30:20:1, by vol.).) Key to Amino Acid Abbreviations: PRO = proline, VAL = valine, LEU = leucine, ILE = isoleucine, HIS = histidine, PHE = phenylalanine, TYR = tyrosine, TRP = tryptophan, MET = methionine, LYS = lysine, ALA = alanine, GLY = glycine, THR = threonine, SER = serine, ASP = aspartic acid, GLU = glutamic acid, ARG = arginine, $CYSO_3$ = cysteine sulphoxide, ME_3LYS = trimethyllysine, DNS = dansyl.

3. Blocked N-terminus

Many proteins have a blocked N-terminus. The coat protein of tobacco mosaic virus was the first shown to have an acetylated amino acid residue at the N-terminus (Narita, 1958). More recently it has been shown that proteins synthesized by a cell-free system from *Escherichia coli* contain N-formylmethionine at the amino end, and this has been proposed as a general mechanism of synthesis on 70S ribosomes, although subsequently the formyl group and, in some cases, several of the amino acid residues at the N-terminal end may be cleaved from the protein (see Smith and Marcker, 1970).

Schmer and Kreil (1969) have developed a micro method for the detection of acetyl and formyl groups in peptides and proteins. In this method the acylhydrazine formed after cleavage of the sample with anhydrous hydrazine is allowed to react with dansyl chloride, and the product, a 1-acyl-2-dansylhydrazine, is identified by two-dimensional thin-layer chromatography on either Kieselgel G or polyamide sheets.

Doolittle and Armentrout (1968) have isolated a pyrrolidonecarboxylyl peptidase from *Pseudomonas fluorescens*, which specifically hydrolyses a terminal pyrrolidone residue from proteins and peptides. However, because of the ease of cyclization of N-terminal glutaminyl residues, the demonstration of pyrrolidone carboxylic acid as a normal constituent in proteins is difficult, and its presence may be a result of the preparation methods used (see Blombäck, 1967).

B. CARBOXY-TERMINUS

The chemical methods which exist for determination of C-terminal amino acids are not comparable in their general usefulness with those described for determination of N-terminal residues. However, good enzymic methods exist which are normally used.

1. Carboxypeptidase Analysis

Three carboxypeptidases (A–C) with different specificities are commercially available (see Table I). With the correct choice of enzyme and conditions, it is usually possible to obtain sequence information for the C-terminal amino acids by following their rates of release. Carboxypeptidase A is often used initially for the determination of the C-terminus. Since it is supplied as a crystalline suspension in water, it is easy to wash the enzyme free of contaminating amino acids before solubilizing (see Ambler, 1967). It releases most rapidly amino acids with aromatic or large aliphatic side chains, but it will not release proline or arginine at all.

Carboxypeptidase B is supplied as a solution which must be kept frozen

TABLE I

Summary of Observed Enzyme specificities

A. ENDOPEPTIDASES[a]

Trypsin	Major Cleavage: Minor Cleavage:	Lys–X Arg–X (see footnote [b])
Chymotrypsin	Major Cleavage: Minor Cleavage:	Tyr–X, Phe–X, Trp–X Met–X, Leu–X, His–X, Asn–X
Thermolysin	Major Cleavage: Minor Cleavage:	X–Leu, Y–Ile, Y–Val, X–Phe X–Ala, X–Tyr, X–Met

B. EXOPEPTIDASES

Carboxypeptidase A	Fast Release: Slow Release: Very slow Release: No release:	Most amino acids Asn, Ser, Lys Asp, Gly, Gly Arg, Pro
Carboxypeptidase B	Fast Release Slow Release No Release:	Arg, Lys Most amino acids Pro
Carboxypeptidase C	Good Release: Slow Release: No Release:	Most amino acids Gly Gly–Pro, Pro–Gly, Pro–Pro

[a] Y represents any amino acid, X represents any amino acid except proline.
[b] Apparent minor cleavage results may be caused by chymotrypsin impurity in the enzyme preparation.

to maintain activity. It is usually necessary to remove the free amino acids present before use and this is best done by gel filtration (see Ambler, 1967). Its specificity is very limited; it releases arginine and lysine very much faster than any of the other amino acids.

Carboxypeptidases C, which occur in a variety of plant tissues (Wells, 1965; Zuber and Matile, 1968) will release all the protein amino acids, including proline. However, the release of glycine is slow and Gly–Pro, Pro–Pro and Pro–Gly bonds appear particularly resistant to hydrolysis (Tschesche and Kupfer, 1972).

The best method of C-terminal analysis using carboxypeptidases is to run a timed series of incubations and follow quantitatively the rates of release of amino acids by using an amino acid analyser (see e.g. Ambler, 1967). However, when the penultimate residue is only released very slowly compared with the C-terminal residue a semi-quantitative method, such as labelling with dansyl chloride, is sufficient to identify the C-terminal residue. Carboxypeptidase analysis does not necessarily give unambiguous results when several amino acids are released very rapidly. For example, distinction between the rates of release of leucine and phenylalanine from jack-bean urease could not be observed after digestion with carboxypeptidase A for 15 sec at 2 °C using an enzyme/substrate ratio of 1:2000 by weight (Bailey and Boulter, 1969). In such cases a chemical method such as hydrazinolysis (Akabori *et al.*, 1952; see also Narita, 1970) or, preferably, tritium labelling (Matsuo *et al.*, 1966) may be used.

2. *Tritium Labelling Method*

Holcomb *et al.* (1968) have indicated that it is possible to identify any *C*-terminal amino acid except proline by using the tritium labelling method. The protein or peptide is dissolved in ^3H_2O (approximately 100 mc/ml) and then 2 vols pyridine and 0·5 vol acetic anhydride are added. Volumes are kept to a minimum to avoid handling excess tritium. After standing for 3 h at room temperature, the sample is dried; it is then redissolved in water and redried six times to remove "washable" isotope. After acid hydrolysis of the samples and separation of the amino acids present, the radioactive *C*-terminal amino acid can be identified. Either a radiochromatogram scanner or liquid scintillation counting may be used, depending on the method used for separating the amino acids.

It has been suggested that labelling occurs via oxazolone formation. Since the procedure is carried out in an aqueous medium in which oxazolones are unstable, it has been proposed that tritiation occurs via the transient formation of an oxazolone in equilibrium with uncyclized amino acid (Matsuo *et al.*, 1966).

Tritiation of non-*C*-terminal glutamic acid and aspartic acid has been observed, and this may give rise to ambiguous results. However, this ambiguity can be avoided if carboxypeptidases are used to release amino acids from the *C*-terminal of the sample (Hsieh *et al.*, 1971).

If the amino acids released by carboxypeptidase hydrolysis are identified as the dansyl derivatives on polyamide sheets and the radioactive amino acid located using a radiochromatogram scanner, it is possible to determine the *C*-terminal residue of proteins and peptides using 1 nmol of sample (Ramshaw *et al.*, 1974a).

IV. DETERMINATION OF AMINO ACID COMPOSITION

Before starting a sequence investigation, it is important to have determined the total amino acid composition of the protein. It is necessary to know the numbers of specific residues present in the protein in order to determine the strategy of sequencing. For example, if the protein contains no arginine or lysine, it is no good using trypsin in an attempt to fragment it. When the proposed sequence is complete, the sum of the amino acid residues in the sequence must equal that determined for the total amino acid composition.

A. PREPARATION OF HYDROLYSATES

All of the methods of determining the total amino acid composition that exist require the prior hydrolysis of the protein into its constituent amino acids. This is carried out under controlled conditions and unless it is carefully carried out excessive breakdown of certain amino acids, which cannot be tolerated, occurs.

A weighed sample of air-dried protein is used for analysis. At the same time, additional samples of the air-dried protein may be taken for moisture, ash and total nitrogen content determinations. The total nitrogen content calculated from the results of the amino acid analysis should be within a few percent of that from the total nitrogen content determination.

The samples for hydrolysis are weighed into heavy-walled Pyrex tubes and suspended in 6 M hydrochloric acid. Oxygen should not be present during the hydrolysis if excessive destruction of certain amino acids is to be avoided. Normally the sample is frozen to avoid excess bumping and oxygen removed under high vacuum. It is important when most of the oxygen has been removed to allow the sample to thaw, otherwise the last traces of oxygen remain. The evacuated tubes are then sealed and hydrolysis carried out at 110 ± 1 °C in a controlled temperature oven. After hydrolysis, the sealed ampoules are removed from the oven and kept in the refrigerator until required. Before analysis the hydrochloric acid is removed by rotary evaporation at 35 °C (Moore and Stein, 1963).

Even with the most careful procedures, the recovery of certain amino acids is not quantitative. Peptide bonds adjacent to isoleucine and valine are hydrolysed more slowly than others and consequently the recovery of these amino acids tends to be low. Asparagine and glutamine are both destroyed and give aspartic acid and glutamic acid respectively. Tryptophan is almost always completely destroyed and serine and threonine are broken down to some extent. Losses of tyrosine and cysteine may also occur, the amount of decomposition depending upon the particular protein. The standard procedure is to hydrolyse samples for 20, 40 and 70 h; zero time values for serine, threonine, tyrosine and cystine are calculated assuming first order kinetics for their destruction, and maximum values are taken for valine and isoleucine (Moore and Stein, 1963).

Tryptophan is determined separately; the most reliable method is that of alkaline hydrolysis of the protein and subsequent analysis of the tryptophan by ion-exchange chromatography (Noltman et al., 1962). The Villegas and Mertz (1971) and Spies and Chambers (1948; 1949) methods both can give variable results with some proteins, whilst that of Horn and Jones (1945) consistently gives lower values than either of these methods. A recent study has shown that if a protein is hydrolysed in 3 M-toluene-*p*-sulphonic acid containing 0·2% 3-(2-aminoethyl)indole in evacuated tubes at 110 °C for up to 72 h, tryptophan is recovering quantitatively, together with the other amino acids (Liu and Chang, 1971). Hydrolysis using Sepharose-bound peptidases has also been suggested for the recovery of all the labile amino acids (Bennett *et al.*, 1971). This method also has the advantage of retaining the amide groups of glutamine and asparagine.

B. AMINO ACID ANALYSIS

The introduction of a paper chromatographic method for the separation of amino acids (Consden *et al.*, 1944) was a major breakthrough in protein chemistry. Concurrent with the development of this technique was the development of column chromatographic techniques, particularly those involving polymeric ion-exchange resins (see, e.g. Moore and Stein, 1951, 1954a). This research resulted with the publication in 1958 of details of an automatic analyser specifically designed for separating and quantitatively determining amino acids (Spackman *et al.*, 1958). This ion-exchange method, with several improvements, is still the major method used in amino acid analysis. Recently, however, both gas–liquid and liquid–liquid chromatography systems have also been developed for this purpose.

1. *Ion-Exchange Chromatography*

Various automatic analysers are now commercially available for the routine separation of amino acids by ion-exchange chromatography; many have automatic loading of samples and automatic integration and calculation of the results. Systems are available which use either a single column method (see Hamilton, 1963) or a double column method (see Spackman *et al.*, 1958). Some current instruments are capable of analysing up to 24 protein hydrolysates per day. The programming used in automatic loading machines requires elution by two or three buffer changes. For machines which run only single analyses, very high resolution can be obtained by using a gradient elution system (Piez and Morris, 1960). This method is particularly useful when the hydrolysate contains unusual amino acids, e.g. trimethyllysine, which may co-chromatograph with the common amino acids using other procedures (DeLange *et al.*, 1969a; Thompson *et al.*, 1970).

By very close attention to reagent purity, an accuracy of ±3% can be obtained in the determination of 10 nmol or less of most amino acids. Cystine and cysteine are best determined as cysteic acid after performic acid oxidation (Hirs, 1967), or as S-carboxymethylcysteine (Bailey and Boulter, 1970); the accuracy of the determination is about ±5–10%. Close attention must be paid to possible variations in the commercially supplied standard amino acid calibration mixtures; in practice, due to a variety of causes, comparative determinations between laboratories may vary by up to $\pm 5\%$.

The detection method normally used on automatic analysers is still the ninhydrin reaction, with absorbance measurements at 570 nm and 440 nm (for proline) (see Moore and Stein, 1954b). Measurement of the ninhydrin absorbance at 405 nm has been suggested for the detection of all amino acids (Ellis and Garcia, 1971). Udenfriend et al. (1972) has recently described a fluorescence detection system using fluorescamine (4-phenylspiro[furan-2(3H),1'-phthalan]-3,3'-dione) as the detection reagent which allows 50 pmol quantities of amino acids to be analysed.

2. Gas–Liquid Chromatography

A technique for the quantitative analysis of the twenty protein amino acids by gas–liquid chromatography has been established (see e.g. Zumwalt et al., 1971; Gehrke et al., 1971). Selection of suitable conditions has been complicated since amino acids do not represent a homologous series. Also, since they are not volatile they must be derivatized before analysis; a suitable derivative should be volatile, stable and readily produced with near quantitative recovery. Of the many derivatives which have been tried (see Linenberg, 1970), the most suitable in these respects are the N-trifluoroacetyl-n-butyl esters and the trimethylsilyl derivatives. The quantitative determination of all twenty amino acids, as their N-trifluoroacetyl-n-butyl esters, can be performed in ca 45 min using two columns (Zumwalt et al., 1971). One column is of 0·65 w/w% EGA on 80/100 mesh AW Chromosorb W (1·5 m × 4 mm I.D. glass) and the other column is of 1·0 w/w% OV-17 on 80/100 mesh H.P. Chromosorb G (1 m × 4 mm I.D. glass).

Stalling et al. (1968) introduced a new silylating agent, bis(trimethylsilyl)-trifluoroacetamide, with which trimethylsilyl derivatives of amino acids can be reproducibly obtained in a single step. Quantitative separation of these derivatives can be achieved on a single column of 10 w/w% OV-11 on 100/120 mesh Supelcoport (6 m × 2 mm I.D.) (Gehrke and Leimer, 1971).

The separation of amino acids as their N-heptafluorobutyryl-n-propyl esters on a single column of 3% OV-1 siloxane on AW-DCMS Chromosorb W may prove suitable for quantitative analysis (Moss et al., 1971).

Gas–liquid chromatographic analyses have now been performed on a wide range of proteins and biological samples. However, in spite of the work of Gehrke and his colleagues, gas–liquid chromatography is still not routinely used to determine all twenty protein amino acids quantitatively, and ion-exchange chromatography remains the preferred method. Currently, gas–liquid chromatography offers advantages in sensitivity, speed and low cost, but it is probable that with future developments, e.g. use of fluorescamine (see above) or high pressure systems, these advantages will be achieved by ion-exchange methods.

V. Fragmentation of Proteins

Most proteins are too large for their sequence to be determined directly. Existing methods can determine a limited number of residues on the protein itself; incomplete yield at each step gives rise to an increasing mixture of polypeptides which at further degradation steps produce ambiguous results. When a protein has been characterized as homogeneous and the amino- and carboxy-groups and total amino acid composition have been determined therefore, the next step is the digestion of the protein into peptides whose composition and sequence can be determined. Ideally, the digestion should be at specific residues only, and peptides occur in high yield. Knowledge of the amino acid composition of the protein then allows the number of bonds which will be cleaved (N), and the number of peptides (N + 1) in the final digestion mixture to be predicted. One of the major difficulties in sequence studies is the problem of the separation of a complex mixture of peptides, generated by fragmentation techniques (see below). A balance has to be struck between the desire to have the smallest number of peptides (ease of separation) against the need to generate peptides whose sequence can subsequently be determined easily. Thus, the normal dansyl-Edman methods can determine about 10–15 residues. However, by using highly purified reagents more than 20 residues can be determined (see Richardson, 1974). Of course, a large peptide can always be separated and then subjected to further digestion, but each separatory step leads to loss in yield. The strategy of fragmentation is, therefore, very important, although owing to the limited number of methods available, the amount of manoeuvre possible is often restricted. Up to now proteins have been sequenced, in the main, manually, but with the development of automated sequences the strategy of fragmentation will change so as to produce a few large peptides.

A. METHODS OF MEASURING HYDROLYSIS

It is useful to follow the course of hydrolysis if high yields of the major peptides are to be obtained without losses due to non-specific cleavage. This is normally done by determining the numbers of new amino groups produced by the hydrolysis using either the ninhydrin method (Moore and Stein, 1954b) or the titrimetric method (see below).

1. Titrimetric Method

In the pH range of 7·4–9·0 each newly liberated carboxyl group released by the breaking of a peptide bond will be fully ionized, whereas the α-amino group also liberated will be only partly charged, leading to a decrease in the pH of the solution. Titration of the unbuffered enzyme mixture with standard alkali directly measures the rate and extent of proteolysis. The alkali uptake is related to the average pK of the α-amino groups and the pH at which the hydrolysis is performed, and is most easily measured with an automatic titrator, i.e. pH stat (Jacobsen *et al.*, 1957). The titration vessel must have temperature regulation, efficient stirring and CO_2 should be excluded by passing through nitrogen gas. The digestion is terminated at the point at which a plateau is reached. By assuming an average pK of liberated amino acids and from the pH of the reaction mixture, the approximate number of bonds broken can be calculated and compared with that expected from the specificity of the enzyme and the amino acid composition of the protein being hydrolysed.

B. ENZYMATIC METHODS OF HYDROLYSIS

Of the various enzymes available, very few exhibit a sufficiently narrow range of specificity to be generally useful in the hydrolysis of proteins for sequence determination. Trypsin and chymotrypsin are by far the most useful endopeptidases. The normal enzyme specificities observed can be varied by the choice of different reaction conditions. Variations may arise from the method of denaturation of the substrate, the enzyme to substrate ratio, temperature, pH and time of digestion.

1. *Trypsin*

Trypsin is the most substrate-specific endopeptidase (see Table I), splitting bonds at the carboxyl group of arginine and lysine. Rates of cleavage of these bonds, however, are affected by the chemical nature of side-chains in their immediate vicinity, the most notable examples being the total resistance to hydrolysis of Lys–Pro or Arg–Pro sequences. Peptide bonds involving modified lysine residues, e.g. ε-N-trimethyllysine, do not appear susceptible to tryptic digestion (Thompson *et al.*, 1970). It is important to remove any chymotrypsin activity from trypsin preparations; satisfactory commercial preparations are now available in which this has been achieved by treatment with L-(1-tosylamido-2-phenyl) ethyl chloromethyl ketone (TPCK), a specific inhibitor for chymotrypsin (Schoellman and Shaw, 1963).

Chemical modifications of proteins, e.g. masking of amino-groups of lysine, will alter the specificity of trypsin to a protein and, similarly, it is possible to expand the specificity by creating new bonds susceptible to cleavage, e.g. S-amino-ethylation of cysteinyl residues.

2. *Chymotrypsin*

The degree of specificity of chymotrypsin is not quite as narrow as that of trypsin; the peptide bonds preferentially cleaved are those involving the carboxyl groups of tyrosine, phenylalanine, tryptophan and leucine. Cleavage may also occur on occasions at a variety of other bonds, but these bonds are split only slowly and do not represent major points of hydrolysis (see Table I). As with trypsin, neighbouring groups have a marked effect on the specificity and rates of hydrolysis. Again, peptide bonds involving proline appear to be resistant to cleavage, although the Phe–Pro bond in α-corticotropin has been reported to be partially hydrolysed (Leonis *et al.*, 1959).

3. *Thermolysin*

The specificity exhibited by thermolysin makes it a useful alternative to chymotrypsin and trypsin. Peptide bonds involving the amino group of hydrophobic residues, particularly isoleucine, leucine and valine, are preferentially cleaved. The resulting peptides are often obtained in high yields. Small peptides are not good substrates for the enzyme. This may be the reason why in certain cases fewer breaks are observed than one would expect from the enzyme specificity. For example, in plant plastocyanins (Ramshaw *et al.*, 1974a; Scawen *et al.*, 1974), in the sequence

$$36 \quad 37 \quad 38 \quad 39 \quad 40 \quad 41 \quad 42$$
$$\text{–Pro–His–Asn–Val–Val–Phe–Asp–}$$

hydrolysis by thermolysin is only observed at valine-39 and there is no activity at either valine-40 or phenylalanine-41.

The use of *N*-terminal analysis alone is insufficient for examining the purity of thermolysin peptides since the amino-group-directed specificity leads to very similar amino-terminal residues for all the peptides. This means that additional peptide purification steps must be included to ensure peptide purity.

4. *Other Enzymes*

Of the many other enzymes available, only papain and pepsin have been used to any significant extent for digestion of proteins for sequence determinations. Their major use, along with other enzymes such as subtilisin and pronase, is for further digestion of large peptides derived from other enzymic digests. Recently, enzymes with novel, and very specific, activities have been described, which may prove useful in future sequence studies, e.g. an enzyme from *Staphylococcus* (Houmard and Drapeau, 1972) which cleaves polypeptide chains on the carboxyl-terminal side of aspartyl and glutamyl residues.

C. CHEMICAL METHODS OF CLEAVAGE

Numerous methods for splitting polypeptide chains by chemical means have now been reported (see Spande *et al.*, 1970). Whilst many of the methods show a very high degree of specificity, in only a few cases are the reaction yields sufficiently high for the methods to be of major practical use.

1. *Cleavage with Cyanogen Bromide*

The most useful chemical cleavage method is that of cyanogen bromide in either 0·1 M HCl or 70% formic acid at methionyl residues (see Fig. 2) (Gross and Witkop, 1961). This reaction is very specific; any non-specific breaks which occur probably result from the acidic reaction conditions which are used. The yields obtained are generally in excess of 80% and may be nearly

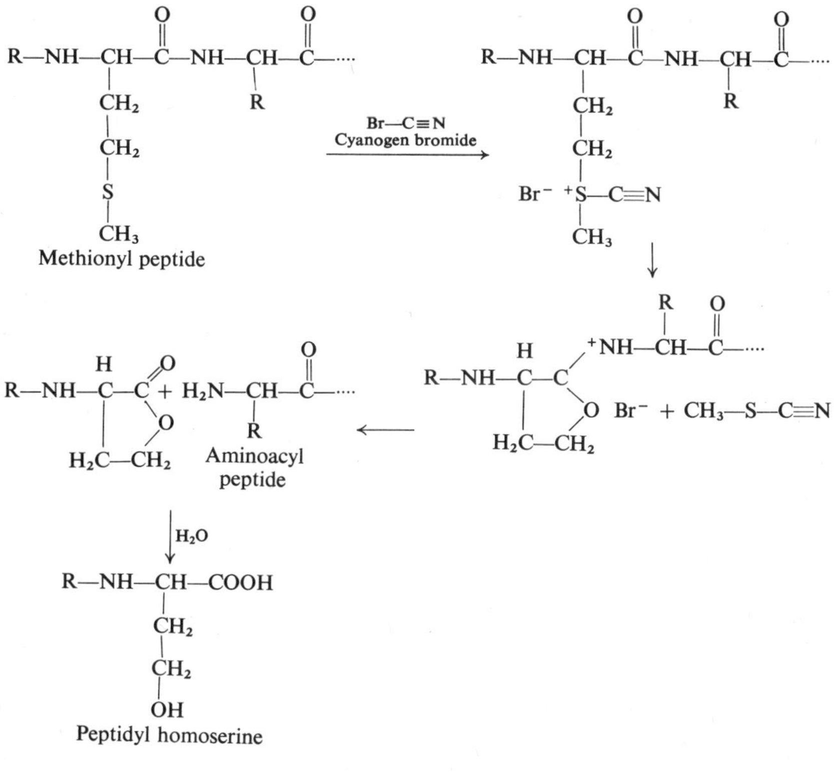

FIG. 2. The mechanism for cleavage of methionyl peptides by cyanogen bromide.

quantitative (Steers *et al.*, 1965). The least susceptible bonds for cleavage seem to be Met–Thr and Met–Ser. The peptides produced all have either homoserine or homoserine lactone as their *C*-terminal residue, with the exception of the *C*-terminal peptide of the protein; a Met–Met sequence or an *N*-terminal methionine gives free homoserine.

Since methionine normally occurs in small amounts in proteins, this increases the probability of obtaining large resultant polypeptide fragments.

2. Cleavage with N-bromosuccinimide

N-bromosuccinimide can be used to cleave proteins primarily at trypto-phanyl, but also at tyrosinyl bonds. If cleavage at both these residues has occurred or has been blocked, histidyl bonds may also be cleaved. The yields from this method are very variable, but generally are fairly low and rarely exceed 50% (see Spande et al., 1970). In addition, side reactions and non-specific cleavage may occur.

3. Other Methods

Dilute acid hydrolysis is occasionally used for hydrolysis of proteins. Treatment of the protein with 0·03 M HCl at 110 °C (Macdowall and Smith, 1965) results in peptide bonds involving aspartic acid being preferentially cleaved at a rate at least 100 times greater than most other bonds (see Schultz, 1967). This method is particularly useful for studies on proteinase inhibitors when normal enzymatic methods cannot be used (see Richardson, 1974).

Specific cleavage of Asn–Gly bonds in good yields by hydroxylamine has been reported (Bornstein and Balian, 1970), but this method is not necessarily generally applicable. Other methods, for example reductive cleavage at proline residues (see Marglin, 1972), do not have sufficiently good yields to be of much use.

VI. SEPARATION OF PEPTIDES

In all separation methods care must be taken to avoid reactions occurring which modify the peptide. Particularly common reactions are cyclization of N-terminal glutamine to a pyrrolidone carboxylyl group and deamidation of amide residues. In addition, volatile buffers should be used so that the purified peptides will be free of contaminating salts.

A. COLUMN CHROMATOGRAPHIC SEPARATION

Either gel filtration or ion-exchange methods may be used. Gel filtration is particularly useful for separating the few large fragments that may result from chemical cleavage, e.g. by cyanogen bromide on plastocyanin (Milne and Wells, 1970). Gels with low exclusion limits, e.g. Sephadex G-15 or G-25 have been used for the preliminary fractionation of mixtures of small peptides (see e.g. Ambler and Wynn, 1973).

Ion-exchange methods are useful when large quantities of material are available, e.g. 100 µmol. They are rarely used with quantities of less than 8–10 µmol (see Kasper, 1970).

B. PAPER ELECTROPHORETIC AND CHROMATOGRAPHIC METHODS

Small peptides (less than twenty residues) can be separated by high-voltage paper electrophoresis at pH 6·5 (pyridine–acetic acid–water, 25:1:225, by volume) on Whatman 3MM paper. A flat-plate apparatus (Gross, 1961) (107 cm × 15 cm; The Locarte Co., London S.W.3, U.K.), at 90 V/cm for 60–150 min, is preferred to electrophoresis in tanks (Michl, 1951) when small quantities of material are to be separated as the longer length of run allows much greater separations. After electrophoresis, the peptides are located on guide strips by the cadmium–ninhydrin reagent of Heilmann et al. (1957) and the starch–iodine method of Pan and Dutcher (1956). The Ehrlich, Sakaguchi and platinic iodide reagents recommended by Easley (1965) can be used to locate specific peptides which contain tryptophan, arginine and methionine and cysteine residues. Recently, phenanthraquinone has been introduced for detection of arginine in a test which is considerably more sensitive than the Sakaguchi test (Yamada and Itano, 1966).

Fluorescamine (see above) can also be used to locate peptides which have primary amino groups (see Vandekerckhove and Van Montagu, 1974).

Peptides are eluted off the paper with water, acetic acid (10% v/v) or pyridine (20% v/v). Their purity can then be estimated by N-terminal analysis by the "dansyl" method (Gray and Hartley, 1963a) and by composition analysis using the dansyl method (Brown and Perham, 1973). Peptides requiring further purification may be separated by electrophoresis at pH 1·9 (acetic acid–formic acid–water, 4:1:45, by volume). If after these two separations the peptides are still not pure, further electrophoresis at pH 3·6 (pyridine–acetic acid–water, 1:10:89, by volume) or paper chromatography using butanol–acetic acid–water–pyridine (15:3:12:10, by volume) is normally sufficient to ensure pure peptides for analysis. Before separating peptides by electrophoresis, care should be taken to ensure that they are soluble in the buffer system to be used. If they are not completely soluble they may streak badly and interfere with the purification of other peptides.

1. *Electrophoretic Mobilities of Peptides*

The mobilities of peptides are measured from the position of the neutral amino acids relative to DNS–Arg–Arg at pH 6·5 and from the position of 1-dimethylaminonaphthalene-5-sulphonic acid relative to DNS–arginine at pH 1·9. The electrophoretic mobility of a peptide at pH 6·5 is directly related to its molecular weight (which is obtained from its amino acid sequence or composition) and its charge (see Fig. 3) (Offord, 1966). A similar relationship also exists for the electrophoretic mobilities of peptides at pH 1·9 (Bailey and Ramshaw, 1973). Consideration of the peptide's molecular weight and mobility is generally used to establish its charge. This allows differentiation

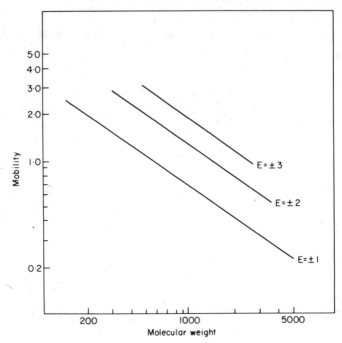

Fig. 3. Peptide mobilities at pH 6.5. (The electrophoretic mobility of peptides from the position of the neutral amino acids relative to dansyl-Arg–Arg is plotted against their molecular weight and charge (E) at pH 6.5. Peptides containing histidine or cysteic acid do not conform to this diagram.)

between asparaginyl and aspartyl residues and glutaminyl and glutamyl residues as the amide residues are both neutral at pH 6·5. This differentiation was not possible using the dansyl method. If a peptide contains a mixture of acid and amide residues, or has a charge greater than ±3, ambiguity may still exist. This can be resolved by following the changes in mobility of the peptides resulting from a series of Edman degradations (see below) on the original peptide.

VII. Sequence Determination

Before starting to sequence the purified peptides it is preferable to have determined their amino acid compositions. This is done by using an automatic amino acid analyser (see above) or by a double-isotope labelling technique using dansyl chloride (Brown and Perham, 1973).

A. The Edman Procedure

Of the various procedures which have been described for the sequential degradation of peptides, only those based on the isothiocyanate degradation, which was first described more than twenty years ago by P. Edman (Edman,

1950), have proved of use in sequencing proteins and peptides (for details, see Edman, 1970).

The reaction can be split into three steps (see Fig. 4). The first step (1) is the formation of a phenylthiocarbamyl (PTC-) derivative of the peptide by coupling the free α-amino terminal group with phenylisothiocyanate (the "coupling" reaction).

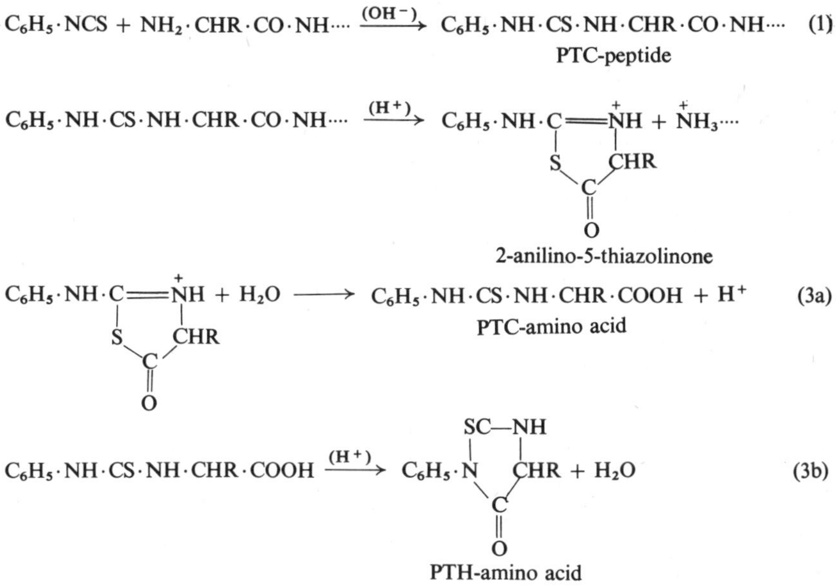

$$C_6H_5 \cdot NCS + NH_2 \cdot CHR \cdot CO \cdot NH \cdots \xrightarrow{(OH^-)} C_6H_5 \cdot NH \cdot CS \cdot NH \cdot CHR \cdot CO \cdot NH \cdots \quad (1)$$
$$\text{PTC-peptide}$$

$$C_6H_5 \cdot NH \cdot CS \cdot NH \cdot CHR \cdot CO \cdot NH \cdots \xrightarrow{(H^+)} C_6H_5 \cdot NH \cdot \overset{+}{C} = \overset{+}{N}H + \overset{+}{N}H_3 \cdots$$

2-anilino-5-thiazolinone

$$C_6H_5 \cdot NH \cdot C = \overset{+}{N}H + H_2O \longrightarrow C_6H_5 \cdot NH \cdot CS \cdot NH \cdot CHR \cdot COOH + H^+ \quad (3a)$$
$$\text{PTC-amino acid}$$

$$C_6H_5 \cdot NH \cdot CS \cdot NH \cdot CHR \cdot COOH \xrightarrow{(H^+)} C_6H_5 \cdot N \quad CHR + H_2O \quad (3b)$$
$$\text{PTH-amino acid}$$

FIG. 4. The mechanism of the Edman degradation procedure.

The next step (2) cleaves the PTC-peptide at the peptide bond nearest to the PTC group. This requires strongly acidic conditions and leads to the formation of a 2-anilino-5-thiazolinone derivative of the first amino acid and a peptide with one amino acid less than the original (the "cleavage" reaction). The shortened peptide has a free α-amino group, and can, therefore, be subjected to a new cycle of degradation.

As the 2-anilino-5-thiazolinone derivative produced from the N-terminal amino acid is unstable, after extraction from the degraded peptide by organic solvent, it is converted in step (3) to the stable 3-phenyl-2-thiohydantoin (PTH) derivative by treatment in aqueous acid (the "conversion" reaction). The conversion reaction proceeds via two steps; first, hydrolysis of the thiazolinone to the PTC-amino acid occurs (3a), and then cyclization of the latter to the corresponding PTH-amino acid (3b).

It is important to separate the cleavage and the conversion reactions, even though both are acid catalysed. The conversion reaction requires an aqueous acid medium and is slow, whereas the cleavage reaction does not require water and is fast. If combined, therefore, the result would be an unnecessarily long exposure of the peptide to hydrolytic conditions.

Three major methods exist for sequence analysis using the Edman degradation. The first, the "direct" method, is to analyse the PTH derivatives (see below) which are obtained on successive steps of degradation (Fraenkel-Conrat, 1954; Blombäck et al., 1966). The other methods both use an "indirect" approach. In the second method, the subtractive method, the amino acid compositions of the peptides before and after a degradation step are compared and the N-terminal amino acid determined by difference (Hirs et al., 1960). The third method, the dansyl-Edman method, is an end-group method using dansyl chloride (see below) (Gray and Hartley, 1963b).

1. The Dansyl-Edman Method

For the dansyl-Edman method (Gray and Hartley, 1963b) the N-terminal residue of the peptide being sequenced is determined by the dansyl method (see above). The peptide is then subjected to one cycle of Edman degradation. A small aliquot of the degraded peptide is then taken and the new N-terminal residue is determined, again by the dansyl method. After identification of the new N-terminal residue the rest of the peptide is subjected to further cycles of Edman degradation and additional residues determined in the same manner (for details, see Gray, 1972). The thiazolinone derivative cleaved off by the Edman degradation is extracted and discarded, i.e. the conversion reaction is not carried out; an exception to this may occur in the identification of amide residues since they cannot be directly determined using the dansyl method. The dansyl-Edman method is very sensitive and requires less than 50 nmol of peptide for the sequence determination of a peptide of up to 8–10 residues. Bruton and Hartley (1970) have devised a micro-procedure of this method which requires only 1 nmol of peptide. The dansyl-Edman method is readily performed on many peptides simultaneously when used in conjunction with polyamide layer chromatography for identification of dansyl amino acids. This makes it ideal for sequence determination studies for which it is currently the preferred manual method.

B. AUTOMATED SEQUENCING METHODS

Two systems for automated sequencing of proteins and peptides have been described. In the one system (Edman and Begg, 1967) the protein/peptide is immobilized as a thin film inside a spinning cup. In the other (Laursen, 1972), the peptide is immobilized by being covalently bound to an insoluble matrix. Now that reliable commercial instruments are available the former method is routinely used in many laboratories; it is this method which will be discussed.

The sequencer automatically performs the coupling and cleavage operations of the Edman sequential degradation delivering the anilino-thiazolinone samples to a refrigerated fraction collector. The conversion step is then carried out manually on a large number of these samples at one time. The reaction steps used are identical to those described for the manual methods, although the instrument design has necessitated small differences in the buffers and reagents which are used, so that they have lower volatilities.

The principle of operation is that all the reaction and extractions are performed as thin films on the inside of a glass cup rotating at high speed. These thin films are very well suited for extraction and drying because of their large surface area and the stabilizing centrifugal force. Each of the operations of the degradation is divided into its individual stages which are programmed sequentially, and once one degradation cycle has been accomplished the whole programme is repeated. Repetitive yields of the order of up to 98% can be obtained, i.e. considerably better than in the manual procedure. These high yields are obtained by using specially prepared high purity chemicals and by careful exclusion of oxygen from the system. Normally, for proteins, the first 30–60 residues can be established before carry-over, loss in yield or increase in background makes interpretation of the results equivocal. In certain cases, however, as many as 80 residues have been determined (Walter et al., 1971). Accumulation of background amino acids probably may result from nonspecific hydrolysis of the sample and this background progressively builds up during sequence analysis. It has been suggested that overlap or carry-over from one cycle to the next results from the fact that the cleavage reaction is not completely quantitative and the products of the reaction are in equilibrium (Edman and Begg, 1967). In the sequencer programme this problem is reduced by repeating the cleavage reaction twice. In the degradation of long polypeptide chains, e.g. greater than 100 residues, the background accumulation is the limiting factor, whereas with short peptide chains, the main difficulty is generally low yields resulting from loss of the peptide in the various organic solvent washes. This loss of yield can be partly avoided either by using a "peptide" programme, which uses volatile buffers and minimizes the solvent washes (see e.g., Hermodsen et al., 1972), or by using a substituted isothiocyanate on the first cycle which subsequently increases the polarity of the peptide (Braunitzer et al., 1971).

1. Identification of PTH Amino Acids

The 2-anilino-5-thiazolinone derivatives formed by the Edman degradation are converted to the PTH derivatives by 0·1 M HCl at 80 °C for 10 min under nitrogen, and then extracted with ethyl acetate (Edman and Begg, 1967). Most PTH amino acids are recovered in good yield. PTH-serine, however, is partially decomposed as a result of a β-elimination reaction. PTH-threonine shows much less tendency for β-elimination; should any decomposition occur, however, the product, PTH-dehydrothreonine, is fairly stable. Both PTH-asparagine and PTH-glutamine are partially deamidated under the conditions used. PTH amino acids with strongly ionized side-chains, e.g. PTH-arginine, PTH-histidine and PTH-cysteic acid, remain in the aqueous phase after the ethyl acetate extraction. If their presence is suspected it is dried down and analysed separately either by electrophoresis or suitable spot tests (see Edman, 1970).

PTH amino acids are identified by thin-layer chromatography on silica plates incorporating a fluorescent indicator and located by their fluorescence quenching (Jeppsson and Sjöquist, 1967). Characteristic colours are given by certain of the PTH amino acids if the plate is stained with ninhydrin after the chromatography. Several solvent systems have been described (see Jeppsson and Sjöquist, 1967; Edman, 1970), which normally allow the identification of all the protein amino acids. However, it is inadvisable to rely solely on any single method of identification.

Gas–liquid chromatography provides a rapid and reliable method of identifying PTH amino acids; it may also be used quantitatively. Pisano and his colleagues (Pisano et al., 1962; 1972; Pisano and Bronzert, 1971) have investigated a variety of systems which use different liquid phases. The result of this work is the use of SP-400 as a liquid phase for the separation of all PTH amino acids directly, or as their trimethylsilyl derivatives when the more polar derivatives of isoleucine and leucine are being analysed.

An alternative method for identifying PTH amino acids is to regenerate the parent amino acid. The amino acid may then be identified using an automatic analyser (see e.g. Smithies et al., 1971), or by converting it to its dansyl derivative and identifying after polyamide sheet chromatography (Ramshaw et al., 1974a). The regeneration is performed using either 6 M HCl at 150 °C for 24 h (Van Orden and Carpenter, 1964), or 0·1 M NaOH at 120 °C for 12 h (Africa and Carpenter, 1966), or 5 M HI at 140 °C for 24 h (Inglis et al., 1971); the latter method is probably the best. None of these methods however, is suitable for regenerating all of the PTH amino acids. However, if the results from the last two methods are used in conjunction, all the PTH-amino acids may be identified (Smithies et al., 1971).

Various alternative isothiocyanate reagents have been suggested for the Edman degradation which are reported to have advantages in the separation of the resulting thiohydantoin derivatives. These include methylisothiocyanate (Richards et al., 1969), pentafluorophenylisothiocyanate (Lequin and Niall, 1972) and fluorescein-isothiocyanate (Kawauchi et al., 1971). None as yet, however, is regularly used as an alternative to phenylisothiocyanate.

VIII. RECONSTRUCTING THE PRIMARY SEQUENCE

Once the protein has been broken into peptides which have been separated and their sequence determined, the complete sequence of the protein is reconstructed. This is accomplished by logical examination and combination of overlapping fragments obtained from the different methods of the hydrolysis of the protein. It is particularly helpful if the *N*- and *C*-terminal peptides have been specifically identified during the sequencing. Various computer programs are available which will perform this reconstruction of the sequence from peptide sequence information.

If a protein being studied is very similar to another protein or group of proteins for which sequence information is already available, alignment of peptides by homology with the other sequence(s) can be of great help in reconstructing its sequence.

If an automatic sequencer has been used to determine the sequence no reconstruction of the sequence from peptides is required; this is one of its major advantages. In sequencer studies, generally, only a few major fragments need placing in the correct order. If, for example, cyanogen bromide fragments have been studied, their order can be determined by isolating specifically the methionine peptides by a diagonal method (Tang and Hartley, 1967) and sequencing these few peptides manually. However, redundant information is still necessary.

IX. Plant Protein Sequence Data

The complete amino acid sequences of numerous plant proteins have now been determined (see Table II). However, it is oñly comparatively recently that many of these sequences have been established and for many years plant proteins were neglected for sequence studies. Thus, it was only in 1967 that the first complete sequences of plant proteins were published: wheat-germ cytochrome *c* (Stevens *et al.*, 1967) and spinach ferredoxin (Matsubara *et al.*, 1967). By 1969 the sequences of only 6 plant proteins had been published compared with over 230 from animal sources (200 of which were mammalian sources) (see Dayhoff, 1969). This large difference is not due to plant proteins being inherently more difficult to sequence. Generally the reason has been that it is considerably more difficult to obtain sufficient quantities of protein for sequence studies from plant sources, because of the small amounts that may be present. For example, at least 200 mg of cytochrome *c* can be purified from 1 kg of horse heart (see Margoliash and Walasek, 1967). To purify a similar quantity of cytochrome *c* from a plant source approximately 250 kg of seeds (Richardson *et al.*, 1971) or 3000 kg of leaves (Asada and Takahashi, 1971) would need to be processed; often as much as 1 g of cytochrome *c* was used to determine the sequence of this protein from various animal sources (see e.g. Chan and Margoliash, 1966; Needleman and Margoliash, 1966).

The availability of only small quantities of material has meant that reliable micro-sequencing methods (e.g. the dansyl-Edman method for peptides purified by high voltage paper electrophoresis), have needed to be established before sequence studies on plant proteins have been attempted. These methods are now available, and using them much valuable sequence information has been determined.

Boulter *et al.* (1972) have used plant cytochrome *c* sequence data to produce

TABLE II

Complete amino acid sequences of plant proteins

Cytochromes c
Triticum aestivum (Wheat)[a]
Phaseolus aureus (Mung bean)[b]
Helianthus annuus (Sunflower)[b]
Ricinus communis (Castor)[b]
Sesamum indicum (Sesame)[b]
Cucurbita maxima (Pumpkin)[b]
Fagopyrum esculentum (Buckwheat)[b]
Brassica oleracea (Cauliflower)[b]
Abutilon theophrasti (Abutilon)[b]
Gossypium barbadense (Cotton)[b]
Ginkgo biloba (Ginkgo)[b]
Brassica napus (Rape)[b]
Guizotia abyssinica (Niger)[b]
Lycopersicon esculentum (Tomato)[c]
Allium porrum (Leek)[d]
Spinacia oleracea (Spinach)[e]
Nigella damascena (Love-in-a-Mist)[f]
Cannabis sativa (Hemp)[g]
Sambucus nigra (Elder)[h]

Acer negundo (Sycamore)[h]

Pastinaca sativa (Parsnip)[h]
Tropaeolum majus (Nasturtium)[h]
Arum maculatum (Arum)[i]
Zea mays (Maize)[i]
Enteromorpha intestinales[j]

Papain
Carica papaya (Papaya)[a]

Leghaemoglobin
Glycine max (Soyabean)[a]

Ferredoxins
Spinacia oleracea (Spinach)[a]
Medicago sativa (Alfalfa)[a]
Scenedesmus sp.[a]
Leucaena glauca[a]
Colocasia esculenta (Taro)[a]

Plastocyanins
Chlorella fusca[k]
Cucurbita pepo (Marrow)[l]
Mercurialis perennis (Dog's Mercury)[m]
Symphytum officinale (Comfrey)[m]
Vicia faba (Broad Bean)[n]
Solanum tuberosum (Potato)[o]
Sambucus nigra (Elder)[p]

Proteinase Inhibitors
Phaseolus lunatus (Lima bean)[a]
Glycine max (Soyabean) Kunitz inhibitor[q]
Glycine max (Soyabean) Bowman inhibitor[r]
Zea mays (Maize) Trypsin inhibitor[s]
Arachis hypogaea (Peanut) Trypsin inhibitor[s]
Solanum tuberosum (Potato) Chymotrypsin inhibitor[t]

Viscotoxins
Viscum album (Mistletoe) Types A2,[a] A3,[a] B[u]

Histone IV
Pisum sativum (Pea)[a]

[a] Cited in Dayhoff (1972).
[b] Cited in Boulter et al. (1972).
[c] Scogin et al. (1972).
[d] Brown and Boulter (1973a).
[e] Brown et al. (1973).
[f] Brown and Boulter (1973b).
[g] Wallace et al. (1973).
[h] Brown and Boulter (1974).
[i] Richardson, D. L. (unpublished).
[j] Meatyard, B. T., unpublished.
[k] Kelly and Ambler (1973).

[l] Scawen and Boulter (1974).
[m] Scawen, M. D., unpublished.
[n] Ramshaw et al. (1974b).
[o] Ramshaw et al. (1974a).
[p] Scawen et al. (1974).
[q] Koide et al. (1972).
[r] Odani and Ikenaka (1972).
[s] Hochstrasser et al. (1970).
[t] Richardson (1973).
[u] Samuelsson and Pettersson (1971).

an affinity tree relating several higher plant sequences and to estimate the times of origin of major plant groups (Ramshaw et al., 1972). Since there is no adequate fossil record for higher plants and the use of morphological and other data has given rise to several conflicting phylogenetic schemes (see Lawrence, 1951), this molecular approach to plant phylogeny may prove exceedingly valuable. Sequence studies on plastocyanins, for which automatic sequencing is particularly useful (Ramshaw et al., 1974a; Scawen et al., 1974), are being used to confirm these results.

The amino acid sequences of various plant proteinase inhibitors have now been reported by several groups (see Table II), and these data have proved useful in studies on the mechanisms of inhibition (Krahn and Stevens, 1972).

Extensive sequence studies on the enzyme papain (see Light et al., 1964; Husain and Lowe, 1969) in conjunction with X-ray crystallographic studies (Drenth et al., 1968), have enabled its complete sequence to be determined (Mitchel et al., 1970). This work has been used to study the mechanism of action of this enzyme.

Many of the other plant protein sequences have been determined for use in comparison with apparently similar proteins obtained from completely different sources. Pea histone IV was compared with bovine histone IV and found to differ in only two positions (Delange et al., 1969b). Soyabean leghaemoglobin has been compared with mammalian haemoglobins and the plant ferredoxins compared with the bacterial ferredoxins (see Dayhoff, 1972).

X. CONCLUSION

Certain trends can be discerned in the study of the primary structure of proteins. Firstly, an increasing number of investigators find it necessary and worthwhile to attempt the elucidation of the primary structure of the major protein involved in their research activity. Secondly, there is the realization that primary structure elucidation is becoming less difficult as techniques improve, and it does not necessarily require excessively expensive equipment. Since the original rather restricted methods of Sanger, a large number of new techniques have been developed, and a new investigator in the field can be fairly confident that the methods exist for the determination of the primary structure of his protein and that these are well documented. The importance of correctly defining a strategy prior to experimentation, and the choice of satisfactory methods from the wealth of those available, are of considerable importance. In many cases now, it is the availability of the pure starting material in sufficient amounts that is the limiting step in primary structure determination. For this reason, methods which use smaller amounts of starting material are at a premium. Thus, separation of peptides on paper rather than by ion-exchange is often to be preferred because of its better resolution (hence less purification steps), greater sensitivity (hence less materials), and its rapidity (speed of a particular operation is also of particular importance in a successful sequence programme). For similar reasons the use of a sensitive indirect rather than direct method of sequence analysis is now preferred, and of the indirect methods there is increasing use of the dansyl-Edman method. Whilst automatic protein/peptide sequencers are not at present being used to determine complete sequences, the manual methods are now being increasingly supported by information obtained with them. Automatic sequencers avoid the most time-consuming part of a sequence determination, namely, the formation and separation of peptides, especially as overlapping peptides need no longer be obtained. Automatic sequencers, whilst expensive to run, are now thoroughly reliable and routinely give 40–60 residues. In addition, the recent developments in the methods available for solid-phase sequencing (Laursen, 1972), should allow this approach to be widely used in the future for the analysis of smaller fragments or peptides, and it may well eventually replace dansyl-Edman analysis. This then is the direction in which we may anticipate the future of sequencing to go, with more and more sequences probably being accomplished by a strategy which involves the sole use of automatic methods. The alternative, mass spectrometry, still only analyses peptides of about ten residues, and whilst temperature control regimes may allow the analysis of mixtures containing small numbers of 2–10 residue peptides (Morris, 1973), the problem of peptide separation cannot be entirely obviated. However, with the further development of high sensitivity techniques, such as the use of field desorption, mass spectrometry may still play a significant role in sequence analysis. It may be particularly useful for studying sets of homologous proteins with few differences between them, and for examining the structures of peptides eluted from fingerprints after fluorescamine staining.

In the future there is, therefore, a need to develop more methods, either enzymic or chemical, which split proteins in high yield at specific residues, particularly those occurring with a low frequency and also to improve the reagents and programmes in the automatic method so as to reduce, still further, side reactions and other losses.

ACKNOWLEDGEMENT

We wish to thank the Nuffield Foundation and the Science Research Council for financial help.

REFERENCES

Africa, B. and Carpenter, F. H. (1966). *Biochem. biophys. Res. Commun.* **24**, 113–119.
Akabori, S., Ohno, K. and Narita, K. (1952). *Bull. chem. Soc. Japan* **25**, 214–218.
Ambler, R. P. (1967). *In* "Methods in Enzymology" (C. H. W. Hirs, ed.), Vol. XI, pp. 155–166 and 436–445. Academic Press, New York and London.

Ambler, R. P. and Wynn, M. (1973). *Biochem. J.* **131**, 485–498.
Asada, K. and Takahashi, M. (1971). *Pl. Cell Physiol.* **12**, 361–375.
Bailey, C. J. and Boulter, D. (1969). *Biochem. J.* **113**, 669–680.
Bailey, C. J. and Boulter, D. (1970). *Eur. J. Biochem.* **17**, 460–466.
Bailey, C. J. and Ramshaw, J. A. M. (1973). *Biochem. J.* **135**, 889–891.
Bennett, H. P. J., Elliott, D. F., Lowry, P. J. and McMartin, C. (1971). *Biochem.* **125**, 80P–81P.
Blombäck, B. (1967). *In* "Methods in Enzymology" (C. H. W. Hirs, ed.), Vol. XI, pp. 398–411. Academic Press, New York and London.
Blombäck, B., Blombäck, M., Edman, P. and Hessel, B. (1966). *Biochim. biophys. Acta* **115**, 371–396.
Bornstein, P. and Balian, G. (1970). *J. biol. Chem.* **245**, 4854–4856.
Boulter, D., Ramshaw, J. A. M., Thompson, E. W., Richardson, M. and Brown, R. H. (1972). *Proc. R. Soc. Lond.* B **181**, 441–455.
Boulton, A. A. and Bush, I. E. (1964). *Biochem. J.* **92**, 11P.
Braunitzer, G., Schrank, B. and Ruhfus, A. (1971). *Hoppe-Seyler's Z. physiol. Chem.* **352**, 1730–1733.
Brown, R. H. and Boulter, D. (1973a). *Biochem. J.* **131**, 247–251.
Brown, R. H. and Boulter, D. (1973b). *Biochem. J.* **133**, 251–254.
Brown, R. H. and Boulter, D. (1974). *Biochem. J.* **137**, 93–100.
Brown, J. P. and Perham, R. N. (1973). *Eur. J. Biochem.* **39**, 69–73.
Brown, R. H., Richardson, M., Scogin, R. and Boulter, D. (1973). *Biochem. J.* **131**, 253–256.
Bruton, C. J. and Hartley, B. S. (1970). *J. molec. Biol.* **52**, 165–178.
Chan, S. K. and Margoliash, E. (1966). *J. biol. Chem.* **241**, 335–348.
Consden, R., Gordon, A. H. and Martin, A. J. P. (1944). *Biochem. J.* **38**, 224–232.
Dayhoff, M. O. (1969). "Atlas of Protein Sequence and Structure", Vol. 4. National Biomedical Research Foundation, Maryland, U.S.A.
Dayhoff, M. O. (1972). "Atlas of Protein Sequence and Structure", Vol. 5. National Biomedical Research Foundation, Maryland, U.S.A.
Delange, R. J., Glazer, A. N. and Smith, E. L. (1969a). *J. biol. Chem.* **244**, 1385–1388.
Delange, R. J., Fambrough, D. M., Smith, E. L. and Bonner, J. (1969b). *J. biol. Chem.* **244**, 5669–5679.
Doolittle, R. F. and Armentrout, R. W. (1968). *Biochemistry* **7**, 516–521.
Drenth, J., Jansonius, J. N., Koekek, R., Swen, H. M. and Wolthers, B. G. (1968). *Nature, Lond.* **218**, 929–932.
Easley, C. W. (1965). *Biochim. biophys. Acta* **107**, 386–388.
Edman, P. (1950). *Acta chem. scand.* **4**, 283–293.
Edman, P. (1970). *In* "Protein Sequence Determination" (S. B. Needleman, ed.), pp. 211–255. Chapman and Hall, London.
Edman, P. and Begg, G. (1967). *Eur. J. Biochem.* **1**, 80–91.
Ellis, Jr. J. P. and Garcia, Jr. J. B. (1971). *J. Chromatogr.* **59**, 321–327.
Fraenkel-Conrat, H. (1954). *J. Am. chem. Soc.* **76**, 3606–3607.
Gehrke, C. W. and Leimer, K. (1971). *J. Chromatogr.* **57**, 219–238.
Gehrke, C. W., Kuo, K. and Zumwalt, R. W. (1971). *J. Chromatogr.* **57**, 209–217.
Gray, W. R. (1967). *In* "Methods in Enzymology" (C. H. W. Hirs, ed.), Vol. XI, pp. 139–151. Academic Press, New York and London.
Gray, W. R. (1972). *In* "Methods in Enzymology" (C. H. W. Hirs, and S. H. Timasheff, eds.), Vol. XXV, pp. 121–138 and 333–344. Academic Press, New York and London.
Gray, W. R. and Hartley, B. S. (1963a). *Biochem. J.* **89**, 59P.
Gray, W. R. and Hartley, B. S. (1963b). *Biochem. J.* **89**, 379–380.

Gros, C. and Labouesse, B. (1969). *Eur. J. Biochem.* **7**, 463–470.

Gross, D. (1961). *J. Chromatogr.* **5**, 194–206.

Gross, E. and Witkop, B. (1961). *J. Am. chem. Soc.* **83**, 1510–1511.

Hamilton, P. B. (1963). *Anal. Chem.* **35**, 2055–2064.

Hartley, B. S. (1970). *Biochem. J.* **119**, 805–822.

Heilmann, J., Barrollier, J. and Watzke, E. (1957). *Hoppe-Seyler's Z. physiol. Chem.* **309**, 219–220.

Hermodson, M. A., Ericsson, L. H., Titani, K., Neurath, H. and Walsh, K. A. (1972). *Biochemistry* **11**, 4493–1502.

Hirs, C. H. W. (1967). In "Methods in Enzymology" (C. H. W. Hirs, ed.), Vol. XI, pp. 197–199. Academic Press, New York and London.

Hirs, C. H. W., Moore, S. and Stein, W. H. (1960). *J. biol. Chem.* **235**, 633–647.

Hochstrasser, K., Illchmann, K. and Werle, E. (1970). *Hoppe-Seyler's Z. physiol. Chem.* **351**, 1503–1512.

Holcomb, G. N., James, S. A. and Ward, D. H. (1968). *Biochemistry* **7**, 1291–1296.

Horn, M. J. and Jones, D. B. (1945). *J. biol. Chem.* **157**, 153–160.

Houmard, J. and Drapeau, G. R. (1972). *Proc. Natn. Acad. Sci. U.S.A.* **69**, 3506–3509.

Hsieh, W. T. Gundersen L. E. and Vestling, C. S. (1971). *Biochem. biophys. Res. Commun.* **43**, 69–75.

Husain, S. S. and Lowe, G. (1969). *Biochem. J.* **114**, 279–288.

Inglis, A. S., Nicholls, P. W. and Roxburgh, C. M. (1971). *Aust. J. biol. Sci.* **24**, 1247–1250.

Jacobsen, C. J., Léonis, J., Linderstrøm-Lang, K., Ottesen, M., (1957). In "Methods of Biochemical Analysis" (D. Glick, ed.), Vol. IV, p. 171–210. Wiley-Interscience, New York.

Jeppsson, J.-O. and Sjöquist, J. (1967). *Analyt. Biochem.* **18**, 264–269.

Kasper, C. B. (1970). In "Protein Sequence Determination" (S. B. Needleman, ed.), pp. 137–184. Chapman and Hall, London.

Kawauchi, H., Kadooka, K., Tanaka, M. and Tuzimura, K. (1971). *Agric. biol. Chem.* **35**, 1720–1726.

Kelly, J. and Ambler, R. P. (1973). *Biochem. Soc. Trans.* **1**, 164–166.

Koide, T., Tsunasawa, S. and Ikenaka, T. (1972). *J. Biochem., Tokyo*, **71**, 165–167.

Krahn, J. and Stevens, F. C. (1972). *Biochemistry* **11**, 1804–1808.

Laursen, R. A. (1972). In "Methods in Enzymology" (C. H. W. Hirs and S. H. Timasheff, eds.), Vol. XXV, pp. 344–359. Academic Press, New York and London.

Lawrence, G. H. M. (1951). "Taxonomy of Vascular Plants". Macmillan, New York.

Leonis, J., Li, C. H. and Chung, D. (1959). *J. Am. chem. Soc.* **81**, 419–423.

Lequin, R. M. and Niall, H. D. (1972). *Biochim. biophys. Acta* **257**, 76–82.

Light, A., Frater, R., Kimmel, J. R. and Smith, E. L. (1964). *Proc. natn. Acad. Sci. U.S.A.* **52**, 1276–1283.

Linenberg, A. (1970). In "Protein Sequence Determination" (S. B. Needleman, ed.), pp. 124–136. Chapman and Hall, London.

Liu, T.-Y. and Chang, Y. H. (1971). *J. biol. Chem.* **246**, 2842–2848.

Macdowall, M. A. and Smith, E. L. (1965). *J. biol. Chem.* **240**, 281–289.

Marglin, A. (1972). *Int. J. Protein Res.* **4**, 47–55.

Margoliash, E. and Walasek, O. F. (1967). In "Methods in Enzymology" (S. P. Colowick and N. O. Kaplan, eds.), Vol. X, pp. 339–348. Academic Press, New York and London.

Matsubara, H., Sasaki, R. M. and Chain, R. K. (1967). *Proc. natn. Acad. Sci. U.S.A.* **57**, 439–445.

Matsuo, H., Fujimoto, Y. and Tatsuno, T. (1966). *Biochem. biophys. Res. commun.* **22**, 69–74.

Michl, H. (1951). *Mschr. Chem.* **82**, 489–493.

Milne, P. R. and Wells, J. R. E. (1970). *J. biol. Chem.* **245**, 1566–1574.

Mitchel, R. E. J., Chaiken I. M. and Smith E. L. (1970). *J. biol. Chem.* **245**, 3485–3492.

Moore, S. and Stein, W. H. (1951). *J. biol. Chem.* **192**, 663–681.

Moore, S. and Stein, W. H. (1954a). *J. biol. Chem.* **211**, 893–906.

Moore, S. and Stein, W. H. (1954b). *J. biol. Chem.* **211**, 907–913.

Moore, S. and Stein, W. H. (1963). *In* "Methods in Enzymology" (S. P. Colowick and N. O. Kaplan, eds), Vol. VI, pp. 817–831. Academic Press, New York and London.

Morris, H. R. (1973). *In* "New Techniques in Biophysics and Cell Biology" (B. J. Smith and R. H. Pain, eds), pp. 149–182. John Wiley, New York.

Moss, C. W., Lambert, M. A. and Diaz, F. J. (1971). *J. Chromatogr.* **60**, 134–136.

Narita, K. (1958). *Biochim. biophys. Acta* **30**, 352–359.

Narita, K. (1970). *In* "Protein Sequence Determination" (S. B. Needleman, ed.), pp. 25–90. Chapman and Hall, London.

Needleman, S. B. and Margoliash, E. (1966). *J. biol. Chem.* **241**, 853–863.

Noltmann, E. A., Mahowald, T. A. and Kuby, S. A. (1962). *J. biol. Chem.* **237**, 1146–1154.

Odani, S. and Ikenaka, T. (1972). *J. Biochem., Tokyo* **71**, 839–848.

Offord, R. E. (1966). *Nature, Lond.* **211**, 591–593.

Pan, S. C. and Dutcher, J. D. (1956). *Analyt. Chem.* **28**, 836–838.

Piez, K. A. and Morris, L. (1960). *Analyt. Biochem.* **1**, 187–201.

Pisano, J. J. and Bronzert, T. J. (1969). *J. biol. Chem.* **244**, 5597–5607.

Pisano, J. J., VandenHeuvel, W. J. A. and Horning, E. C. (1962). *Biochem. biophys. Res. Comm.* **7**, 72–86.

Pisano, J. J., Bronzert T. J. and Brewer H. B. (1972). *Analyt. Biochem.* **45**, 43–59.

Ramshaw, J. A. M., Thompson, E. W. and Boulter, D. (1970). *Biochem. J.* **119**, 535–539.

Ramshaw, J. A. M., Richardson, D. L., Meatyard, B. T., Brown, R. H., Richardson, M., Thompson, E. W. and Boulter, D. (1972). *New Phytol.* **71**, 773–779.

Ramshaw, J. A. M., Scawen, M. D., Bailey, C. J. and Boulter, D. (1974a). *Biochem. J.* **139**, 583–592.

Ramshaw, J. A. M., Scawen, M. D. and Boulter, D. (1974b). *Biochem. J.* **141**, 835–843.

Richards, F. F., Barnes, W. T., Lovins, R. E., Salomone, R. and Waterfield, M. D. (1969). *Nature, Lond.* **221**, 1241–1244.

Richardson, M. (1974). *Biochem. J.* **137**, 101–112.

Richardson, M., Richardson, D. L., Ramshaw, J. A. M., Thompson, E. W. and Boulter, D. (1971). *J. Biochem. Tokyo* **69**, 811–813.

Samuelsson, G. and Pettersson, B. M. (1971). *Eur. J. Biochem.* **21**, 86–89.

Sanger, F. (1945). *Biochem. J.* **39**, 507–515.

Scawen, M. D., Ramshaw, J. A. M., Brown, R. H. and Boulter, D. (1974). *Eur. J. Biochem.* **44**, 299–303.

Scawen, M. D. and Boulter, D. (1974). *Biochem. J.* **143**, 257–264.

Schmer, G. and Kreil, G. (1969). *Analyt. Biochem.* **29**, 186–192.

Schoellman, N. G. and Shaw, E. (1963). *Biochemistry* **2**, 202–255.

Schultz, J. (1967). *In* "Methods in Enzymology" (C. H. W. Hirs, ed.), Vol. XI, pp. 255–263. Academic Press, New York and London.

Scogin, R., Richardson, M. and Boulter, D. (1972). *Archs Biochem. Biophys.* **150**, 489–492.

Seiler, N. (1970). *In* "Methods of Biochemical Analysis" (D. Glick, ed.), Vol. 8, pp. 259–337. Interscience Publishers.

Seiler, N. and Wiechmann, M. (1966). *Z. Anal. Chem.* **220**, 109–127.

Smith, A. E. and Marcker, K. A. (1970). *Nature, Lond.* **226**, 607–610.

Smithies, O., Gibson, D., Fanning, E. M., Goodfliesh, R. M., Gilman, J. G. and Ballantyne, D. L. (1971). *Biochemistry* **10**, 4912–4921.

Spackman, D. H., Stein, W. H. and Moore, S. (1958). *Anal. Chem.* **30**, 1190–1206.

Spande, T. F., Witkop, B., Degani, Y. and Patchornik, A. (1970). *Adv. Protein Chem.* **24**, 96–260.

Spies, J. R. and Chambers, D. C. (1948). *Anal. Chem.* **20**, 30–39.

Spies, J. R. and Chambers, D. C. (1949). *Anal. Chem.* **21**, 1249–1266.

Stalling, D. L., Gehrke, C. W. and Zumwalt, R. W. (1968). *Biochem. biophys. Res. Commun.* **31**, 616–622.

Stark, G. R. (1972). *In* "Methods in Enzymology" (C. H. W. Hirs and S. N. Timasheff, eds.), Vol. XXV, pp. 103–120. Academic Press, New York and London.

Steers, Jr. E., Craven, G. R., Anfinsen, C. B. and Bethune, J. L. (1965). *J. biol. Chem.* **240**, 2478–2484.

Stevens, F. C., Glazer, A. N. and Smith, E. L. (1967). *J. biol. Chem.* **242**, 2764–2779.

Tang, J. and Hartley, B. S. (1967). *Biochem. J.* **102**, 593–599.

Thompson, E. W., Laycock, M. V., Ramshaw, J. A. M. and Boulter, D. (1970). *Biochem. J.* **117**, 183–192.

Tschesche, H. and Kupfer, S. (1972). *Eur. J. Biochem.* **26**, 33–36.

Udenfriend, S., Stein, S., Böhlen, P. Dairman, W., Leimgruber, W. and Weigele, M. (1972). *Science, N.Y.* **178**, 871–872.

Vanderkerckhove, J. and Van Montagu, M. (1974). *Eur. J. Biochem.* **44**, 279–288.

Van Holde, K. E. (1970). *In* "Protein Sequence Determination" (S. B. Needleman, ed.), pp. 4–24. Chapman and Hall, London.

Van Orden, H. O. and Carpenter, F. H. (1964). *Biochem. biophys. Res. Commun.* **14**, 399–403.

Villeagas, E. and Mertz, E. T. (1971). "Chemical screening methods for maize protein quality at CIMMYT". *Res. Bull.* No. 20, CIMMYT, Mexico.

Wallace, D. G., Brown, R. H. and Boulter, D. (1973). *Phytochemistry* **12**, 2617–2622.

Walter, R., Schlesinger, D. H., Schwartz, I. L. and Capra, J. D. (1971). *Biochem. biophys. Res. Commun.* **44**, 293–298.

Weiner, A. M., Platt, T. and Weber, K. (1972). *J. biol. Chem.* **247**, 3242–3251.

Wells, J. R. E. (1965). *Biochem. J.* **97**, 228–235.

Woods, K. R. and Wang, K. T. (1967). *Biochim. biophys. Acta* **133**, 369–370.

Yamada, S. and Itano, H. A. (1966). *Biochim. biophys. Acta* **130**, 538–540.

Zuber, H. and Matile, Ph. (1968). *Z. Naturforschung* **23**, 663–665.

Zumwalt, R. W., Kuo, K. and Gehrke, C. W. (1971). *J. Chromatog.* **55**, 267–280.

CHAPTER 2

Immunochemical Investigations of Plant Proteins

JEAN DAUSSANT

Physiologie des Organes Végétaux, C.N.R.S., 92-Bellevue, France

I. Introduction

The specificity of immunochemical reactions was already used by plant taxonomists at the beginning of the century (Magnus, 1908; Wells and Osborne, 1913). However, only in the past decade have they been applied to plant physiology and plant protein biochemistry, mainly due to the methods introduced by Oudin (1946), Ouchterlony (1949) and Grabar and Williams (1953).

This chapter does not pretend to be an exhaustive review on the results obtained by these methods. Its main concern is to deal with two aspects of the application of immunochemical methods in plant protein research. Firstly, descriptions of the methods including particular modifications and remarks of more general interest are reported; difficulties due to plant material are mentioned. Secondly, the usefulness of these methods in various fields of plant protein research are outlined and illustrated by examples.

II. IMMUNOCHEMICAL METHODS

A. EXTRACTION

A number of immunochemical methods imply the presence of antigen in solutions. Proteins which cannot be extracted into buffered solutions for the most part escape direct immunochemical investigation, and this is a limitation in the practicability of immunochemical methods for study of plant proteins. This happens with certain organelle proteins and also with certain seed storage proteins, which are barely soluble in water or saline solutions and which may constitute a high percentage of the seed protein content. Nevertheless, this difficulty is sometimes overcome by using reducing agents or urea for solubilizing these proteins (Benhamou-Glynn *et al.*, 1965; Escribano *et al.*, 1966; Tronier and Ory, 1970). Proteins may be extracted from leaves, roots, stems, flower parts, fruit, galls or seeds. This material often needs to be taken at different stages of physiological evolution such as development, germination, maturation, ageing. One commonly used way of fixing plant material consists of soaking it in liquid nitrogen and freeze drying it.

No general method exists for extracting all the protein; optimal conditions have to be determined for each plant organ, even for each physiological stage in growth. Nonetheless, certain precautions can be used for preventing difficulties encountered in extraction of plant proteins. The suspension obtained by grinding the plant material in a buffer brings together various molecular types which can interact with each other. This may result in insolubilization of proteins or modification of their physicochemical properties (Grabar and Daussant, 1968; Young, 1965). Fats, organic acids, phenols, quinones, pectins and various pigments may interfere with protein extractions. In order to eliminate fats, the freeze dried material or the seed flour can be treated with organic solvents at low temperature and dried before being homogenized in buffer solution. The considerable amount of organic acid in certain tissues requires checking in order to make sure that the pH of the extraction buffer remains constant. Polyvinylpyrrolidone or polyethyleneglycol are often used to complex with polyphenols; this prevents them from binding with the proteins (Badran and Jones, 1965; Loomis, 1969; Loomis and Battaille, 1966). Alternatively raising the pH of the extraction buffer reduces the hydrogen bonding between polyphenols and proteins. Quinone formation is more likely at these higher pHs but can be checked by using reducing agents (McCown *et al.*, 1968; Tucker and Fairbrothers, 1970). Inhibitors of polyphenol oxidases are also added to the extraction buffer (Anderson, 1968; Loomis, 1969). Pectins are eliminated by using $CaCl_2$ (Frenkel *et al.*, 1969). The extraction procedure is often performed between 0 and 4 °C in order to prevent enzymic action. Nevertheless, these temperature conditions are inadequate for plant material, e.g. seed tissues, containing cryoprotein (Ghetie and Buzila, 1962).

After elimination of insoluble material by centrifugation, proteins are separated from other constituents by using ammonium sulphate precipitation, dialysis and exclusion chromatography. The protein content is often low and extracts can be concentrated by ammonium sulphate precipitation, pervaporation or freeze drying.

B. PREPARATION OF IMMUNE SERA

Immunization of animals, usually rabbits, with plant proteins remains a rather empirical process. The specificity of the reactions has to be checked with the sera taken before immunization. Non-specific reactions between immune sera and extracts of plant material have been reported (Rohringer and Stahmann, 1958). Examples of immunization schedules for plant proteins are reported in Table I.

Two types of immune sera can be distinguished: there are those which are specific for several proteins contained in extracts of the whole plant, the plant organ, the plant tissues or the plant subcellular particles; and there are those which are specific for one protein only. The usefulness of each of these immune sera will be outlined in the next section. For obtaining immune sera of the first type, the animals are immunized with complex extracts or with the protein fraction of these extracts. The preparation of immune sera of the second type requires a previous purification of the protein used for immunization. The combination of classical methods such as ammonium sulphate precipitation, ion exchange chromatography, exclusion chromatography and preparative electrophoresis have been used in the preparations of serum specific for storage protein in seeds, for the major protein of leaves and for enzymes. For example β-amylases from various seeds have been used for obtaining a serum specific for β-amylases only, in certain seeds extracts; as mentioned in section IIIB, part of the antibodies of anti-wheat β-amylase immune sera also react with the barley enzyme and inversely, part of the antibodies of immune sera specific for barley β-amylase also react with the wheat enzyme. This makes possible the use of heterologous immune serum for detecting β-amylase. The first anti-barley β-amylase and anti-wheat β-amylase immune sera obtained were not strictly monospecific for these enzymes when tested with homologous extracts (i.e. barley extract with anti-barley β-amylase, and wheat extract with anti-wheat β-amylase (Fig. 1)). Fortunately, the antibodies against the impurities of β-amylase from either barley or wheat are species specific and the immune serum used with heterologous extracts becomes monospecific for β-amylase (Fig. 1). Needless to say, monospecific immune sera can sometimes be obtained by absorption of immune sera: a serum specific for α-amylase of barley malt was obtained by absorption of anti-barley malt serum with barley proteins (Daussant, 1966; Grabar and Daussant, 1964a).

An important problem arises in the preparation of monospecific immune

TABLE I

Schedule for immunization of rabbits

Proteins	Injections
Phaseolus cotyledon proteins	Intravenous injections 2–3 times a week over a period of 3–6 weeks. Doses: starting with 1–3 mg, finishing with 6–10 mg. Animals are bled 6–10 days after the last injection.[a]
SOYABEAN 2% solution from whole soyabean meal extracts 2% solution of purified 11S constituent 2% solution of purified 7S constituent	Intraperitoneal injections of the solutions emulsified with equal volume of Freund's complete adjuvant at weekly intervals: 1 ml is injected the first week; 2 ml are injected the second week; 5 ml are injected the third week. After 1 month rest period, the animals are given 5 ml booster injection by the same route.[bc]
PEANUT Proteins from cotyledon extract: 30 mg/ml; purified arachin: 10 mg/ml; conarachin: 10 mg/ml	Subcutaneous injections at 4 locations on the rabbit's back at monthly intervals of the solutions emulsified with equal volume of Freund's complete adjuvant. Animals are bled 2 weeks after the third injection.[d]
BARLEY Freeze dried protein preparations	Seven subcutaneous injections of 1 ml solution between scapulae at weekly intervals. After a 6 months' rest, animals are sensitized by fresh injections and a fortnight later are bled.[e]
BARLEY AND WHEAT Soluble proteins 20 mg/ml	Subcutaneous injections at 4 locations on the rabbit's back with 1 ml of protein solution emulsified with Freund's complete adjuvant at monthly intervals. Animals were bled 2 weeks after the last injection. 3–5 injections are usually performed.
Purified enzymes: β, α amylases: 2–3 mg/ml	Same procedure—2 or 3 injections are performed.[f g]
WHEAT "Insoluble" proteins Gliadin: 20 mg in 1 ml aluminium hydroxide suspension	Four intramuscular injections of 1 ml at 10 day intervals. Animals are bled 10 days after the last injection.[h]
Gluten: 10 mg aqueous solutions	Six intramuscular injections with 1 ml every fourth day. Animals are bled 10 days after the last injection.[i]

Proteins	Injections
Acetic extracts (10 mg/ml) adjusted at pH between 5 and 6. Phosphate extracts containing urea 3 M (10 mg/ml)	Intradermal injections at weekly intervals under the foot pad of the solutions emulsified with an equal volume of Freund's complete adjuvant: (1) one injection of 1 ml; (2) four injections of 1 ml; (3) four injections of 2 ml. After previous tests of the immune sera, eventually three further injections with 1, 1·5 and 2 ml.[j]
WHEAT LEAVES Purified fraction I protein 0·2–0·3 mg/ml	Two subcutaneous injections at four locations on the rabbit's back of 1 ml emulsified with 1 ml Freund's complete adjuvant at monthly intervals. Animals are bled six weeks after the last injection.[k]
RED KIDNEY BEAN LEAVES Crude fraction I protein 5 mg/ml	One intravenous injection with 1 ml. Five further intramuscular injections at weekly intervals with 1 ml of the protein solution emulsified with 1 ml of Freund's complete adjuvant.[l]
CALLUS AND TUMOUR Tobacco tissues	Intramuscular injections of 2–2·5 mg proteins emulsified with Freund's incomplete adjuvant. Five to six injections at 3 day intervals, the final injection being given 7 days after the preceding one. Animals are bled every 7 days.[m]
PREPARATIONS OF BEER CHILL HAZE 10 mg/ml in veronal buffer or in phosphate buffer pH 6·6 +containing 5% polyvinylpyrrolidone	First intramuscular injection of 2 ml of the solution emulsified with an equal volume of Freund's complete adjuvant. Three weeks later, four injections with 1 ml of the solution successively intramuscularly and intravenously at 1 day intervals. Animals are bled 10 days after the last injection.[n]
FERREDOXIN OF SPINACH 5 mg/ml in 0·85% NaCl solution	First injection of 1 ml solution + 1 ml complete Freund's adjuvant subcutaneously in the hind paw. Four further injections at weekly intervals of 2 ml in the ear vein. Serum taken 1 day before the last injection contained antibodies. Two weeks later 2 ml of the solution, then 2 days later 4 ml of the solution were injected in the ear vein. One week later 4 blood samples are withdrawn at weekly intervals. All these sera contained antibodies against ferredoxin.[o]

[a] Kloz, 1971.   [h] Ewart, 1966.
[b] Catsimpoolas et al., 1968b   [i] Elton and Ewart, 1963.
[c] Catsimpoolas and Meyer, 1968a.   [j] Benhamou-Glynn et al., 1965.
[d] Daussant et al., 1969a, b.   [k] Bourcelier and Daussant, 1973.
[e] Hill and Djurtoft, 1964.   [l] Falk and Bogorad, 1969.
[f] Daussant, 1966.   [m] Chadha and Srivastava, 1971.
[g] Daussant and Abbott, 1969; and unpublished.   [n] Grabar and Daussant, 1968.
                    [o] Hiedemann-Van Wyk and Kannangara, 1971.

sera when isoenzymes are studied; if antibodies are required against all proteins bearing a given enzymic activity, preparative procedures separating this
activity into different fractions would not fulfil the aim of the study unless
immune sera against each purified fraction are prepared. Affinity chromatography based on enzymic properties of the enzymes would be particularly
suitable for such preparations. For example, in order to prepare anti-malt
α-amylase immune sera, the enzyme was purified from malt (Daussant, 1966)
with a procedure based on properties of α-amylase which in certain conditions
of solvent and temperature forms insoluble complexes with glycogen (Loyter
and Schramm, 1962).

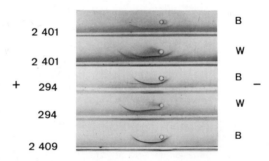

FIG. 1. Immunoelectrophoretic analysis of barley (B) and wheat (W) extracts using different
anti-β-amylase immune sera:
        2401: immune serum prepared with purified wheat β-amylase
        294: immune serum prepared with purified barley β-amylase
        2409: immune serum prepared with crystallized barley β-amylase.

The serum specific for a particular enzyme needs to be tested, first in order
to check that all antigens detected with this serum bear the enzymic activity
studied; secondly in order to see if the antibodies produced react with all or
only with part of the enzymes studied. The first control may succeed by using
amido black staining and enzymic characterization reactions on the immunoelectrophoregram obtained with the extract from which the enzyme was
purified. The second control can be made by measuring the enzymatic
activity remaining after absorption of the enzymes in the extract.

1. Precipitin Reaction

The precipitin test tube reaction is used, mostly by taxonomists, for evaluating the degree of reactivity between an immune serum prepared with proteins extracted from one plant species and the proteins extracted from a second plant species. Serial concentrations of an antigen solution are added to a constant amount of immune serum. After standing in conditions which favour the precipitin reaction (for example, 1 h at 37 °C, then 1–2 days at 4 °C), the precipitates are washed several times with 0·85 M sodium chloride buffered to pH 7–8. Then, their protein content is measured. Results plotted against the corresponding antigen quantities added to the serum provide the precipitin curve for this antigen solution. Curves corresponding to the other antigen solutions allow a comparison of the reactivity of the immune serum with protein extracts from the homologous and from the various heterologous species; areas below the quantitative precipitin curves provide the terms of comparison. Another method consists in measuring the turbidity which occurs when the antigen solution is mixed with the immune serum. The results are expressed in "correspondence values" which are the ratios between the surface of the curves given by the heterologous antigens and the surface of the curve given by the homologous antigens. These curves actually reflect the sum of several reactions: reactions between the different antigens and the corresponding antibodies and, for one antigen, the reactions between the different determinants and their corresponding antibodies (Grabar, 1961). Since the responses of animals to immunization are qualitatively and quantitatively different, the correspondence values obtained are dependent on the serum used. These results have therefore no absolute value. The results obtained with one immune serum are used for constructing series indicating relative serological similarities on the antigen sources compared (Hillebrand and Fairbrothers, 1969).

The classical ring-test technique is also used in taxonomic studies. The immune serum and the antigen solution are overlayered in a glass tube of small diameter. The extent of the reaction is evaluated by the time taken for a precipitate to appear at the interface or by the greatest dilution at which the precipitin reaction is still visible at the interface.

2. Qualitative Methods based on the Specific Precipitation in Gels

(a) The gels. Agar, ionagar and agarose gels at 1 or 1·2% concentration are mostly used for double diffusion according to Ouchterlony (1949) or immuno-electrophoresis according to Grabar and Williams (1953). For the preparation of 1% agarose gel, 1 g of agarose is heated in 50 ml distilled water until it is completely dissolved and 50 ml of warmed 0·05 M veronal buffer pH 8·2 is then added. The liquid is filtered through cheesecloth before being poured onto a microscope slide previously washed with alcohol and on which a drop of hot agarose gel is spread out in order to fix the gel layer on the glass slide. For experiments with proteins insoluble in water, urea can be included in the gels; urea is added to the warmed veronal and a final concentration in the gels as high as 3 M can be obtained. Nevertheless, the gel hardens very slowly and is not as firm as the gel without urea. For a gel including 3 M urea, a 1·4% ionagar concentration is used.

(b) *Double diffusion and immunoelectrophoresis* (*I.E.A.*). Wells and troughs are cut in the gels according to classical designs such as those indicated in Fig. 1 for I.E.A. or in Fig. 10A for double diffusion. When only dilute protein solutions are available, the wells can be filled several times over, allowing short intervals between applications. For electrophoresis, the ends of the gels are covered with filter papers soaked in the veronal buffer, which ensure the connection with the electrode baths. A voltage ranging from 4 to 6 V/cm length of the gel is applied for 1–3 h, the timing depending on the proteins to be studied. A drop of blue bromophenol can be added in a well in order to locate the electrophoretic migration front.

After electrophoresis, the troughs are cut and filled with the immune serum. The time of diffusion usually varies between 2 and 3 days. When urea is added to the gel, these gels have to be washed (three times with veronal buffer, for 20 min each time) after electrophoresis in order to eliminate urea from the gel before the immune serum is applied.

(c) *Positioning of reactants.* Questions concerning antigen identification or serological comparison of proteins from different species may be answered by special positioning of the reactants in the gel for double diffusion or immunoelectrophoretic analysis. Examples of these dispositions for plant protein analysis are given in Fig. 2. In Fig. 2A, a variant of the double diffusion according to Abelev (1960), the disposition allows one to check the absence of one barley malt antigen, the α-amylase, in barley extract. Enzymatic characterization of α-amylase on precipitin bands was performed on this diffusion pattern. The presence of α-amylase in a 100-fold diluted malt extract (well D) was detected by the deviation of the precipitin band in this area. No deviation of the precipitin band occurred in the neighbourhood of well E which received seven applications of barley extract. This result indicates that if small amounts of the antigen exist in the barley extract its concentration was at least 700-fold smaller than in malt extract. Similar experiments suggested that if the malt amylase antigen existed in barley extract, it occurred in quantities at least 2800-fold smaller than in the malt extract (Daussant, 1966).

Using the classical interrupted trough technique (Fig. 2B), it was possible to identify the same antigen in albumin (A) and globulin (G) fractions of barley. This technique was systematically used for comparing antigens occurring at different places on the immunoelectrophoregram of different wheat species (Nimmo and O'Sullivan, 1967). In Fig. 2C, another disposition (Grabar and Daussant, 1968) was used for comparing antigens, detected in barley extract (B) and beer chill haze (H), which had different electrophoretic

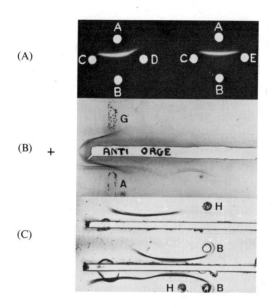

FIG. 2. (A) Double diffusion according to Abelev (1960); α-amylase characterization—A: anti-barley malt immune serum; B: barley malt extract; C: veronal buffer; D: 100-fold diluted barley malt extract; E: 7 successive applications of barley extract. (B) Immuno-electrophoretic analysis. A: barley albumins; G: barley globulins. Immune serum anti-barley proteins (Grabar and Daussant, 1964b). (C) Immunoelectrophoretic analysis. B: barley extract; H: beer chill haze proteins. Anti-chill haze immune serum (Grabar and Daussant, 1968).

mobilities. By placing the wells containing the two solutions on the same side of the trough at a convenient distance from each other, it was possible to observe an identity reaction between the two antigens with the immune serum used.

(d) *Discontinuous immunoelectrophoresis.* Another technique consists of combining electrophoresis in polyacrylamide gel, giving electrophoretic separation with antigen–antibody reactions in agar gel (Catsimpoolas and Meyer, 1968a, b). After electrophoresis in polyacrylamide gel, the slabs of polyacrylamide are completely imbedded in 1% agarose gel, troughs are cut in this gel at a distance of about 7 mm of the slab and the immune serum is placed in the troughs. Cutting the polyacrylamide slab lengthwise makes it possible to pour a thinner layer of agarose gel (see Fig. 3).

Fig. 3. Polyacrylamide gel electrophoresis (1) and immunoelectrophoresis (2, 3) of purified wheat β-amylase with an anti-wheat protein immune serum. A: amido black staining; B: β-amylase characterization.

(e) *Characterization of enzymes on specific precipitin bands.* Using the activity which frequently remains on the enzyme–antibody precipitin bands in gel, it is possible to apply various enzymic characterization procedures to immuno-techniques in gels. Such methods allow the identification of an antigenic enzyme in an antigen mixture or antigenic comparisons between enzymes with close substrate specificity. From a methodological point of view, one may consider, in certain cases, that these characterizations make possible the use of a non-monospecific immune serum as a monospecific immune serum.

The reactions used for enzymic characterization are well documented and those used after double diffusion or I.E.A. have been recently reviewed (Uriel, 1971). Before being used for characterization reactions, the agar plates have to be washed in order to eliminate the proteins which do not enter the specific precipitate. The washing is usually performed by constant stirring with 8·5 g/l sodium chloride, buffered with the veronal solution used for electrophoresis for 3–5 days. The characterization reactions are performed on the washed gels or on the gels previously covered with a sheet of filter paper and dried at room temperature.

3. *Quantitative Methods founded on the Specific Precipitation in Gels*

Immunochemical methods derived from single diffusion, double diffusion or immunoelectrophoresis are particularly suitable for quantification of one antigen in various preparations such as extracts of different plant organs or in extracts of organs at different stages of ontogenesis. In such work, it must be assumed that the antigen in the different solutions presents the same reactivity with the immune serum. These methods involve the use of monospecific immune sera. In certain cases, however, enzymic characterization on the precipitin bands or special technical conditions allow one to use non-specific immune sera. If the protein is available in a pure state, protein amounts in the different solutions can be evaluated; a reference scale is determined with serial dilutions of a known amount of the protein and the results obtained with the experimental solutions under the same conditions are reported on the same scale. If a pure protein preparation is not available, the relative amounts of the antigen in various solutions can be obtained by taking one of these solutions as a reference.

In single diffusion and immunoelectrophoretic techniques, the antigen moves by diffusion or electrophoresis in a gel containing antibodies. A relationship is established between the distance the precipitin band migrates and the antigen concentration. In these experiments, preliminary assays must be made in order to find the best range of antiserum and antigen concentrations for optimal results. The immune serum is added to the gel, kept at 50 °C and carefully mixed before being poured into glass tubes or on glass plates. The final concentration of the immune serum in the gel may range from less than 1% to 20%.

(a) *Single diffusion* (Oudin, 1946). This method was used for quantification of the 11S component and trypsin inhibitor of soyabean (Catsimpoolas and Leuthner, 1969; Catsimpoolas and Meyer, 1968b; Catsimpoolas et al., 1969). A series of glass tubes (diameter 2 mm, length 100 mm) are first coated with 0·1% agar gel, dried at 70 °C and half filled with 0·6% ionagar in veronal buffer containing the immune serum. Other gels containing various quantities of the antigen instead of the immune serum are placed above the first gel. The migration distance of the antigen–antibody precipitin band is measured after 20 h diffusion at 20 °C. There is a straight line relationship between the log of the soyabean inhibitor concentration and the migration distance.

(b) *Radial diffusion* (Mancini et al., 1965; Ryan, 1967; Fig. 4B). An agar immune serum gel is poured on a glass plate and holes are punched in the gel. Different concentrations of the antigen are filled into the holes. The diameter of the precipitin zones is measured after diffusion for at least one day at 20 °C. There is a straight line relationship between the antigen concentration and the diameter of the precipitin zone (Fig. 4C).

(c) *Double diffusion* (Ouchterlony, 1949). This technique used with agar gels similar to those prepared for qualitative double diffusion or immunoelectrophoresis allows semi-quantitative evaluations (Daussant et al., 1969b; Neucere and Ory, 1970). A central hole surrounded at equal distances by six other holes are punched into the gel. The central well is filled with the immune serum, the surrounding wells with serial dilutions of the antigen. After a given time of diffusion, a certain amount of the antigen is determined with the last dilutions for which the precipitin band is visible.

(d) *Quantitative immunoelectrophoresis* (Laurell, 1966, Fig. 4A). Holes of 3 mm in diameter are punched into the agarose immune serum gel prepared in veronal buffer 0·25 M, pH 8·2; they are then filled with the samples. Electrophoresis (at 2·5–7 V/cm) is carried out at 20 °C or at 4 °C for 5–15 h. There is a straight line relationship between the peak height and the amount of protein precipitated (Fig. 4C).

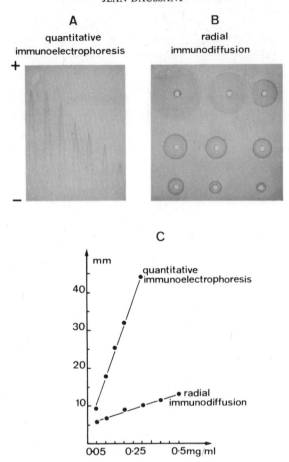

Fig. 4. Two immunochemical techniques in gel for quantitation of one protein in a protein
mixture using a monospecific immune serum. Protein mixture: crude extract of sunflower
protein. Monospecific immune serum: rabbit serum immune specific for one globulin
purified from sunflower. For both techniques, the same gel was used: 1·2% agarose gel in
veronal buffer (0·025 M pH 8·6 containing 1% anti-sunflower globulin immune serum (see
text p. 41); equal volumes of solution with different protein concentrations were deposited
in the wells. (A) From the left to the right, the wells were filled with solutions at 0·5; 0·4;
0·3; 0·2; 0·15; 0·1; 0·05 mg/ml. Electrophoresis: 3 V/cm for 5 h at 20 °C. (B) Starting from
the top of the figure, and for each line from the left to the right, the wells were filled with
solutions at 15; 5; 1·66; 0·5; 0·4; 0·3; 0·2; 0·1; and 0·05 mg/ml. Diffusion: 3 days at 20 °C.
(C) Relation between peak heights (for A), or diameter size (for B) and protein concentration deposited in the corresponding wells.

(e) *Quantitative immunoelectrophoresis with bidimensional electrophoresis*
(Ressler, 1960, Fig. 5). Proteins are first separated by agar electrophoresis.
The gel in which the electrophoresis was performed is imbedded in another
agar gel containing the immune serum; the electrophoresis is performed in a

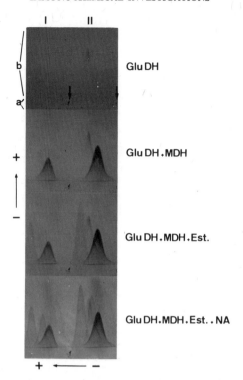

FIG. 5. Quantitative immunoelectrophoresis with bi-dimensional electrophoresis followed by four successive characterizations. Top of the figure—(a) 1·2% agarose gel in 0·025 M veronal buffer pH 8·6. Protein extract of tulip petals were deposited in the two wells indicated by the arrows. For the pattern below I the well was filled twice with the extract. For the pattern below II the well was filled fourfold with the extract. A first electrophoresis was then performed for 1 h at 4 °C with 6 V/cm. (b) 1·2% agarose gel in veronal buffer 0·025 M pH 8·6 containing 15% immune serum anti-tulip proteins in which the gel (a) has been imbedded for a second electrophoresis. The second electrophoresis was performed for 6 h at 4 °C with 2·5 V/cm. After 1 h washing of the gels a first characterization (glutamate dehydrogenase) was performed and a photo recorded. Other parts of the figures: Three other characterizations were successively performed on these gels, a photo was recorded after each additional characterization. GluDH: glutamate dehydrogenase characterization; MDH: malate dehydrogenase characterization; Est: esterase characterization; NA: amido black staining.

direction at right angles to the direction of the first electrophoresis. For each pattern several peaks are obtained, the surface of each peak being proportional to the quantity of corresponding antigen.

(f) *Remarks.*

1. Preliminary quantitative methods in gels consist of determining for each immune serum used the smallest concentration in the gel for which the precipitin band is still visible; this gives maximum sensitivity. The determina-

tion of the smallest concentration of the serum in the gel presents other advantages; (a) the consumption of the immune serum will be smaller and (b) if the serum is not monospecific but if the precipitin line corresponding to the protein studied is the most important, the use of the diluted immune serum results in the occurrence of the principal precipitin band only. That was the case for the immune serum used in the experiments reported in Fig. 4.

2. Radial immunodiffusion was shown to be more sensitive than single immunodiffusion, using trypsin inhibitor as the test protein (Catsimpoolas and Leuthner, 1969). Results obtained with radial immunodiffusion and with mono-dimensional quantitative immunoelectrophoresis using a sunflower protein are compared in Fig. 4C. Experiments have been conducted with the same immune serum concentrations as those which corresponded to the optima for each of these techniques. From Fig. 4C, it is apparent that the quantitative immunoelectrophoresis is more sensitive than the radial immunodiffusion. It is worth noticing that the results were obtained more quickly with quantitative immunoelectrophoresis (5 h) than with radial immunodiffusion (3 days). With the 5 h period of electrophoresis, the relationship between peak length and protein concentration is linear up to 3 mg/ml; a longer time of electrophoresis would be required for obtaining linearity up to 5 mg/ml.

3. Enzymic characterization can be performed on radial immunodiffusion and quantitative immunoelectrophoresis patterns. It is necessary to wash the gels beforehand, in order to eliminate the proteins which do not participate in the specific precipitates. A solution of 8·5 g sodium chloride per litre buffered with the veronal to pH 8–8·6 is generally used. However, in quantitative immunoelectrophoresis, the relatively long duration of electrophoresis already removes a number of proteins from the gel, and the washing can therefore often be limited to one or a few hours only. Here lies one advantage of quantitative immunoelectrophoresis over the radial immunodiffusion; the first technique is more suitable for studying unstable enzymes which rapidly lose their activity than radial immunodiffusion which takes longer.

As for qualitative immunochemical methods, enzymic characterization permits the use of organo-specific immune sera as monospecific immune sera. Several enzymes can be characterized on the same gel; for example, in Fig. 5, three enzyme reactions and amido black staining have been successively performed on the same pattern of a bidimensional immunoelectrophoresis. Radial immunodiffusion as well as unidimensional quantitative immunoelectrophoresis can be used for the same purpose (Daussant et al., 1971; Lanzerotti and Gullino, 1972).

In the case of isoenzyme investigations, however, quantitative immunoelectrophoresis with bidimensional electrophoresis remains the only suitable technique of investigation. In this case, several antigens have the same enzymic activity; with radial immunodiffusion and monodimensional quantitative immunoelectrophoresis, the migration distances of the precipitin lines bearing the enzymatic activity depend on the amount of the different antigens and there is generally no means of identifying the different isoenzymes. Quantitative immunoelectrophoresis with bidimensional electrophoresis, however, provides a means, with the first electrophoresis in agarose gel, for identifying the different isoenzymes from their electrophoretic mobilities.

4. Absorption Methods

The technique of immune serum absorption is used for eliminating from an immune serum only those antibodies that react with certain antigens. Conversely the absorption techniques can be used for eliminating from a solution only those antigens that correspond to the antibodies present in a given immune serum. Both types of absorption are important in immunochemical investigation on plant proteins.

(a) *Antibody absorption.* Absorption can be obtained by adding small amounts of the antigen preparation to the immune serum (0·2 ml preparation to 1 ml serum). The mixture, carefully shaken, is kept in conditions which favour the specific precipitation (1 h at 37 °C, then one or two days at 4 °C). The supernatant is tested by the ring-test technique; if the result is positive, the absorption has to be repeated.

In immunoelectrophoretic or double diffusion experiments, absorption can be performed more quickly and without diluting the immune serum (Hill and Djurtoft, 1964). After electrophoresis of the protein extract, the antigen preparation is deposited in the trough used for the immune serum. When the antigen solution is sucked in by the gel, the immune serum is poured into this trough. The antibodies absorbed form a precipitate with the absorbing antigen all along the trough and are prevented from reacting with the proteins first submitted to electrophoresis.

The efficiency of the absorption is tested by double diffusion using the absorbed immune sera and several concentrations of the absorbing antigen. When the first technique is used, an aliquot of the immune serum is treated like the absorbed portion but using buffer instead of antigen preparation. This technique results in the dilution of the immune serum. Nevertheless, the use of glutaraldehyde for polymerizing the proteins into soluble active antigens avoids any dilution (Avrameas and Ternynck, 1969). The second technique does not imply any dilution.

(b) *Enzyme absorption.* The extract containing enzymes is diluted in order to provide a solution with the right range of activity. Increasing amounts of the immune sera are added to a constant amount of the diluted extract and all samples are adjusted to the same volume. After standing in conditions which favour the specific precipitation (1 h at 37 °C and then 2 days at 4 °C), the samples are centrifuged and the enzymic activity of the supernatant is measured. Parallel experiments are performed with non-specific immune sera in order to take into account the non-specific action of the serum on the enzymic activity. Results of such absorption for wheat β-amylase are shown in Fig. 6. While it is clear that the sera increase the activity non-specifically, the anti-β-amylase is specific for the bulk of the activity extracted from wheat.

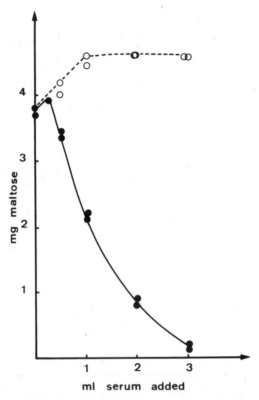

FIG. 6. Absorption of β-amylase activity from wheat extract with different immune sera. ●—●—● Anti-β-amylase immune serum; ○--○--○ immune serum not specific for β-amylase. For each absorption increasing amounts of immune sera were added to 2 ml extract and samples were adjusted to 5 ml (see text). The activity is expressed in mg/maltose/ml sample supernatant/30 min incubation.

Another technique recently developed combines in the same gel the immunochemical absorption and the evaluation of the activity of the remaining enzymes (Daussant and Skakoun, 1974). An example of this technique concerns α-amylase; its activity can be evaluated according to a diffusion method in agar gel containing soluble starch pre-incubated with β-amylase (Möttönen, 1970). 3% Starch in 0·2 M acetate buffer pH 5·7 is incubated at 37 °C for 16 h with 10 mg of a commercial β-amylase preparation per g starch. With this solution, a 1·2% agarose–1·5% incubated starch gel is prepared in the same buffer and poured on glass slides. Wells of 3 mm diameter are cut in the gel and filled with different concentrations of the amylase solution. After 24 h incubation at 20 °C, the gels are stained with an iodine solution. The diameter of the white circles occurring on a red background is proportional to the log concentration of α-amylase. For absorption experiments, the anti-α-amylase immune serum is first poured into the holes. When the serum is sucked in by the gel, the holes are filled with the antigen solutions. After diffusion and staining, the remaining α-amylase activity detected indicates the action of enzyme which has not been precipitated by the antibodies previously deposited in the wells (Fig. 12A, absorption). Parallel experiments are performed with sera other than the anti-α-amylase one in order to control the specific action of the anti-α-amylase immune serum (Fig. 12A, control). This simplified procedure for enzyme absorption and subsequent measurement of activity could also be applied for other enzymes under certain conditions; the pH of antigen–antibodies reaction should be of the same range as the pH of enzyme activity and substrate and co-enzyme have to be soluble in the gel.

5. *Micro-complement Fixation*

In this method (Levine, 1967; Wasserman and Levine, 1961), the antigen–antibody reaction is reflected by the amount of complement fixed on the antigen–antibody complex. This method, details of which were given by Levine (1967), consists of adding increasing amounts of antigen to samples containing constant amounts of decomplemented immune serum and of guinea pig complement. Blanks with serum only, antigen only, complement only are run in parallel. After 24 h at 4 °C, the amount of complement not fixed in the first reaction is measured. Equal amounts of sensitized sheep red cells are added to each sample at 37 °C. The action of the complement on the sensitized red cells results in lysis of the red cells and the amount of haemoglobin released is measured spectrophotometrically at 413 nm. This gives the quantity of complement remaining in each sample. The curve of complement fixed at different dilutions of the antigen for constant amounts of antibodies and complement is bell shaped. Maximum amount of complement is fixed for a certain ratio between antigen and antibodies, the equivalent point. For excess of either antigen or antibodies, less complement is fixed.

This method needs monospecific immune sera when applied to the study of one protein in a mixture. It is particularly suitable for detecting small differences in protein structures. For example, β-amylase in barley and barley malt extracts were compared both by double diffusion and micro-complement fixation using the same anti-barley β-amylase immune serum (Fig. 7). The malt extract was diluted to yield a β-amylase activity equal to the activity of the barley extract. These extracts and the undiluted immune serum were used for double diffusion. Malt and barley extracts diluted 40-fold were used as starting solutions for the micro-complement fixation method with 900-fold diluted immune serum. No antigenic differences were detected by

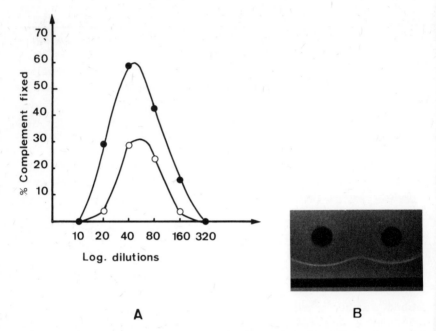

A   B

FIG. 7. Comparison by micro-complement fixation and double diffusion between β-amylase from ungerminated and germinated barley seeds. The same monospecific immune serum anti-ungerminated barley β-amylase was used in both experiments. Extracts were dialysed and diluted in order to provide the same β-amylase activity (see text). (A) Micro-complement fixation: ●—●—● with extract of ungerminated seeds; ○—○—○ with extract of germinated seeds. (B) Double diffusion: left well—extract of ungerminated seeds; right well—extract of germinated seeds.

double diffusion; however, a clear cut difference in the reactivity of the two enzymes appeared with the micro-complement fixation technique. The greater sensitivity of micro-complement fixation on double diffusion for detecting differences between antigens was reported in studies concerning phytochrome conformational changes (Hopkins and Butler, 1970) and phytochrome from different plants (Pratt, 1973).

III. Application of Immunochemical Methods in Plant Protein Studies

The object of this section is to summarize the contributions immunochemical methods have made in research on plant proteins. A few typical examples are cited to illustrate different aspects of these applications. The following aspects are being dealt with: identification and detection; taxonomic studies; localization; and physiological studies.

A. IDENTIFICATION AND DETECTION

Detection and quantification of proteins is one of the first steps in taxonomic, physiological and biochemical studies. Often, the aim is the identification of proteins common or specific to different plants, organs, tissues or to a plant at different stages of development. Plant sources of protein are of interest for basic as well as for applied research.

1. *Nomenclature and Identification of Functions of Antigens*

Immunochemical methods based on precipitation in the gel and immuno-electrophoretic analysis have been used for detecting and determining the number of proteins in plant extracts. Recently, it became necessary to establish a system of protein nomenclature for those seed proteins which are commonly investigated by laboratories and used by industries. A commission of the European Brewery Convention (1967) stipulated the characteristics required for a system of barley protein nomenclature; the constituents were to be identified according to the greatest number of criteria based on independent properties. Moreover, the nomenclature had to be sufficiently flexible to identify a constituent even while it was undergoing modifications in the course of natural or industrial processes such as germination or beer making. Immunoelectrophoretic analysis seemed to fulfil these tasks, since constituents are always characterized by mobility and antigenicity, often by enzymic activity. A standard horse immune serum specific for barley and barley malt proteins was specified by the European Brewery Convention (1971). Based on immunochemical investigations on soyabean proteins (Catsimpoolas *et al.*, 1968a, b; 1969; Catsimpoolas and Ekenstam, 1969; Catsimpoolas and Meyer, 1968a, b), the Soybean Protein Nomenclature Committee sponsored by the Oil Seed Division of the American Association of Cereal Chemists proposed a more elaborate reference system (Catsimpoolas, 1969b). Monospecific immune sera for certain soyabean proteins were provided as well as a serum specific for water soluble proteins. It became apparent that the classical definition of proteins according to their solubility in different solvents was insufficient to classify a number of cereal proteins. For example, albumin and globulin fractions of barley contain identical antigens, in addition to constituents specific for each fraction (Grabar *et al.*, 1962; Grabar and Daussant, 1964b). A major constituent of gliadin was identified in the glutenin of wheat (Beckwith and Heiner, 1966; Benhamou-Glynn *et al.*, 1965; Escribano *et al.*, 1966) and this suggested that the main glutenin fraction has structural features common to some gliadin constituents

(Ewart, 1966). Based on immunodiffusion experiments, it was even suggested
that glutenin consists of an association of several molecular types including
gliadin, globulin and albumin constituents (Konarev and Gavriluk, 1970). The
barley globulins defined according to their sedimentation coefficients and
named, α, β, ρ, γ, δ (Quensel, 1942) each contain several antigenic constituents, and the presence of the same antigen in several fractions suggested that
these constituents could exist with varying degrees of aggregation (Nummi,
1963a, b).

Immunochemical studies have shown that soyabean lipoxidase is immuno-
chemically distinct from haemagglutinin, trypsin inhibitor and storage pro-
teins glycinin and α-, β-, γ-, δ-conglycinin (Catsimpoolas, 1969a). Cryo-
precipitability, a property of proteins in numerous seeds (Ghetie and Buzila,
1962) has been attributed to the major storage proteins in groundnuts
(Daussant et al., 1969b; Neucere, 1969) and in pea (Buzila, 1966). Physico-
chemical differences concerning molecular size and absorption properties
furthermore suggested that vicillin (pea storage protein) and the cryo-protein
are different forms of the same antigen (Buzila, 1967). Several enzymes have
been identified by their antigenic activities. For example, three amylolytic
enzymes extracted from germinated barley, namely α- and β-amylases and
x-enzyme, are distinct antigens (Niku-Paavola and Nummi, 1971). The
characterization of two enzymic activities on the same precipitin band (Fig. 8)
has provided immunochemical confirmation (Daussant et al., 1971) that
indole acetic acid oxidases are peroxidases (Siegel and Galston, 1967).

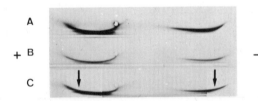

FIG. 8. Immunoelectrophoretic analysis of crown gall tissue extract using an anti-crown
gall immune serum. Peroxidase characterization: (A) and extreme parts of immunoelectrophoregram of (C) arrowed. Auxin oxidase characterization: (B) and central part of immunoelectrophoregram of (C) (Daussant et al., 1971).

The distinction between 1,5-diphosphate carboxylase (fraction I protein)
and protochlorophyllide holochrome has been based on immunochemical
methods. The first of these proteins is a major leaf protein (Kawashima and
Wildman, 1970) while the second is converted under the influence of white
light into chlorophyllide holochrome. The two proteins have different
absorption spectra with maximum absorption respectively at 637 nm and

670 nm (Falk and Bogorad, 1969). A normal serum and an anti-ribulose diphosphate carboxylase immune serum were used in absorption experiments on crude preparations of protochlorophyllide holochrome. Evidence was obtained that these proteins are distinct (Falk and Bogorad, 1969). The specific immune serum precipitates the 1,5-ribulose diphosphate carboxylase from the crude preparation of protochlorophyllide holochrome but does not interfere either qualitatively or quantitatively with the *in vitro* photoconversion of this latter protein.

2. Origin of Proteins

The origin of proteins is particularly important in cancer studies. The transformation of the normal plant cell into tumour cell induced by the presence of *Agrobacterium tumefaciens* in a plant wound raises the question of whether it is possible to introduce a self-replicating foreign genome into the host genome. Decisive evidence showing that bacteria-free tumour tissues contain an integrated bacterial genome has been provided by immunochemical methods, which have shown that bacteria-specific proteins occur in these tissues (Chadha and Srivastava, 1971; Shilperoort *et al.*, 1969). Serology has similarly been used in studies of plant diseases caused by viruses (Vuittenez, 1966) or fungi (Tempel, 1959).

3. Molecular Heterogeneity of Enzymes

Enzymes are heterogenous if constituents from the same origin catalysing the same reaction differ in some of their physico-chemical properties. According to the recommendations of the IUPAC-IUB Commission on Biochemical Nomenclature (1971), the term "multiple forms" should be used for enzymes occurring naturally in a single species. The term "isoenzyme" should be applied when this heterogeneity is due to genetically determined differences in primary structure. The antigenic specificity of proteins revealed by immunochemical methods provides comparative data on the structure of enzymic constituents present in a protein mixture. These methods can therefore offer suitable criteria for determining enzymic heterogeneity.

Using these methods, several distinct molecular species of peroxidase have been distinguished in crown gall tumour extracts of *Datura stramonium* (Daussant *et al.*, 1971; Roussaux *et al.*, 1971) and in leaf extracts of wheat, rye, maize and oats (Alexandrescu and Hagima, 1973). Antigenic heterogeneity was also detected for acid phosphatase extracted from the leaves of these cereals (Alexandrescu and Povov, 1973). A serum prepared with purified α-amylase from barley malt showed the presence of two isoenzymes in an extract of 7-day-germinated barley seeds (Fig. 9). The spur occurring at the intersection of the precipitin bands shows up antigenic differences between these two constituents, which indicates that they correspond to different molecular species. The semi-identity reaction, nevertheless, reflects antigenic similarities between these constituents. The structural relationship may be discussed in terms of hybridization between different polypeptide chains but needs further investigation. The results concerning the absorption of the extract of 7-day-germinated seeds with the immune serum indicates that this

FIG. 9. α-amylases isoenzymes in germinated barley. Immunoelectrophoretic analysis of an extract of 7 days germinated barley seeds. Immune serum anti-α-amylase of germinated barley. (A) α-amylase characterization; (B) amido black staining. Arrows indicate the spur.

immune serum reacts with the bulk of the enzyme in this extract (J. Daussant, unpublished data). A similar heterogeneity due to the presence of several molecular species for α-amylase has been found in germinated seeds of wheat (Daussant and Corvazier, 1970), oat and maize (Alexandrescu and Mihailescu, 1973). The polymorphism of barley β-amylase is well known; the enzyme extracted with water or saline solution is called free β-amylase while the remaining enzyme, extractable using reducing agents, is named bound β-amylase. The free β-amylase is heterogeneous in its molecular size (Nummi, 1967). Some of these constituents are inter-convertible by reduction or by oxidation. A further structural relationship between these different constituents and the bound β-amylase was provided by using several immune sera; antigenic identity reactions occurred between all these constituents. These results suggest that the enzyme exists in several different states of aggregation either within itself or with other constituents. These results do not exclude the possibility that several molecular species of the enzyme may be involved in these aggregates; however, with the immune sera used and with the methods of precipitation in gel, these two possibilities could not be distinguished from each other (Daussant et al., 1966; Nummi, 1967).

Molecular heterogeneity exists in proteins other than enzymes; soyabean trypsin inhibitor is poly-disperse and immunochemical investigations on this inhibitor are expected to indicate considerable heterogeneity (Catsimpoolas et al., 1969). The immunochemical identity between the different forms of soyabean haemagglutinins has been discussed in terms of common antigenic sites situated on the protein or on the carbohydrate moieties (Catsimpoolas and Meyer, 1969). Immunochemical comparison between large and small phytochromes purified from oats indicated that the small phytochrome is not a product of disaggregation or hydrolysis of the large phytochrome into identical subunits (Cundiff and Pratt, 1973).

IMMUNOCHEMICAL INVESTIGATIONS

4. *Immunochemical Identification in Applied Research*

The specificity of immunochemical identification provides a method of quality control for plant products. Thus, with methods based on the specific precipitation in gel, it is possible to detect as little as 5% of barley flour in wheat flour (Liuzzi and Angeletti, 1969) and as little as 5% contamination of *Triticum vulgare* flour in *Triticum durum* flour (Piazzi and Cantagalli, 1969; Piazzi *et al.*, 1972). Assays for detecting soya protein manufactured food products are difficult to carry out because of the effect of heating on the proteins (Krüger and Grossklaus, 1970). For this purpose, it is probably better to select and purify those seed proteins which are the most resistant to heat and to prepare the corresponding immune sera. Proteins vary in their capacity for retaining both their antigenic properties and their solubility when exposed to high temperatures. Heating of groundnuts results in the loss of the majority of antigens detected (Neucere *et al.*, 1969) but the main storage protein, α-arachin, resists 1 h exposure at 155 °C. This resistance may be due to the location of this protein in particles surrounded by a membrane, the aleurone grains (Neucere, 1972). Nevertheless, this protein is particularly resistant to heat; when the flour, instead of the whole nuts, is exposed for 1 h at 110 °C, the α-arachin is still easily detectable by immuno-electrophoresis (Neucere *et al.*, 1972). Other examples of heat resistance have been noted during studies on barley, barley malt and beer proteins (Djurtoft and Hill, 1966; Donhauser, 1967; Grabar, 1960; Grabar and Daussant, 1964b, 1968, 1971; Grabar *et al.*, 1969; Hill and Djurtoft, 1964; Loisa *et al.*, 1971). Of all the constituents detected in barley with an anti-barley protein immune serum, only a few are still detectable in beer with the same immune serum. Certain proteins of maize and rice are also heat resistant: this property has been used to see whether maize and rice have been employed for beer-making (Schuster and Donhauser, 1967). Some proteins of orange juice also retain part of their original antigenic specificity after pasteurization (Cantagalli *et al.*, 1972).

Studies on thermal denaturation of the antigenicity of purified glycinin showed that the protein still reacts with the immune serum after 1 h exposure at 90 °C (Catsimpoolas *et al.*, 1969). Nevertheless, radial immunodiffusion and micro-complement fixation showed changes in the reactivity of the protein with the immune serum after exposure for 30 min at 70 °C and 80 °C respectively (Catsimpoolas *et al.*, 1971).

The antigenic specificity of proteins can also be used in food technology as one criterion for evaluating the effect of industrial processes on protein denaturation (Neucere and Hensarling, 1973).

B. TAXONOMIC STUDIES

Serology is used in taxonomy for detecting differences or similarities between organisms which cannot be revealed by other methods (Moritz, 1964). Actually, the best comparison at the protein level would be the amino acid sequences of individual proteins from different plants (see Chapter 1). However, this is not yet generally practicable because of the number of samples needing to be investigated. Immunochemical methods, therefore, still provide a valuable tool for this purpose (Kloz, 1971). Serology has been widely used in plant taxonomy and several reviews are available (Fairbrothers, 1968; Hillebrand and Fairbrothers, 1969, 1970; International Phytoserologists, 1968; Vaughan, 1968). We therefore limit this section to some special problems encountered in this type of research.

For systematic studies, only proteins from the same organ taken at the same stage of ontogeny are strictly comparable. Mature seeds which represent well defined physiologically stable material constitute the material most often used in these studies. Experiments with *Phaseolus* show that the influence of age on mature seeds does not significantly affect the immunoelectrophoretic pattern of these seed proteins. The quality of the immune sera used is of great importance; there are immune sera of both high and low specificities. The first type provide low correspondence values and are therefore useful for separating closely related taxa such as species within the same genus. The second type give high correspondence values and permit a comparison of distantly related taxa such as representatives of different families.

Two techniques are mainly applied in taxonomic studies: turbidimetric analysis and the methods of precipitation in gels. Turbidimetric analysis measures the extent of reaction between an antigen mixture and the corresponding serum and, furthermore, the reaction between antigen mixtures from other origins and the same immune serum. From these data, one obtains the relative correspondence of each sample with the reference. As already mentioned, this method does not distinguish between qualitative and quantitative data. This may cause some unexpected results, because the reaction of an immune serum is stronger with the antigens of an heterologous source than with the antigens of the homologous source. Such results could be due to a common antigen, present in greater amounts in the heterologous than in the homologous source. However, use of the absorption technique combined with this method eliminates such anomalies (Moritz, 1964).

The double diffusion and immunoelectrophoretic analyses provide another way of determining the number of individual constituents, detectable with a given immune serum, in extracts from different species. These methods permit a more detailed comparison of specific proteins and provide a considerable insight into taxonomic relationships (Kloz, 1971). Even deeper insight can be expected if some of these constituents are compared quantitatively according to their origin. Quantitative comparison of individual constituents in different species can be obtained by single diffusion using monospecific sera or quantitative immunoelectrophoresis. Thus, comparison of soluble proteins from different strains of blue-green algae has been reported (Lindblom, 1972). Additional data can be expected by antigenic comparison of a given functional protein originating from different species; for example, one enzyme (Alexandrescu and Hagima, 1973; Codd and Schmid, 1972; Daussant, 1968) or even one enzyme subunit (Gray and Kewick, 1973), or a trypsin inhibitor (Gurvsiddaiah *et al.*, 1972) or phytochrome (Rice and Briggs, 1973). Monospecific sera for the protein of one species fulfil the task when used in double diffusion (Fig. 10A). For enzymes which can be characterized in the gel, organo-specific

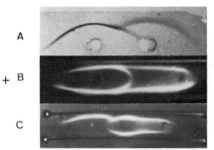

FIG. 10. Antigenic comparison between β-amylases from barley and wheat. (A) Double diffusion of barley extract (left well) and of wheat extract (right well) with a monospecific anti-barley β-amylase immune serum. Amido black staining. (B and C) Immunoelectrophoretic analysis of barley extract (left wells) and of wheat extracts (right wells) with anti-barley immune sera (B) and with anti-wheat immune sera (C). β-amylase characterization—Note the inhibition of barley β-amylase with one of the heterologous immune sera (Nummi *et al.*, 1970).

immune sera may also be used. A characterization reaction is performed on the pattern (Fig. 10B). Thus immunochemical comparison of β-amylase from barley and wheat indicated the presence of three types of antibodies: those reacting with the enzymes from both plants, those reacting with the wheat enzyme only and those reacting with the barley enzyme only (Daussant, 1968; Nummi *et al*, 1970).

The distinction between closely related species or varieties is a difficult task. The crude extracts normally used for immunization may contain some antigens specific for one of these species: nevertheless, the bulk of the antigenic material of these extracts is very probably constituted of antigens common to all of these species. Consequently, one cannot expect to obtain easily an immune serum suitable for distinguishing between these species. These difficulties have been underlined in studies concerning immunochemical comparisons between several wheat species and cultivars (Daussant and Abbott, 1969; Elton and Ewart, 1963; Grabar *et al.*, 1962; Nimmo and O'Sullivan, 1967). They may be overcome if the immune sera are prepared with a reduced number of proteins, chosen from those for which physico-chemical differences can be detected according to origin. A previous electrophoretic or chromatographic comparison between the extracts would probably indicate differences in certain fractions and these fractions should be used for immunization. An example of this approach has been given in studies concerning the differentiation between *Triticum aestivum* and *Triticum durum*. An absorption of the immune serum may be necessary to eliminate antibodies still reacting with protein common to all cultivars (Piazzi and Cantagalli, 1969; Piazzi *et al.*, 1972).

C. LOCALIZATION

A direct method here is the localization *in situ* of antigens with labelled antibodies and microscope or electron microscope analysis (Arrameas, 1970; Nairn, 1962). For example, using fluorescent antibodies specific for soluble wheat proteins, these proteins were observed in high concentration around each starch granule (Barkow, 1973).

The immunochemical analysis of subcellular preparations is another means of studying protein localization. Nevertheless, the direct analysis is restricted to those antigens which can be solubilized. A study of the localization of peanut storage proteins in protein bodies and in cytoplasm involved three techniques: (1) preparation of protein bodies (aleurones) by differential and gradient centrifugation; (2) electron microscopy analysis of the preparations; and (3) immunoelectrophoresis of proteins extracted from the protein bodies. The study confirmed that α-arachin is an aleurin and showed that one antigenic constituent of α-conarachin is also particle bound, another constituent appearing more as a cytoplasmic protein (Daussant *et al.*, 1969b; Neucere and Ory, 1970). Similar techniques were used in studies of barley protein bodies. It was shown that the bound barley β-amylase was associated with these protein bodies and released from them with reducing agents (Tronier and Ory, 1970).

A special immunochemical approach for elucidating the structure of organelles and assigning functions on these substructures has been provided by studies on chloroplast lamellae. These internal chloroplast membranes, densely packed in the grana or loosely organized in the stroma, are asymmetrical and little is known about the distribution of function within them. In immunochemical studies (Menke, 1972; Park and Sane, 1971), antibodies against whole chloroplasts, whole subparticles of chloroplasts (Berzborn *et al.*, 1966; Briantais and Picard, 1972) or purified proteins (Hiedemann-Van Wyk and Kannangara 1971; Kannangara *et al.*, 1970) were used. The antibodies can react with the antigens on the particles when the antigenic sites are accessible and situated on the outer surface of the lamellae. One technique which can be used with the immune sera and with the particle preparations is agglutination. Agglutination becomes visible when the bivalent antibodies react with antigenic sites situated on two different particles. Because of steric hindrance, the antibodies may also react with the antigenic site of one particle only. This probably happens when the antigen stays in a hole of the outer surface. In this case, the antibodies are absorbed on the particles but there is no agglutination. Agglutination becomes visible after addition of an anti-γ-globulin immune serum to the preparation (Menke, 1972). One may also use the antibodies by observing their influence on the photosynthetic activities of isolated chloroplasts. Thus, for example, using serum monospecific for the coupling factor of phosphorylation, for ferredoxin TPN reductase or for ferredoxin, these antigens were found to be close together on the outer surface of the lamellae. However, the molecular structure of these membranes seems to depend on the procedure used for their preparation.

D. PHYSIOLOGICAL STUDIES

1. *Descriptive Aspects*

Comparisons of antigenic material in different organs have shown that there are antigens specific for one organ and others common to several organs or even to the whole plant. In *Vicia faba*, antigens common to seeds, roots, stems and leaves were found, in addition to antigens specific for each of these organs (Ghetie and Buzila, 1963a, b). Analogous results were obtained for groundnut cotyledons and roots (Daussant et al., 1969a). Immunochemical studies conducted on tissue culture of tobacco stem (Boutenko and Volodarsky, 1968) at different stages of differentiation, including callus organogenesis and regenerated plant, fully confirmed these results. An antigenic chymotrypsin inhibitor present in potato tuber was also found under certain conditions in the plant leaflets; nevertheless, there is a semi-identity reaction between these tuber and leaflet antigens (Ryan, 1968).

Studies on developing and germinating beans (Kloz et al., 1966) and wheat (Rainey and Abbott, 1971) have indicated that antigens are also produced at different stages in the development of the plant. This was also found in studies on the embryo axis of maize during germination (Khavkin et al., 1971) and on groundnut storage organs (Daussant et al., 1969a) during germination. The antigenic chymotrypsin inhibitor detected in different organs of potato was shown to be transitory; its presence in tissues coincides with establishment or maintenance of meristematic regions in the plant (Ryan et al., 1968).

Storage proteins have been especially studied during seed germination. The different molecular species of storage proteins vanish as germination progresses but not all at the same rate. Some of these storage proteins, e.g. phaseolin in *Phaseolus*, 11S and 7S constituents in soyabean, α-arachin in peanut, undergo particular modifications during seed germination. By using the precipitation reaction in gels no modification in their antigenic specificity can be detected. Nevertheless, the antigens form a population with their electrophoretic mobilities shifted towards the anode. These modifications could be explained by a progressive deamidation of asparagine and glutamine residues in the proteins; in this case, the deamidation would not modify their antigenic specificity but result in an increase in negative charges by exposure of carboxyl groups, causing the observed electrophoretic shift (Catsimpoolas *et al.*, 1968a; Daussant *et al.*, 1969a; Kloz *et al.*, 1966).

Diseases in plants also seem to be associated with changes in protein patterns. The particular problem of plant tumours where bacterial proteins have been detected in tumour cells has already been mentioned (p. 51). Black root infection as well as sterile cutting of potatoes is accompanied by the appearance of a new antigen having peroxidase activity (Uritani and Stahmann, 1961a, b). Crown gall infection as well as a sterile wound on *Datura stramonium* stem considerably increases the amount of at least two antigens bearing peroxidase and indoleacetic acid oxidase activities (Daussant *et al.*, 1971). Insect or mechanical wounding of the leaves of potato and tomato plants results in a rapid accumulation of proteinase inhibitor. The accumulation of this protein was measured in different parts of the plant using radial immunodiffusion with a corresponding monospecific immune serum (Ryan, 1967). This antigen increases in amount within a few hours in damaged and adjacent leaves, and the response is both light and temperature dependent (Green and Ryan, 1972, 1973). An antigen specific for the leaf of a female plant of *Mercurialis annua* has been detected. The action of cytokinins on the synthesis of this specific protein has been studied (Durand-Rivieres, 1969; Durand and Durand-Rivieres, 1969).

2. De novo *Synthesis and Precursor Activation*

Qualitative or quantitative changes in enzyme activity, which occur during plant growth can be ascribed to *de novo* synthesis of the enzyme, precursor activation or a decrease in the rate of enzyme degradation. Immunochemical methods offer one experimental approach of many that are available (Filner *et al.*, 1969).

An inactive precursor should be detectable with a serum specific for the active enzyme because of its structural similarity to the enzyme (Barett and Thompson, 1965; Lehrer and van Vunakis, 1965); a monospecific anti-enzyme immune serum could, therefore, be used to detect such a precursor. α-Amylase activity, which is absent or very low in mature barley seeds, drastically increases during germination (Kneen, 1944). The lack of reaction between a serum specific for α-amylase from germinated seeds and the proteins of ungerminated seeds was revealed by double diffusion, and by absorption of the serum with mature seed extracts and absorption with the flour itself. These results indicate that no precursor of α-amylase exists in the ungerminated seeds, while strongly suggesting that the enzyme was synthesized during germination (Grabar and Daussant, 1964a).

Direct evidence for *de novo* synthesis can be obtained by combining immunochemical techniques with *in vivo* labelled amino acid incorporation into proteins. An illustration of this approach is in studies on wheat proteins during germination (Daussant and Corvazier, 1970). Seeds were germinated for 7 days in water containing labelled amino acids. The extracts of these seeds and of ungerminated seeds were compared by immunoelectrophoresis using immune sera specific for proteins present in ungerminated seeds, for β-amylase and for α-amylases. Immunoelectrophoretic and radio-immuno-electrophoretic patterns are shown in Fig. 11. The radioactivity detected on the precipitin bands corresponding to the α-amylase isoenzymes proved that at least some of the molecules of both α-amylase isoenzymes have been synthesized. Moreover, as in the case of barley, the anti-α-amylase immune serum did not react with proteins from ungerminated wheat seeds. These results indicate that the increase of α-amylase activity during wheat germination for most of the α-amylase antigens is due to *de novo* synthesis. Since more than 97% of the α-amylase activity in 7-day-germinated wheat is specifically eliminated by the immune serum (Daussant and Renard, 1972) the conclusion is that the bulk of the α-amylase activity found in wheat at this stage of germination results from *de novo* synthesis.

The quantitative application of immunochemical techniques provides a means of determining whether an increase in one enzymic activity corresponds to an increase in the amount of antigenic enzyme present. Such an approach showed that the increase in ribulose diphosphate carboxylase activity in etiolated barley leaves after exposure to light is due to the synthesis of new enzyme molecules (Kleinkopf *et al.*, 1970). This method has also shown that the increase in certain peroxidases following crown gall infection and wounding in *Datura stramonium* was due to an increase in the amount of the

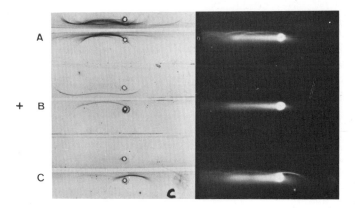

FIG. 11. Immunoelectrophoregrams of proteins extracted from ungerminated wheat seeds (wells above the canals) and from germinated wheat seeds (wells under the canals). (A) Anti-protein extracted from ungerminated wheat seeds immune serum. (B) Anti-β-amylase immune serum. (C) Anti-α-amylase immune serum. Left: protein stained pattern. Right: autoradiogram (Daussant and Corvazier, 1970).

antigenic enzymes (Daussant *et al.*, 1971). It also indicated that an increase in RNase activity following potato tuber damage was not correlated with an increase in RNase protein (Pitt, 1971).

3. Changes in Enzymes due to in vivo Protein Modification or to Substitution in Enzyme Molecules

Differences in physico-chemical properties in one enzyme or in one iso-enzyme system at different developmental stages of a plant organ reflect events which may occur at quite distinct levels. Existing protein may itself undergo modifications, as reported for storage proteins in soyabean, ground-nuts and *Phaseolus* (see p. 57). Alternatively, the genome expression may be differently oriented, a new enzyme population with its own characteristics occurring additionally or replacing the first enzyme population. Examples are β-amylase in germinating wheat and α-amylase in both developing and germinating wheat.

It has been mentioned already that β-amylases exist in mature seeds in several "free" forms, readily extractable with water and saline solutions, and in a "bound" form not extractable with these solutions but with reducing agents. The different forms of the free and bound β-amylases are antigenically identical. When the free forms are also treated with reducing agents, they all behave electrophoretically like the bound β-amylase. During germination, the amount of free β-amylase increases, while the amount of bound β-amylase decreases. Nevertheless, the electrophoretic mobility of the enzyme moves to-wards the cathode. The experiments reported for wheat β-amylase, with *in vivo* incorporation of labelled amino acids into proteins, did not show any radio-activity in the β-amylase (Fig. 11). This result strongly indicates that the bulk of the antigenic β-amylases with the new electrophoretic mobility is due to modification of the existing β-amylase rather than to synthesis of new mole-cules (Daussant and Corvazier, 1970). Similar features have been observed for barley β-amylase during germination (Daussant, 1966; Daussant *et al.*, 1966; Grabar and Daussant, 1964a).

The activity of α-amylase present at an early stage of development in wheat decreases and practically disappears as maturation proceeds (Olered, 1964). This activity again significantly increases during germination (Kneen, 1944). This raises the question of whether the enzymes involved at these two distinct periods in the life of the seeds are structurally identical or different. The α-amylase reactivity of extracts of developing and germinating seeds with a serum specific for α-amylase of germinating seeds were compared by absorp-tion techniques (Fig. 12). The serum absorbed more than 97% of the enzyme activity from the germinated seeds but did not noticeably affect the activity of developing seeds. Thus, the bulk of the activity in germinating seeds is due to antigens distinct from those bearing the bulk of the activity of the developing seeds. Consequently, the first population of α-amylase characteristic of

developing seeds has been replaced by a population of α-amylase corresponding to different molecular species (Daussant and Renard, 1972).

These events probably involve the expression of different loci of the genome.
The results call for further studies concerning the disappearance of the first
α-amylase population. It remains to be seen whether it corresponds to inactivation or degradation of the enzyme. One might expect that the use of
serum specific for these enzymes could give an answer to this question.

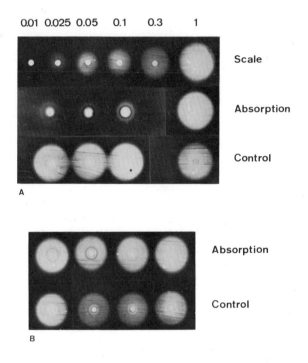

FIG. 12. (A) α-amylase activity in extracts of 7-day-germinating seeds before and after
immunoabsorption. Scale: α-amylase activity of different dilutions of this extract. Aliquots
of this extract were diluted in order to get a scale ranging from this concentration (1), on the
right, to a 100-fold smaller concentration (0·01) on the left. Absorption: α-amylase activity
after absorption with the anti-α-amylase immune serum. From the left to the right, the 4
wells were first filled with, respectively, undiluted, twice diluted, 4-fold diluted immune
serum, and veronal buffer before being filled again with the extract (concentration 1 on the
scale). Control: α-amylase activity after absorption with an immune serum not specific for
α-amylase. The dispositions of the serum and the extract are identical to the disposition
depicted for the preceding part of the figure. (B) α-amylase activity in extracts of developing
seeds picked 19 days after anthesis, before and after immunoabsorption. Absorption:
α-amylase activity after absorption with the anti-α-amylase immune serum. From the left
to the right, the 4 wells were first filled with, respectively, undiluted, twice diluted, 4-fold
diluted immune serum and veronal buffer before to be filled again with the extract. Control:
α-amylase activity after absorption with an immune serum not specific for α-amylase. The
dispositions of the immune serum and the extract are identical to the disposition depicted
for the preceding part of the figure (Daussant and Renard, 1972).

62 JEAN DAUSSANT

IV. CONCLUSION

Studies on plant protein in fundamental as well as in applied research are extending. Highly specific tools are needed in order to detect a constituent in a mixture, to recognize it in varying environmental conditions, to locate it in a cell or in subcellular particles and eventually to characterize modifications in its structure. It has been the aim here to illustrate how the immunochemical methods alone or in combination with other methods can fulfil this task in a wide range of applications.

ACKNOWLEDGEMENTS

The author wishes to thank Professor P. Grabar, Institut Pasteur, Paris and Doctor P. Benda, Collège de France, Paris, for valuable discussions and for reviewing this paper. This work was supported in part by a grant from the Délégation Générale à la Recherche Scientifique et Technique (No. 7270143).

REFERENCES

Abelev, G. I. (1960). Modification of the agar-precipitin method for comparing two antigen-antiserum systems. *Folia Biologica* **6**, 56–59.
Alexandrescu, V. and Hagima, I. (1973). Peroxidase in leaves of some cereals, immunochemical study. *Rev. roum. Biochim.* **10**, 15–21.
Alexandrescu, V. and Mihailescu, F. (1973). Immunochemical investigation on germinated seeds endosperms: α-amylase of some cereals. *Rev. roum. Biochim.* **10**, 89–94.
Alexandrescu, V. and Povov, D. (1973). Immunochemical study of acid phosphatases in leaves of some cereals. *Rev. roum. Biochim.* **10**, 7–13.
Anderson, J. W. 1968. Extraction of enzymes and subcellular organelles from plant tissues. *Phytochemistry* **7**, 1973–1988.
Avrameas, S. (1970). Immunoenzyme techniques. Enzymes as markers for the localization of antigens and antibodies. *Int. Rev. Cytol.* **27**, 349–385.
Avrameas, S. and Ternynck, T. (1969). The cross-linking of proteins with glutaraldehyde and its use for the preparation of immunoadsorbents. *Immunochemistry* **6**, 53–56.
Badran, A. M. and Jones, D. E. (1965). Polyethylene glycol–tannins interaction in extracting enzymes. *Nature Lond.* **206**, 622–624.
Barett, J. T., Thompson, L. D. (1965). Immunochemical studies with chymotrypsinogen A. *Immunology* **8**, 136–143.
Barkow, K. K. (1973). The localization of water soluble proteins in the wheat endosperm as revealed by fluorescent antibody techniques. *Experientia* **29**, 222–231.
Beckwith, A. C. and Heiner, D. C. (1966). An immunological study of wheat gluten proteins and derivatives. *Archs Biochem. Biophys.* **117**, 239–247.
Benhamou-Glynn, N., Escribano, M. J. and Grabar, P. (1965). Étude des protéines du gluten à l'aide de méthodes immunochimiques. *Bull. Soc. Chim. biol.* **47**, 141–156.

Berzborn, R., Menke, W., Trebst, A. and Pistorius, E. (1966). Über die Hemmung photosynthetischer Reaktionen isolierter Chloroplasten durch Chloroplasten Antikörper. *Z. Naturf.* **21**b, 1057–1059.

Bourcelier, C. and Daussant, J. (1973). La fraction I. protéique du limbe de blé: purification et caractérisation immunochimique préliminaire. *C.r. hebd. Séanc. Acad. Sci., Paris* D **276**, 2525–2528.

Boutenko, G. and Volodarsky, A. D. (1968). Analyse immunochimique de la différenciation cellulaire dans les tissus de culture de tabac. *Physiol. veg.* **6**, 299–309.

Briantais, J. M. and Picard, M. (1972). Immunological evidences for a localization of system I on the outside face and of system II on the inside face of the chloroplast lamella. *FEBS Letters* **20**, 100–104.

Buzila, L. (1966). Purificarea Crioproteinei din Mazare sub control Immunochimic. *St. Cerc. Biochim.* **9**, 21–24.

Buzila, L. (1967). Physiochemical and immunochemical correspondences between the cryoprotein and vicilin of pea seeds. *Rev. roum. Biochim.* **4**, 103–108.

Cantagalli, P., Forconi, F., Cagnoni, G. and Pieri, J. (1972). Immunochemical behaviour of the proteins of the Orange. *J. Sci. Fd. Agric.* **23**, 905–910.

Catsimpoolas, N. (1969a). Isolation of Soy-bean lipoxidase by isoelectric focusing. *Archs Biochem. Biophys.* **131**, 185–190.

Catsimpoolas, N. (1969b). A note on the proposal of an immunochemical system of reference and nomenclature for the major soy-bean proteins. *Cereal Chem.* **46**, 369–371.

Catsimpoolas, N., Campbell, T. G. and Meyer, E. W. (1968a). Immunochemical study of changes in reserve proteins of germinating soyabean seeds. *Pl. Physiol., Lancaster* **43**, 799–805.

Catsimpoolas, N., Campbell, T. G. and Meyer, E. W. (1969). Association dissociation phenomena in glucinin. *Archs Biochem. Biophys.* **131**, 577–586.

Catsimpoolas, N. and Ekenstam, C. (1969). Isolation of alpha, beta and gamma Conglycinins. *Archs Biochem. Biophys.* **129**, 490–497.

Catsimpoolas, N., Kenney, J. and Meyer, E. W. (1971). The effect of thermal denaturation on the antigenicity of glycinin. *Biochim. biophys. Acta* **229**, 451–458.

Catsimpoolas, N. and Leuthner, E. (1969). Immunochemical methods for detection and quantitation of Kunitz soybean trypsin inhibitor. *Analyt. Biochem.* **31**, 437–447.

Catsimpoolas, N., Leuthner, E. and Meyer, E. W. (1968b). Studies on the characterization of soybean proteins by immunoelectrophoresis. *Archs Biochem. Biophys.* **127**, 338–345.

Catsimpoolas, N. and Meyer, E. W. (1968a). Immunochemical study of soyabean proteins. *Agric. Fd Chem.* **16**, 128–131.

Catsimpoolas, N. and Meyer, E. W. (1968b). Immunochemical properties of the 11 S component of soybean proteins. *Archs Biochem. Biophys.* **125**, 742–750.

Catsimpoolas, N. and Meyer, E. W. (1969). Isolation of soybean hemagglutinin and demonstration of multiple forms by isoelectric focusing. *Archs Biochem. Biophys.* **132**, 279–285.

Catsimpoolas, N., Rogers, D. A. and Meyer, E. W. (1969). Immunochemical and disc electrophoresis study of soybean trypsin inhibitor. *Cereal Chem.* **46**, 136–144.

Chadha, K. C. and Srivastava, B. I. S. (1971). Evidence for the presence of bacteria specific proteins in sterile Crown Gall tumor tissue. *Pl. Physiol., Lancaster* **48**, 125–129.

Codd, G. A. and Schmid, G. H. (1972). Serological characterization of the glycolate-oxidizing enzymes from tobacco, *Euglena gracilis* and of a yellow mutant of *Chlorella vulgaris*. *Pl. Physiol.*, *Lancaster* **50**, 769–773.

Cundiff, S. C. and Pratt, L. H. (1973). Immunological determination of the relationship between large and small sizes of phytochrome. *Pl. Physiol.*, *Lancaster* **51**, 210–213.

Daussant, J. (1966). Étude des protéines de l'orge et du malt par des méthodes immunochimiques. *Biotechnique* **4, 5, 6**, 1–40.

Daussant, J. (1968). Examples of qualitative immunochemical comparison between isoenzymes in different seed species. *Serol. Mus. Bull.* **40**, 7.

Daussant, J. and Abbott, D. C. (1969). Changes in soluble proteins of wheat during germination. *J. Sci. Fd. Agric.* **20**, 633–637.

Daussant, J. and Corvazier, P. (1970). Biosynthesis and modification of alpha and beta-amylases in germinating wheat seeds. *FEBS Letters* **7**, 191–194.

Daussant, J. and Grabar, P. (1966). Comparaison immunologique des alpha-amylases extraites de céréales. *Annls. Inst. Pasteur, Paris* (suppl.) **110**, 79–83.

Daussant, J., Grabar, P. and Nummi, M. (1966). Beta-amylase. II. Identification des différentes beta-amylases de l'orge et du malt. *Proc. Europ. Brew. Conv.* **10**, 52–69 (Elsevier, Amsterdam).

Daussant, J., Neucere, N. J. and Conkerton, E. J. (1969a). Immunochemical studies on *Arachis hypogaea* proteins with particular reference to reserve proteins. II. Protein modification during germination. *Pl. Physiol.*, *Lancaster* **44**, 480–484.

Daussant, J., Neucere, N. J. and Yatsu, L. Y. (1969b). Immunochemical studies on *Arachis hypogaea* proteins with particular reference to the reserve proteins. I. Characterization, distribution and properties of alpha-arachin and alpha-conarachin. *Pl. Physiol.*, *Lancaster* **44**, 471–479.

Daussant, J. and Renard, M. (1972). Immunochemical comparisons of alpha-amylases in developing and germinating wheat seeds. *FEBS Letters* **22**, 301–304.

Daussant, J., Roussaux, J. and Manigault, P. (1971). Caractérisations immuno-chimiques de deux auxine oxydases extraites de tumeurs végétales. *FEBS Letters* **14**, 245–250

Daussant, J. and Skakoun, A. (1974). Combination of absorption technique and alpha-amylase activity determination in the same gel medium. *J. immunol. Methods* **4**, 127–133.

Djurtoft, R. and Hill, R. J. (1966). Immunoelectrophoretic studies of proteins in barley, malt, beer and beer haze preparations. *Proc. Europ. Brew. Conv.* **10**, 137–146 (Elsevier, Amsterdam).

Donhauser, S. (1967). Immunologische Untersuchungen über die Veränderung der salzöslichen Eiweissfraktionen von der Gärste bis zum Bier unter Variationen des Malzungsverfahrens sowie Untersuchungen über Rochfruchtbier. *Proc. Europ. Brew. Conv.* **11**, 323–335 (Elsevier, Amsterdam).

Durand-Rivieres, R. (1969). Mise en évidence d'une protéine antigénique spécifique dans les méristèmes et les feuilles femelles de *Mercurialis annua* L. *C.r. hebd. Séanc. Acad. Sci., Paris* Ser. D. **268**, 2046–2048.

Durand, B. and Durand-Rivieres, R. (1969). Cytokinines et régulation de la synthèse d'une protéine antigénique spécifique du sexe femelle chez une plante dioïque *Mercurialis annua* L. *C.r. hebd. Séanc. Acad. Sci., Paris* Ser. D. **269**, 1639–1641.

Elton, G. A. H. and Ewart, J. A. D. (1963). Immunological comparison of cereal proteins. *J. Sci. Fd Agric.* **14**, 750–758.

Escribano, M. J., Keilova, H. and Grabar, P. (1966). Étude de la gliadine et de la gluténine après réduction ou oxydation. *Biochim. biophys. Acta* **127**, 94–100.

European Brewery Convention (Barley Protein Committee) 1967. The E.B.C. system of reference for barley proteins *J. Inst. Brew.* **73**, 381–386.

European Brewery Convention (Barley Protein Sub-Committee) (1971). The second E.B.C. reference serum for barley and malt proteins. *News Letter* **2**, 2.

Ewart, J. A. D. (1966). Cereal proteins: Immunological studies. *J. Sci. Fd Agric.* **17**, 279–284.

Fairbrothers, D. E. (1968). Chemosystematics with emphasis on systematic serology. *In* "Modern Methods in Plant Taxonomy" (V. H. Heywood, ed.), pp. 141–174. Academic Press, London and New York.

Falk, R. H. and Bogorad, L. (1969). Immunological distinction between fraction I protein and protochlorophyllide holochrome. *Pl. Physiol.*, *Lancaster* **44**, 1669–1671.

Filner, P. Wray, J. L. and Varner, J. E. (1969). Enzyme induction in higher plants. *Science, N.Y.* **165**, 358–366.

Frenkel, C., Klein, I. and Dilley, D. R. (1969). Methods for the study of ripening and protein synthesis in intact pome fruits. *Phytochemistry* **8**, 945–955.

Ghetie, V. and Buzila, L. (1962). Crioproteinele Vegetale. *St. Cerc. Biochim.* **5**, 65–75.

Ghetie, V. and Buzila, L. (1963a). Immunochemical investigations on the germination of bean. *St. Cerc. Biochim.* **6**, 51–57.

Ghetie, V. and Buzila, L. (1963b). Electrophoretic and immunoelectrophoretic studies of cytoplasmic proteins during the development of *Vicia faba* leaves. *St. Cerc. Biochim.* **6**, 49–59.

Grabar, P. (1960). Étude du "trouble au froid" par des méthodes immunochimiques. *Brasserie* **15**, 121–128.

Grabar, P. (1961). Use of immunoelectrophoretic analysis in the study of specific precipitation. *In* "Immunochemical Approaches to Problems of Microbiology" (M. Heidelberger and O. Plescia, eds), pp. 20–29. Rutgers Univ. Press.

Grabar, P., Benhamou, N. and Daussant, J. (1962). Études électrophorétiques et immunochimiques de quelques protéines de l'orge et du blé. *Archs Biochem. Biophys.* (suppl.) **1**, 187–189.

Grabar, P. and Daussant, J. (1964a). Study of barley and malt amylases by immunochemical methods. *Cereal Chem.* **41**, 523–532.

Grabar, P. and Daussant, J. (1964b). Recherches immunochimiques sur les protéines de l'orge et du malt. *Proc. Europ. Brew. Conv.* **8**, 45–58 (Elsevier, Amsterdam).

Grabar, P. and Daussant, J. (1968). Identification des constituants protéiques du trouble au froid de la bière. *Proc. Europ. Brew. Conv.* **10**, 147–155 (Elsevier, Amsterdam).

Grabar, P. and Daussant, J. (1971). Immunochemical studies on the nitrogenous constituents of beer foam. *J. Inst. Brew.* **77**, 544–546.

Grabar, P., Daussant, J., Enari, T. M. and Nummi, M. (1969). L'origine des troubles au froid. *Proc. Europ. Brew. Conv.* **11**, 379–386 (Elsevier, Amsterdam).

Grabar, P. and Williams, C. A. (1953). Méthode permettant l'étude conjuguée des propriétés électrophorétiques et immunochimiques d'un mélange de protéines. Application au sérum sanguin. *Biochim. biophys. Acta* **10**, 193–194.

Gray, J. C., and Kewick, R. G. O. (1973). A serological investigation of ribulose 1·5-diphosphate carboxylase and its subunits. *Biochem. Soc. Trans.* **I**, 455–458.

Green, T. R. and Ryan, C. A. (1972). Wound induced proteinase inhibitor in plant leaves; a possible defense mechanism against insects. *Science, N.Y.* **175**, 776–778.

Green, T. R. and Ryan, C. A. (1973). Wound induced proteinase inhibitor in tomato leaves. Some effects of light and temperature on the wound response. *Pl. Physiol., Lancaster* **51**, 19–21.

Gurvsiddaiah, S., Kuo, T. and Ryan, C. A. (1972). Immunological comparison of chymotrypsin inhibitor I among several genera of the Solanaceae. *Pl. Physiol., Lancaster* **50**, 627–631.

Hiedemann-Van Wyk, D. and Kannangara, G. C. (1971). Localization of ferredoxin in the thylakoid membrane with immunological methods. *Z. Naturf.* **26** b, 46–50.

Hill, R. J. and Djurtoft, R. (1964). Some immunoelectrophoretic studies on barley proteins. *J. Inst. Brew.* **70**, 416–424.

Hillebrand, G. R. and Fairbrothers, D. E. (1969). A serological investigation on intrageneric relationships in *Viburnum* (Caprifoliaceae). *Bull. Torrey bot. Club* **96**, 556–567.

Hillebrand, G. R. and Fairbrothers, D. E. (1970). Phytoserological systematic survey of the Caprifoliaceae. *Brittonia* **22**, 125–133.

Hopkins, D. W. and Butler, W. L. (1970). Immunochemical and spectroscopic evidence for protein conformational changes in phytochrome transformations. *Pl. Physiol., Lancaster* **45**, 567–570.

International Working Session of Phytoserologists, Kiel (Germany) (1968). *Serol. Mus. Bull.* **40**, 1–8.

I.U.P.A.C.—I.U.B. Commission on Biochemical Nomenclature (CBN) recommendations (1971). *Europ. J. Biochem.* **24**, 1–3.

Kannangara, G. C., Van Wyk, D. and Menke, W. (1970). Immunological evidence for the presence of latent Ca^{++} dependent ATPase and carboxy dismutase on the thylakoid surface. *Z. Naturf.* **25** b, 613–618

Kawashima, N. and Wildman, S. G. (1970). Fraction I protein. *A. Rev. Pl. Physiol.* **21**, 325–358.

Khavkin, E. E., Antipina, A. I. and Misharin, S. I. (1971). [Immunochemical study on proteins of the growing part of maize roots]. *Dokl. Akad. Nauk. SSSR* **199**, 972–975.

Kleinkopf, G. E., Huffaker, R. C. and Matheson, A. (1970). Light induced *de novo* synthesis of ribulose 1–5 diphosphate carboxylase in greening leaves of barley. *Pl. Physiol., Lancaster* **46**, 416–418.

Kloz, J. (1971). Serology of the Leguminosae. *In* "Chemotaxonomy of the Leguminosae" (J. B. Harborne, D. Boulter and B. L. Turner, eds), pp. 309–365. Academic Press, London and New York.

Kloz, J., Turkova, V. and Klozova, E. (1966). Proteins found during maturation and germination of seeds of *Phaseolus vulgaris*. *Biologia Pl.* **8**, 164–173.

Kneen, E. (1944). A comparative study of the development of amylases in germinating cereals. *Cereal Chem.* **21**, 304–314.

Konarev, V. G. and Gavriluk, J. P. (1970). [Immunochemical study on the composition of glutenin]. *Dokladi* **7**, 16–18.

Krüger, H. and Grossklaus, D. (1970). Untersuchungen zum serologischen Nachweis von Soja-protein in erhitzten Fleischerzeugnissen. *Fleischwirtschaft* **11**, 1529–1532.

Lanzerotti, R. H. and Gullino, P. M. (1972). Immunochemical quantitation of enzymes using multispecific antisera. *Analyt. Biochem.* **50**, 344–353.

Laurell, C. B. (1966). Quantitative estimation of proteins by electrophoresis in agarose gel containing antibodies. *Analyt. Biochem.* **15**, 45–52.

Lehrer, H. I. and Van Vunakis, H. (1965). Immunochemical studies on carboxypeptidase A. *Immunochemistry* **2**, 255–262.

Levine, L. (1967). Micro-complement fixation. In "Handbook of Experimental Immunology" (D. Weir, ed.), pp. 707–719. Blackwell, Oxford.

Lindblom, M. (1972). Antigen-antibody crossed electrophoresis of soluble proteins in different strains of Spirulina. Physiologia Pl. 26, 318–320.

Liuzzi, A. and Angeletti, P. U. (1969). Application of immunodiffusion in detecting the presence of barley in wheat flour. J. Sci. Fd Agric. 20, 207–209.

Loisa, M., Nummi, M. and Daussant, J. (1971). Quantitative Erfassung von bestimmten Bier-Eiweisskomponenten nach einem immunologischen Verfahren. Brauwissenschaft 24, 366–368.

Loomis, W. D. (1969). Removal of phenolic compounds during the isolation of plant enzymes. Methods in Enzymology XIII, 555–563.

Loomis, W. D. and Bataille, J. (1966). Plant phenolic compounds and the isolation of plant enzymes. Phytochemistry 5, 423–438.

Loyter, A. and Schramm, M. (1962). The glycogen-amylase complex as a means of obtaining highly purified alpha-amylases. Biochim. biophys. Acta 65, 200–206.

Magnus, W. (1908). Weitere Ergebnisse der Serum Diagnostik für theoretische und angewandte Botanik. Ber. Deut. Bot. Gessell. 26 a, 532–539.

Mancini, G., Carbonara, A. O. and Heremans, J. F. (1965). Immunochemical quantitation of antigens by single radial immunodiffusion. In "Immunochemistry", pp. 235–254. Pergamon Press, Oxford.

McCown, B. H., Beck, G. E. and Hall, T. C. (1968). Plant leaf and stem proteins. I. Extraction and electrophoretic separation of the basic water-soluble fractions. Pl. Physiol., Lancaster 43, 578–582.

Menke, W. (1972). 40 Jahre Versuche zur Aufklarung der molekularen Struktur der Chloroplasten. "Jahrbuch der Max Planck Gesellschaft zur Förderung der Wissenschaft", pp. 132–155.

Moritz, O. (1964). Some special features of serobotanical work. In 'Taxonomic Biochemistry and Serology" (C. A. Leone, ed.), pp. 275–290. Ronald Press, New York.

Möttönen, K. (1970). Comparison of the agar-plate diffusion method for alpha-amylase assay with the colorimetric stationary-stage method. J. Sci. Fd Agric. 21, 261–272.

Nairn, R. C. (1962). "Fluorescent Protein Tracing". E. and S. Livingstone, Edinburgh.

Neucere, N. J. (1969). Isolation of arachin, the major peanut globulin. Analyt. Biochem. 27, 15–24.

Neucere, N. J. (1972). Effect of heat on peanut proteins. I. Solubility properties and immunochemical electrophoretic modifications. J. agric. Fd Chem. 20, 252–255.

Neucere, N. J., Conkerton, E. J. and Booth, A. N. (1972). Effect of heat on peanut proteins. II. Variations in nutritional quality of the meals. J. agric. Fd Chem. 20, 256–259.

Neucere, N. J. and Hensarling, T. (1973). Immunochemical-cytological study of proteins from partially defatted peanuts. J. agric. Fd Chem. 21, 192–195.

Neucere, N. J. and Ory, R. L. (1970). Physicochemical studies on the proteins of the peanut cotyledon and embryonic axis. Pl. Physiol, Lancaster, 45, 616–619.

Neucere, N. J., Ory, R. L. and Carney, W. B. (1969). Effect of roasting on the stability of peanut proteins. J. agric. Fd Chem. 17, 25–28.

Niku-Paavola, M. L. and Nummi, M. (1971). A new amylolytic enzyme. Acta chem. scand. 25, 1492–1493.

Nimmo, C. C. and O'Sullivan, M. T. (1967). Immunochemical comparisons of antigenic proteins of Durum and Hard Spring wheat. Cereal Chem. 44, 584–591.

Nummi, M. (1963a). Fractionation of barley globulins on dextran gel columns. *Acta chem. scand.* **17**, 527–529.

Nummi, M. (1963b). Exclusion chromatography, ultracentrifugation and salting out of barley globulins. *Suomen Kemistilehti* B **36**, 112–114.

Nummi, M. (1967). "Studies on the heterogeneity of soluble barley proteins with particular reference to beta-amylase." Thesis, Helsinki, The State Institute for Technical Research.

Nummi, M., Daussant, J., Niku-Paavola, M. L., Kalsta, H. and Enari, T. M. (1970). Comparative immunological and chromatographic study of some plant beta-amylases. *J. Sci. Fd Agric.* **21**, 258–260.

Olered, R. (1964). Studies on the development of alpha-amylase activity in ripening wheat. *Arkiv. Kemi* **22**, 175–184.

Ouchterlony, O. (1949). Antigen–antibody reactions in gels. *Acta path. microbiol. scand.* **26**, 507–515.

Oudin, J. (1946). Methode d'analyse immunochimique par précipitation spécifique en milieu gélifié. *C.r. hebd. Séanc. Acad. Sci., Paris* **222**, 115–116.

Park, R. B. and Sane, P. V. (1971). Distribution of function and structure in chloroplast lamellae. *A. Rev. Pl. Physiol.* **22**, 395–430.

Piazzi, S. E. and Cantagalli, P. (1969). Immunochemical analysis on soluble proteins of wheat. *Cereal Chem.* **46**, 642–646.

Piazzi, S. E., Riparbelli, G., Sordi, S., Cantagalli, P., Pocchiari, F. and Silano, V. (1972). Immunochemical characterization of specific albumins of bread wheat. *Cereal Chem.* **49**, 72–78.

Pitt, D. (1971). Purification of a ribonuclease from potato tubers and its use as an antigen in the immunochemical assay of this protein following tuber damage. *Planta* **101**, 333–351.

Pratt, L. H. (1973). Comparative immunochemistry of phytochrome. *Pl. Physiol., Lancaster,* **51**, 203–209.

Quensel, O. (1942). "Untersuchungen über die Gerstenglobuline." Inaugural dissertation, Uppsala (Sweden).

Rainey, D. W. and Abbott, D. C. (1971). Changes in the buffer-soluble proteins during maturation of the wheat kernel. *J. Sci. Fd Agric.* **22**, 607–610.

Ressler, N. (1960). Two dimensional electrophoresis of protein antigens with antibody containing buffer. *Clin. chim. Acta* **5**, 795–800.

Rice, H. V. and Briggs, W. R. (1973). Immunochemistry of phytochrome. *Pl. Physiol., Lancaster* **51**, 939–945.

Rohringer, R. and Stahmann, M. A. (1958). Immunochemical studies with tomato leaf proteins. *Science, N.Y.* **127**, 1336–1337.

Roussaux, J., Daussant, J. and Manigault, P. (1971). Mise en évidence de certaines peroxidases dans les tissus sains et tumoraux de *Datura stramonium* L. Colloques Internationaux du CNRS, n° 193 "Les cultures de tissus de plantes", pp. 437–442.

Ryan, C. A. (1967). Quantitative determination of soluble cellular proteins by radial diffusion in agar gels containing antibodies. *Analyt. Biochem.* **19**, 430–440.

Ryan, C. A. (1968). Synthesis of chymotrypsin inhibitor I. Protein in potato leaflets induced by detachment. *Pl. Physiol., Lancaster* **43**, 1859–1865.

Ryan, C. A., Huisman, O. C., Van Denburgh, R. W. (1968). Transitory aspects of a single protein in tissues of *Solanum tuberosum* and its coincidence with the establishment of new growth. *Pl. Physiol., Lancaster* **43**, 589–596.

Schuster, K., and Donhauser, S. (1967). Separation and characterization by immunological and physico-chemical methods of the salt soluble proteins of barley and adjuncts occurring in the brewing process. *Brauwiss* **20**(4), 135–144; **20**(5), 209–214; **20**(6), 234–247.

Shilperoort, R. A., Meijs, W. H., Pippel, G. M. and Veldstra, H. (1969). *Agrobacterium tumefaciens* cross-reacting antigens in sterile Crown Gall tumours. *FEBS Letters* **3**, 173–176.

Siegel, B. Z. and Galston, A. W. (1967). Indole acetic acid oxidase activity of apoperoxidase. *Science, N.Y.* **157**, 1557–1559.

Tempel, A. (1959). Serologisch onderzoek bij *Fusarium oxysporum*. *Meded. Landb.-Hoogesch. Wageningen* **59**(7), 1–60.

Tronier, B. and Ory, R. (1970). Association of bound beta-amylase with protein bodies in barley. *Cereal Chem.* **47**, 464–471.

Tucker, A. O. and Fairbrothers, D. E. (1970). Extraction of leaf proteins of *Mentha* (Labiatae) for disc electrophoresis. *Phytochemistry* **9**, 1399 1403.

Uriel, J. (1971). Color reactions for the identification of antigen-antibody precipitates in gel. *In* "Methods in Immunology and Immunochemistry", Vol. 3 (C. A. Williams and M. W. Chase, eds), pp. 294–321. Academic Press, New York.

Uritani, I. and Stahmann, M. A. (1961a). The relationship between antigenic compounds produced by sweet potato in response to black rot infection and the magnitude of disease resistance. *Agric. biol. Chem.* **25**, 479–486.

Uritani, I. and Stahmann, M. A. (1961b). Changes in nitrogen metabolism in sweet potato with black rot. *Pl. Physiol., Lancaster* **36**, 770–782.

Vaughan, J. G. (1968). Serology and other protein separation methods in studies of angiosperm taxonomy. *Sci. Prog. Oxf.* **56**, 205–222.

Vuittenez, A. (1966). Utilisation de la sérologie dans l'étude de certaines maladies à virus des plantes fruitières ligneuses. *Bull. Soc. Physiol. veget.* **12**, 355–375.

Wasserman, E. and Levine, L. (1961). Quantitative micro-complement fixation and its use in the study of antigenic structure by specific antigen-antibody inhibition. *J. Immunol.* **87**, 290–295.

Wells, H. G. and Osborne, T. B. (1913). The specificity of the anaphylaxis reaction depends on the chemical constitution of proteins or on their biological relations. The biological reactions of the vegetable proteins. *J. infect. Dis.* **12**, 341–358.

Young, R. E. (1965). Extraction of enzymes from tannin bearing tissue. *Archs Biochem. Biophys.* **111**, 174–180.

CHAPTER 3

Properties of and Physiological Changes in Storage Proteins

HERMANN STEGEMANN

*Biochemical Institute, Messeweg 11, D33 Brunswick,
West Germany*

I. INTRODUCTION

Dealing with storage proteins is a difficult task. First of all, there is no clear definition of what storage proteins are. Do they store nitrogen, ready-made amino acids or even proteins to be assembled into glycoproteins by the Golgi apparatus as in the submaxillary gland? What are storage proteins synthesized for? Are they needed for lowering the osmotic pressure of the amino acid pool, or for binding ammonia as amide without change of the overall acidity? It is well known that most of the excess nitrogen goes first into the three amides asparagine, glutamine and urea. But if one does not take the word "storage" too literally, there may well be more behind the scenes. May they not serve as macromolecules shielding other compounds from enzymic actions, which would give them prime importance in metabolic regulations?

There are a number of hints, but no proof, that storage proteins have such a function. Although knowledge concerning their physiology, degradation and synthesis is scarce, there are indications for their having a regulatory function. Pain and Clemens (1973) have reviewed the role of soluble proteins, in plants and animals, in the control of protein synthesis. They conclude that initiation and elongation factors are the same in all eukaryotic cells, that exchange takes place between proteins in cell sap and ribosomes to regulate ribosomal activity. Protein inhibitors for ribonucleases respond to various changes in cellular physiology. The balance between opposing factors, for instance enzyme and inhibitor, is not understood in spite of many gross observations of changes brought about by changing nutritional state, hormone levels and so on. The functions of the well known inhibitors of proteinases in potatoes have still not been elucidated and little is known about the activation of zymogens or the method of release of insulin from proinsulin. There are only a very few examples where the interplay between macromolecules can be partially explained. There is the protein–protein interaction in the case of the cofactor–protein for the activation of actin–myosin–ATPase in muscle contraction (Pollard and Korn, 1973). A protein–carbohydrate interaction occurs between the inhibitor protein from beans and the pectinase from *Fusarium* (Fisher *et al.*, 1973). This involves the globulin from jack beans (*Canavalia ensiformis*), the haemagglutinin concanavaline A, which interacts specifically with multiple α-D-gluco(-or manno)syl groups, e.g. glycogen (and fructosyl) residues. This globulin of a plant, which could be considered as a storage protein too (2–3% of the total protein) initiates differentiation and cell division in animals and is very much in focus because it provides a possible means for distinguishing between normal and tumour cells. (For a review of recent literature see Agrawal and Goldstein, 1972; Lis and Sharon, 1973; Ryan, 1973.)

II. General Properties of Storage Proteins

In the plant the storage proteins are present in seeds, tubers and bulbs. They all serve the same purpose and are utilized in the first stages of growth, whatever this means. Their differentiation into globulins, albumins, acid- and alkali-soluble proteins (Osborne's fractionation) is quite arbitrary, but still has some importance. Most of these proteins are not very soluble in water and their isolation is hindered by tanning due to phenolic substances and oxidizing enzymes. Rapid separation after extraction is essential. Gels are very useful for such separations since they can tolerate considerable amounts of the various impurities present in crude plant extracts. Only following the application of starch and polyacrylamide gel electrophoresis has detailed information become available on protein patterns in storage tissues.

One general feature is that proteins from seeds and tubers are highly amidated. This is especially so in rape seeds (Lönnerdal and Janson, 1972) where they have an isolectric point of 11 and almost all the aspartic and glutamic acid residues are amidated. The amide group is not very stable in glutamine (Stegemann, 1958) and one should take this into consideration, since many variations in behaviour on electrophoregrams could be due to partial hydrolysis of the proteins in question, during extraction and isolation procedures. On the other hand, Midelfort and Mehler (1972) have also shown that deamidation can also occur *in vivo*.

Another general phenomenon is the decrease in the molecular weight of storage proteins during senescence and this seems to apply to proteins in the leaf as well (Parups, 1971). But it is not just a degradation; it consists first of a dissociation of larger protein complexes as the rate-limiting factor, with a concomitant increase of urea in the tissue (Catsimpoolas et al., 1968). In potatoes, it is possible to reverse this trend if old cells are induced to regenerate new cells, e.g. by slicing old tissue. The mol. wt of the protein subunits in the potato tissue changes from about 18 000 to 36 000 (Stegemann et al., 1973) which is the same range as found in the gliadins (Bietz and Wall, 1972; Ewart, 1973).

There is a third very important general property of these proteins. The more soluble storage proteins give different electrophoretic patterns between species and also between varieties. This was first recognized by us in 1961 (Stegemann and Loeschcke, 1961), and independently, it was later found to be true for many mono- and dicotyledonous plants. Therefore, it is possible to identify varieties by gel electrophoresis or, since the main difference between these proteins is in their charge, by electrofocusing (Macko and Stegemann, 1969). The knowledge we have accumulated about potato varieties will be discussed in more detail below. A brief mention will first be made of seed proteins (see Altschul et al., 1966 and also chapters 2 and 4 in this book).

III. Cereal Proteins

In wheat, gluten is the principle storage protein formed in the proteoplasts and this consists of gliadin and glutenin (80–85% of the total endosperm protein), together with albumin and globulin. Gliadin and glutenin are acetic acid-soluble and accumulate in visible protein bodies common in the endosperm of all cereals. The water-soluble proteins of homogenized endosperm with different functions are identical with the cytoplasmic ground substance and more specifically with the endoplasmic reticulum. They are synthesized here and sometimes called adherent protein, in contrast to the former "interstitial" protein (Kent, 1969). The water-soluble proteins—albumins—can be used for taxonomic purposes.

Most proteins in the gliadin-fraction are single polypeptide chains, have mol. wt near 36 500 and a similar amino acid sequence. Glutenin consists of polypeptides of at least 15 distinct mol. wts ranging from 12 000 to 133 000 (Osborne fractionation), not formed by association of lower mol. wt components. No ester cross-linkage could be found and a test for glycosidic linkages (treatment with borohydride in alkaline medium) destroyed all intact subunits. Albumin and globulin polypeptides (mol. wt 11 400) remain unchanged following reduction with mercaptoethanol (Bietz and Wall, 1972). Reduced gliadin components resemble reduced glutenin subunits in electrophoretic mobilities; differences among varieties exist. Their changes during sprouting have not yet been investigated in great detail.

IV. Potato Tuber Proteins

This section is restricted to results obtained with potato tubers. Many of the publications in this field are referred to in a recent paper (Stegemann *et al.*, 1973). The tuber contains up to 4% proteins and amino acids, but only half of the nitrogen is in a macromolecular state. Twenty-five years ago, two main proteins were thought to occur in the potato tuber; these were called tuberin and tuberinin. Today, in a sample of 20 µl of raw sap with about 200 µg protein, it is possible to see between 10 and 30 or 40 main bands after gel

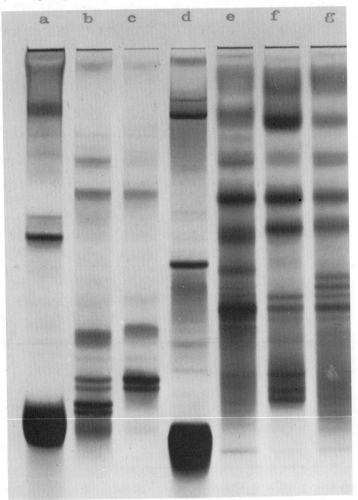

Fig. 1. Electrophoresis of proteins from serum: human (a), two strains of trout (b, c), rabbit (d) and from sap of three potato varieties: "Lerche" (e), "Lori" (f), "Rosa" (g). 5% cyanogum in Tris-borate-buffer pH 8·9.

electrophoresis and staining with Supranolcyanin (Bayer) or Coomassie-Blue (ICI). Only a few of the proteins detectable by the Coomassie Blue stain are enzymes, their scale being 10^{-1} to 10^{-12} below the level of the storage proteins. Thus, some 70–80% of the extractable protein in the potato tuber belongs to so-called storage protein with unknown functions.

By following changes in protein patterns during growth, storage and sprouting, we hoped to gain some insights into the variation with physiological conditions. Since protein patterns are used for variety identification, it was also necessary to learn in which period in the life cycle the pattern is stable enough for this purpose. Genetic variations have to be separated from physiological and pathological variations. Sap as it came from the tuber, by pressing, was used either from fresh or frozen tissue. The sap was never freeze dried and never fractionated. Buffers high in borate were used because they did not interfere with constituents of the raw sap, if certain precautions were observed. For reliable comparison of many samples, the slab gel technique was necessary. Polyacrylamide and methylene-bisacrylamide are the supporting gel media. In Figs. 1–11, the anode is at the bottom and the sample is applied to the upper part of the gel. The apparatus used is described elsewhere (Stegemann, 1972a, b).

Protein patterns have been in use for more than 10 years as a tool for variety identification. In the dormant tuber, there is a well defined relationship among the stainable proteins, not influenced by climate, soil or treatment with growth regulators; Fig. 1 shows the patterns of three different potato varieties, compared with the better known patterns of sera from animals. Different varieties are well defined and we have not yet found two varieties with an identical protein pattern. Much work has also been done with enzymes (phosphorylases, amylases, peroxidases, esterases), their patterns being more or less related to the genetic origin of the plants (Siepmann and Stegemann, 1967). The position of some enzymes relative to the Coomassie Blue-stainable proteins are given in Fig. 2.

In recent years it has become obvious that some of the potato proteins in any one variety disappear as sprouting begins, whereas most of them stay unchanged. The disappearing band is located in the same position in all varieties and is therefore a general feature (Fig. 3). The mobility of some other bands changes in the stored tuber due to a decrease of the mol. wt of some subunits. However, most of the characteristic patterns of the variety are still detectable.

The electrophoretic pattern of immature tubers is also of interest. In this case, the characteristic pattern of each variety is not completely developed and a predominant band is present in all potato varieties. This band has the unusual property that its mobility is the same in gels of pH 8·9 and 7·9, in contrast to all other bands (Fig. 4). It is possible that this band represents a protein linked to an acidic polysaccharide. Under climatic conditions in central Europe, it disappears in August, when the characteristic patterns of

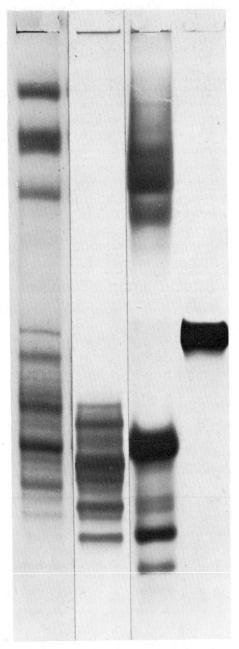

FIG. 2. Comparison of constituents from tubers, variety "Maritta", stained for proteins (Supranolcyanin, Bayer), esterases (α-naphthyl acetate/echtblausalz RR), peroxidases (H_2O_2/o-dianisidine) and phosphorylases (primer technique, G-1-P incubation, iodine staining). Electrophoresis at pH 8·9 from the same gel slab.

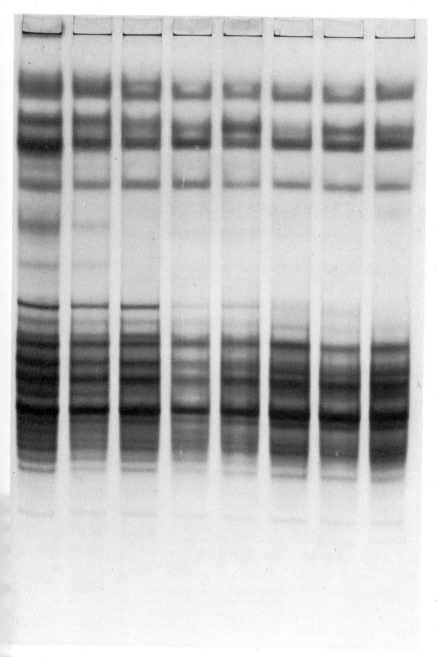

FIG. 3. Proteins from mature and sprouting tubers, respectively (variety "Maritta") taken from September until May (except December), gel electrophoresis at pH 7·9.

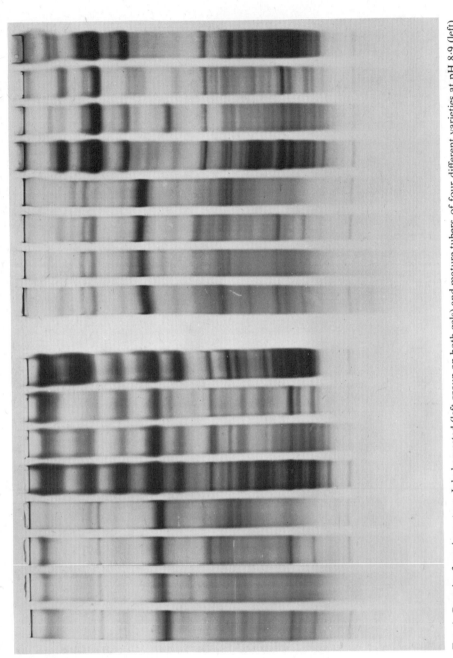

Fig. 4. Proteins from immature, July-harvested (left group on both gels) and mature tubers of four different varieties at pH 8·9 (left) and 7·9 (right)

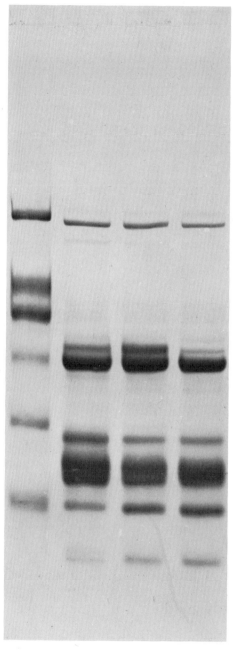

FIG. 5. Size distribution of tuber proteins from three different varieties (mol. wt markers and electrophoretic conditions as in Fig. 8) after incubation with Na-dodecylsulphate (SDS) at 100 °C for 3 min.

the different varieties develop. This protein is as typical of an immature state as is the protein which disappears as sprouting begins. This latter one can be seen in any variety tested and is therefore a general physiological marker for the mature tuber. Experiments are under way to correlate its disappearance with "ripeness" from the processer's point of view. Botanists have no accurate way of describing maturity in a potato tuber.

There is very little difference in the mol. wt of the major proteins among the varieties (Fig. 5), which means that the genetically distinct pattern of a variety

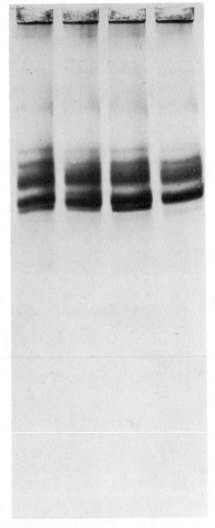

Fig. 6. Protein patterns of four different potato varieties after gel-electrophoresis in 20% ·acetic acid and 5 M urea. Here, the electrodes have been reversed, the anode is on top.

is mainly charge-dependent. The same is true for a gel electrophoresis run in 5 M urea and 20% acetic acid (Fig. 6). Dr Hoff from Purdue University in West Lafayette (U.S.A.) supplied some crystals of proteins isolated from the outer layer of the potato. These crystals have been known for a long time to have a higher mol. wt than the other proteins; they can be split into smaller units by reduction (Fig. 7). As mentioned earlier, the mol. wt changes during ageing. In mature tubers, the main group of proteins have subunit mol. wt of 16 800, 18 000 and 19 500. The proteins of mol. wt 16 800 are altered by treatment with mercaptoethanol and give two faster migrating bands of 13 800 and 10 200. Smaller mol. wt bands are not detectable by the SDS-method, under our experimental conditions. Old potato tissue can be induced to form meristematic cells by slicing; the mol. wts of the main proteins increase after this treatment (Fig. 8).

A very good separation of tuber proteins is achieved first by electro-focusing the sample and then by electrophoresis of the separated proteins in the second dimension (Fig. 9). An even better survey of the properties of the main proteins comes from a 3-step technique and a 2-dimensional separation. First, the proteins are separated according to their isoelectric points by electrofocusing in gels with ampholines, e.g. pH 5–7. Then the separated proteins are reduced within the gel rod with mercapto-ethanol and they are loaded with SDS (sodium dodecylsulphate). Then, as the third step, the rod is put on top of a rectangular gel slab and the proteins in the rod are forced to migrate into a SDS-containing gel by electrophoresis. Thus, the mol. wt of any of the proteins with a definite isoelectric point can be determined (Fig. 10). Experimental details are given in Stegemann et al. (1973). It is surprising that each of the three proteins with distinctly different isoelectric points gives one (and if reduced, three) proteins of the same mol. wt. We have not yet determined the physiological importance of this observation, but it may be that these 3 proteins come from a common ancestor and differ only in their charge distribution.

This mapping technique reveals more of the composition of a complex mixture than any other method and is a true check for purity. However, what does the distribution of the spots mean? Are these all storage proteins with a few common primary structures and differing mainly in charge? The charge difference could often be due to different amounts of amides in the chain, and this could be caused either by bad conditions during preparation as revealed by combined preparative and analytical electrophoresis (Fig. 11) or by natural circumstances (Midelfort and Mehler, 1972). Is the amide content a mirror of planned or built-in obsolescence, as Robinson et al. (1973) think? Has a planned degradation a regulating function?

A still more open question is: why are the patterns dependent on the genetic background? Are they synthesized in the usual manner on ribosomes and glutamine and asparagine coded for in the usual way? This would mean different DNA for coding. Or is only one protein synthesized and this parent

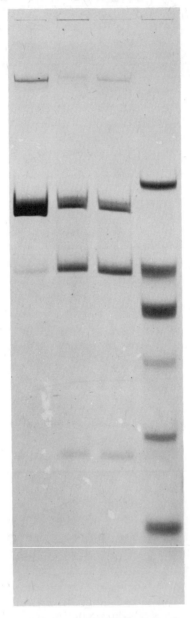

FIG. 7. Gel electrophoresis of protein crystals from tubers (gift from Dr J. E. Hoff, Purdue University, Lafayette, U.S.A.) after pretreatment with SDS in the presence of mercapto-ethanol (ME) for 3 and 20 h and without ME for 20 h at 50 °C (left and right). Electrophoretic conditions as in Fig. 8.

protein changed later by a gene- or sequence-directed amidation–deamidation reaction, like the hydroxylation in procollagen? We hope to get some of the answers to these questions in our future programme of research, by finger-printing the major components, by end group analysis, by immunochemical means and by radioactive labelling.

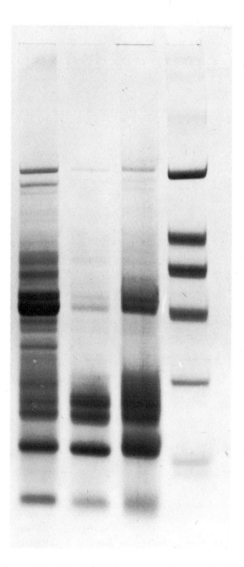

FIG. 8. Tuber proteins, variety "Lerche", from young tissue on the left. Same tissue after storage for 9 months (middle) and 1 day after cutting it into 1 mm thick slices. Mol. wt markers (right) 925, 670, 482, 370, 250 and 124 × 10². SDS-electrophoresis in SDS-phosphate buffer pH 7·1.

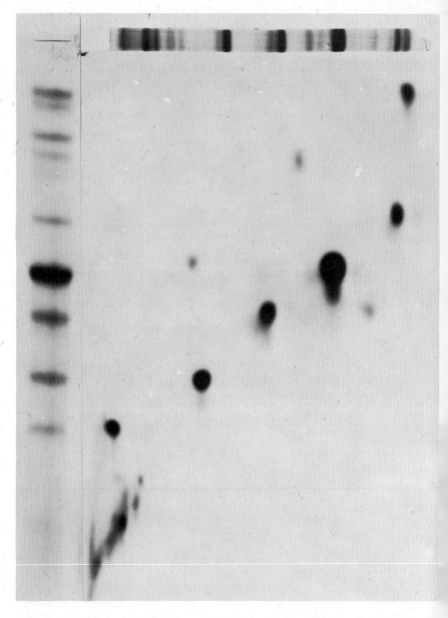

FIG. 9. Protein mapping of tuber proteins. First dimension (left to right) focusing in ampholine pH 5–7; second dimension (top to bottom) electrophoresis at pH 8·9. Upper gel cylinder inserted later for photography, somewhat shrunken.

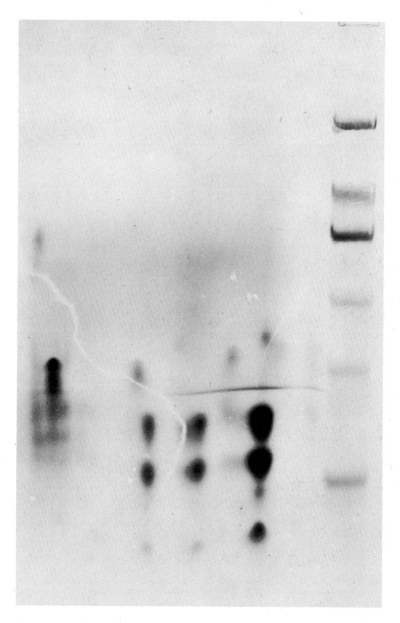

FIG. 10. Same as Fig. 9, but after electrofocusing gel rod incubated with SDS/mercapto-
ethanol and electrophoresis performed in SDS-phosphate buffer.

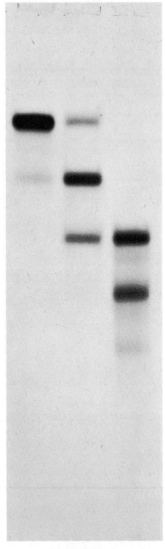

Fig. 11. One of the main proteins in potato tubers isolated by preparative eluting gel electrophoresis and after incubation at pH 10 at 50 °C for 2 and 20 h respectively (left to right).

REFERENCES

Agrawal, B. B. L. and Goldstein, I. J. (1972). Concanavalin A, the jack bean phyto-hemaglutinin. *In* "Methods in Enzymology" (Colowick-Kaplan, ed.), Vol. **28**, Part B, pp. 313–318. Academic Press, New York.

Altschul, A. M., Yatsu, L. Y., Ory, R. L. and Engleman, E. M. (1966). Seed Proteins. *A. Rev. Pl. Physiol.* **17**, 113–136.

Bietz, J. A. and Wall, J. S. (1972). Wheat Gluten Subunits: Molecular Weights Determined by Sodium Dodecyl Sulfate-Polyacrylamide Gel Electrophoresis. *Cereal Chem.* **49**, 416–430.

Catsimpoolas, N., Ekenstam, C., Rogers, D. A. and Meyer, E. W. (1968). Protein Subunits in Dormant and Germinating Soybean Seeds. *Biochim. biophys. Acta* **168**, 122–131.

Ewart, J. A. D. (1973). Sodium Dodecyl Sulfate Electrophoresis of Wheat Gliadins. *J. Sci. Fd Agric.* **24**, 685–689.

Fisher, M. L., Anderson, A. J. and Albersheim, P. (1973). Host-Pathogen Interactions. VI. A Single Plant Protein Efficiently Inhibits Endopolygalacturonases Secreted by *Colletotrichum lindemuthianum* and *Aspergillus niger. Pl. Physiol., Lancaster* **51**, 489–491.

Kent, N. L. (1969). *In* "Structural and Nutritional Properties of Cereal Proteins" (R. A. Lawrie, ed.), pp. 280–299. Proceedings of the 16th Easter School. University of Nottingham.

Lis, H. and Sharon, N. (1973). The biochemistry of plant lectins. *A. Rev. Biochem.* **42**, 541–574.

Lönnerdal, B. and Janson, J. Ch. (1972). The Low Molecular Weight Proteins in Rapeseed. Isolation and Characterization. *Biochim. biophys. Acta* **278**, 175–183.

Macko, V. and Stegemann, H. (1969). Mapping of Potato Proteins by Combined Electrofocusing and Electrophoresis. Identification of Varieties. *Hoppe-Seyler's Z. physiol. Chem.* **350**, 917–919.

Midelfort, C. F. and Mehler, A. H. (1972). Deamidation *in vivo* of an Asparagine Residue of Rabbit Muscle Aldolase. *Proc. natn. Acad. Sci., U.S.A.* **69**, 1816–1819.

Pain, V. M. and Clemens, M. J. (1973). The Role of Soluble Protein Factors in the Translation Control of Protein Synthesis in Eukaryotic Cells. *FEBS Letters* **32** 205–212.

Parups, E. V. (1971). Disc Electrophoresis of Proteins of Senescing and Fresh Leaves and Petals of Certain Ornamental Plants. *J. Am. Soc. hort. Sci.* **96**, 168–171.

Pollard, Th.D. and Korn, E. D. (1973). Acanthamoeba Myosin. *J. biol. Chem.* **248**, 4691–4697.

Robinson, A. B., Irving, K. and McCrea, M. (1973). Acceleration of the Rate of Deamidation of GlyArgAsnArgGly and of Human Transferrin by Addition of 1-Ascorbic Acid. *Proc. natn. Acad. Sci., U.S.A.* **70**, 2122–2123.

Ryan, C. A. (1973). Proteolytic Enzymes and their Inhibitors in Plants. *A. Rev. Pl. Physiol.* **24**, 173–196.

Siepmann, R. and Stegemann, H. (1967). Enzym-Elektrophorese in Einschluß-Polymerisaten des Acrylamids. A. Amylasen und Phosphorylasen. *Z. Naturf.* **22b**, 949–955.

Stegemann, H. (1972). Bau, Einrichtungen und Geräte des Instituts für Biochemie. *Glas- u Instrumententechnik* **16**, 925.

Stegemann, H. (1958). Eine Mikrobestimmung von Amid-Stickstoff, speziell in Proteinen. *Hoppe-Seyler's Z. physiol. Chem.* **312**, 255–263.

Stegemann, H. (1972). Apparatur zur thermokonstanten Elektrophorese oder Fokussierung und ihre Zusatzteile. *Z. analyt. Chem.* **261**, 388–391.

Stegemann, H., Francksen, H. and Macko, V. (1973). Potato Proteins: Genetic and Physiological Changes, Evaluated by One- and Two-Dimensional PAA-Gel-Techniques. *Z. Natur.* **28c**, 722–732.

Stegemann, H. and Loeschcke, V. (1961). Über die Proteine der Kartoffelknolle. *Landw. Forschung* **14**, 259–261; Stegemann, H. and Loeschcke, V. (1965). Proteinmuster der Kartoffelknolle—ein Sortencharakteristikum. *Jahresberichte der Biolog. Bundesanstalt,* **A52.**

CHAPTER 4

The Proteins of Barley

GISÈLE PRÉAUX AND R. LONTIE

Laboratorium voor Biochemie, Katholieke Universiteit te Leuven,
Louvain, Belgium

I. THE PIONEER WORK OF OSBORNE

One of the first classifications of the barley proteins, and still the most common, goes back to Osborne at the end of the last century (1895). He showed that the seeds of barley, like those of other cereals, contain "proteid matters soluble in water, in sodium chloride solutions, and in alcohol, and that after complete extraction with all these reagents there remains a considerable quantity of proteid which can be partly extracted by dilute potash solutions, but the greater part of which is insoluble in any reagent hitherto applied". He further pointed out, however, that as an "aqueous extract of any seed is in reality a dilute saline solution, owing to the salt extracted from the seed, and as the proteid matter soluble in alcohol dissolves to a slight extent in very dilute saline solution, it was preferable to obtain the proteids properly soluble in water by extracting the meal with sodium chloride solutions, dialysing away the salts and filtering off the proteid that thereby precipitated.

In this way the protein matter, soluble in pure water, which had been extracted from the meal was obtained in solution by itself." He therefore always followed this procedure.

As solvents for the extractions, Osborne (1895) successively employed 10% salt and 75% alcohol solutions, the latter usually with heating. The proteins were then separated from the extracts either by dialysis, heat coagulation or precipitation, further purified, dried and their amount determined by weight. The fraction partly soluble in aqueous potash could not be purified. The glutelin fraction was therefore determined by difference from the total protein content. The latter was deduced from the nitrogen content of the barley flour, assuming that this all consisted of protein with a 17% nitrogen content.

Osborne (1895) called the albumin, the protein fraction soluble in water, "leucosin" from the identity of its elementary composition with the fraction isolated from wheat and rye kernel. In a similar way he called the globulin, the protein only solubilized by the presence of salt, "edestin" from the similarity in composition with the globulin found in a large number of other seeds. The latter name fell very rapidly into disuse, however (Osborne and Harris, 1903), and was only retained for the globulin of hemp seed and castor bean as the globulin fraction of the other seeds did not yield the same proportion of nitrogenous decomposition products. Bishop (1928) abandoned both terms leucosin and edestin on account of variations in the relative amounts of the constituent amino acids in different samples.

The fraction soluble in alcohol, although similar in many respects, appeared to differ slightly in elementary composition from those fractions extracted under the same conditions from other seeds and thus seemed to be characteristic of barley. Osborne (1895) therefore proposed to adopt for it the then disused name "hordein" which had been introduced by Proust (1817) and applied later on by Hermbstädt (1831) to extracts of barley grain.

Hordein was, from the beginning of this century on, by far the best characterized of the barley proteins. Like the alcohol-soluble proteins of the other cereals, it was found to contain a large amount of nitrogen as ammonia and a very low amount of basic nitrogen (Osborne and Harris, 1903). On acid hydrolysis it yielded much glutamic acid, proline and ammonia (Kleinschmitt, 1907; Osborne and Clapp, 1907). Referring to this relatively large proportion of proline and amide nitrogen, Osborne (1908) then proposed to designate the "alcohol-soluble" proteins of the cereals by the group name "prolamin".

The fraction insoluble in water, in salt solution and in alcoholic solution and for which the group name glutelin was used (Osborne, 1895) in analogy with the similar fraction from other seeds, was clearly quantitatively the most important (Fig. 1). Osborne (1895) considered that the insolubility in alkali of the larger part of it "could have arisen from the treatment for removal of the other proteids if this fraction were not already insoluble". The small part that could be solubilized into alkali, on the contrary, was considered either as

Albumin (*Leucosin*)	H_2O	0·3%
Globulin (*Edestin*)	Salt solution	1·95
Prolamin *Hordein*	Alcohol–water	4·0
Glutelin *Insoluble proteins*	Alkaline solution	4·5

FIG. 1. Percentage of the four protein fractions in the barley kernel as determined by Osborne (1895).

"a protein of the properties of glutelin or as a portion of the proteins which failed to be extracted by neutral solvents, either because they were contained in unruptured cells, which were afterwards destroyed by the alkaline solution, or were retained in the meal residue in combination with other substances, such as nucleic acid or tannin, which rendered it insoluble in neutral solution" (Osborne, 1919).

II. Factors Influencing the Yield of Protein Extracted

A. QUANTITATIVE ASPECTS

Bishop (1928) was one of the first to consider quantitative aspects of protein extraction from barley. Following Osborne (1895), he always extracted the albumin and globulin fractions together using neutral salt solutions, but determined the yield of the extraction from the nitrogen content of the solutions as such, without any previous separation or purification of the proteins. This was then referred to as the salt-soluble nitrogen content, as the extracts also contained amino acids, proteoses and some non-amino acid nitrogen. In the same way, the total nitrogen of the alcoholic extracts was taken as a measure of the hordein fraction present, while the glutelin fraction was determined by difference from the total nitrogen content of the barley meal. For the extraction of the salt-soluble proteins, Bishop (1928) preferred potassium sulphate to sodium chloride for convenience in the subsequent Kjeldahl process, after he observed that a 5% potassium sulphate solution yielded (by two successive extractions) the same maximum amount as sodium chloride at a concentration between 5 and 10%.

He used centrifugation instead of filtration in order to separate the supernatants. He found that at least three extractions with the same solvent were necessary, and even five in practice, owing to the slowness of the diffusion of the proteins from the particles of barley. He performed the extractions under continuous shaking as this was found to increase the efficiency. He also standardized the grinding method since the solubility was affected by the fineness and evenness of the meal. For the extraction of the hordein fraction, Bishop (1928) used hot 70% (v/v) alcohol. He found this necessary for achieving complete extraction, since hordein is not very soluble in cold alcohol when the barley has been previously extracted with salt solution.

B. TOTAL NITROGEN CONTENT OF BARLEY

Bishop (1928) first applied his quantitative method to a series of barleys of the two-row Plumage-Archer variety differing in their total nitrogen content. He made the interesting observation that the glutelin fraction was increasing in the same way as the total nitrogen while the hordein was increasing proportionally more and the salt-soluble fraction considerably less. When the percentages of the three nitrogen fractions are considered as a function of the total nitrogen content of barley (Fig. 2), it appears that the percentage of glutelin does not vary, while that of the hordein is increasing and that of the salt-soluble fraction decreasing with the total protein content of the barley meal. Bishop (1928) further reported that other barleys gave curves of apparently similar type but with somewhat different numerical values and later on he also found it to be true for six-row barley varieties (Bishop, 1930, 1939).

This regular relationship in amounts of different proteins first observed by Bishop was further confirmed by other workers in Denmark (Hofman-Bang, 1931), Germany (Fink and Kunisch, 1937) and France (Urion and Golovtchenko, 1940; Urion et al., 1944a). Urion and Golovtchenko (1940) further suggested the use of the percentage of glutelin, as "a varietal constant" for

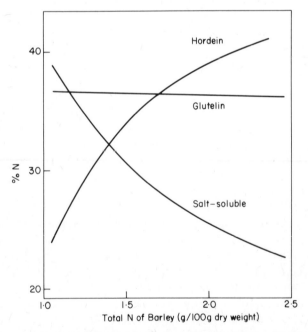

Fig. 2. Percentage of the three nitrogen fractions as defined by Bishop (1928) as a function of the nitrogen content of barley.

characterizing a pure variety. Urion *et al.* (1944a), however, pointed out the need for caution in doing so, since the percentage of glutelin was found to be a function of the degree of grinding.

C. FINENESS OF GRINDING

Bishop (1928) used a large adjustable cone coffee mill, which reduced most of the material to a finer state than most other mills. This coffee mill, however, had the disadvantage of leaving behind some large particles. In order to eliminate these, and, thus to ensure evenness, Bishop (1929) proposed to grind the barley first as finely as possible in the coffee mill and then to regrind it in the "Wiley mill" using a 0·5 mm sieve. This latter mill ground the material without heating by the cutting action of knives. This method seemed necessary to him, and was always used in his later work, since he made the important observation that the yield of the salt-soluble, of the alcohol-soluble and of the glutelin fraction depended on the fineness of grinding. The first two fractions increased when it was used, while the third one necessarily decreased since it was determined by difference from the total nitrogen content of the barley.

A very similar observation was made a year later by Hofman-Bang (1930). He also extended his study, however, to a finer grinding than that used by Bishop (1929), since he thought "that this might increase the amount of soluble proteins found, owing to the fact that when the cell walls were disintegrated, the salt solution had more ready access to the proteins. Moreover the total surface of the barley after the attack by the potassium sulphate was greatly increased by finer grinding". Indeed this finer grinding (Fig. 3) allowed him to increase still more the yield of the salt-soluble proteins but not, or only to a lesser extent, that of the alcohol-soluble proteins The highest and most constant values were reached with the ball mill. Hofman-Bang (1930) thus proposed to adopt it as a standard method of grinding; the frictional heat effect that occurred in its use was found to have only a negligible influence on the analytical results.

Urion and Golovtchenko (1940) followed the fineness of grinding by a microscopic study and found that no further increase in the yields of the salt-soluble and alcohol-soluble nitrogen fraction was obtained when all the cellular organization had been destroyed. This led at the same time to a minimum value for the varietal constant (% glutelin). These authors pointed out that the cellular membranes, as long as they remain unbroken, must prevent free diffusion of the large molecules of proteins; the extraction is then necessarily incomplete, since for all the cells from which the envelopes were not destroyed, it is limited to a diffusion through a semi-permeable membrane.

In order to facilitate the comparison of results obtained in different brewing laboratories it was decided to use the same mill (Bishop, 1961). The Casella version of the Wiley type mill which works by a combined cutting and sieving action on a 1 mm sieve was chosen for this purpose. It was used by

Enari and Mikola (1961) and by Ory and Henningsen (1969) among others. But as barley was heated to some extent with this mill, Enari *et al.* (1962) therefore preferred, at least for their later research work on barley, to use the Miag-Seck vertical axis cone mill which gave a meal very similar to that obtained with the Casella mill, while heating effects were avoided. By measuring the yield of the albumin fraction, they also tested the effect of a Miag-Seck roller mill which gives a much coarser grist than the other two mills; as expected, it only gave a very low yield.

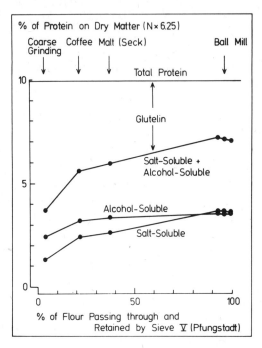

FIG. 3. Percentage of the three nitrogen fractions expressed as per cent of protein of the dry barley, as a function of the fineness of the meal (Hofman-Bang, 1930).

In this laboratory a Raymond Hammer Mill, provided with a 0·5 mm sieve, has always been used. In order to ensure as reproducible results as possible and to avoid heating, the material is always ground twice and each time in the presence of an equal volume of dry ice (Lontie and Voets, 1959; Préaux *et al.*, 1963).

D. WATER CONTENT OF THE GRAIN

The water content of the barley was also found to influence the degree of grinding. The grain should preferably be milled dry (10% moisture or less) as done by Bishop (1929), since it ensures a more complete and comparable fineness of grinding (Bishop, 1930) and reduces the time required for milling (Hofman-Bang, 1931). Drying of the barley to 10% moisture or less also has the advantage of rendering the salt-soluble proteins more easily available as shown by the slight increase in yield (Hofman-Bang, 1931) and the corresponding decrease of the glutelin fraction.

E. STORAGE OF THE BARLEY GRAIN

Storage of the grain for a long time has the opposite effect of drying and thus causes a decrease in the salt-soluble proteins (Bishop, 1930). The decrease is always more pronounced for undried than for dried barley; furthermore the temperature of storage has an effect as can be seen (Fig. 4)

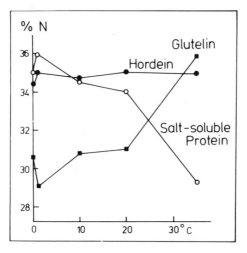

Fig. 4. Percentage of the three nitrogen fractions from undried barley stored for 1 month at different temperatures (Hofman-Bang, 1931).

in the case of undried barley kept for one month at different temperatures (Hofman-Bang, 1931). The decrease in yield of the salt-soluble fraction is small on going from barley kept in the cold to barley kept at room temperature but is greater with barley stored at 30 °C. With this decrease, there is a corresponding increase in the glutelin fraction, since the yield of hordein fraction remains the same, even after a storage period of 2 months. After 4 years, however, a slight decrease of the alcohol-soluble fraction has been reported (Urion et al., 1944b).

F. CHOICE OF THE SOLVENTS

In order to obtain the maximum yield of a given fraction, a proper choice of solvent has to be made; otherwise there is an artificial increase in the glutelin fraction, and also possibly in the hordein fraction. This is best illustrated with the salt-soluble proteins, where for a given meal the temperature (Urion *et al.*, 1944b), the concentration of the salt solution (Fig. 5) (Urion *et al.*, 1944c) and the pH (Fig. 6) (Urion *et al.*, 1944d) were found to influence the yield. The duration of the extraction must also be long enough (Enari *et al.*, 1962) and the extraction with each solvent be repeated at least 3 to 5 times (Bishop, 1928; Lontie and Voets, 1959; Enari *et al.*, 1962; Préaux *et al.*, 1963) in order to ensure as complete an extraction as possible.

When potassium sulphate is added to water in order to solubilize the globulin fraction, the yield of the salt-soluble nitrogen increases up to a salt concentration of 6·5%, when a constant value is reached (Fig. 5). This increase is completely at the expense of the glutelin nitrogen since the yield of hordein remains constant and is thus apparently not influenced by the salt concentration of the previous extractions. This constancy in yield of the hordein fraction, whether the globulins have or have not first been removed, is shown in the work of Rose and Anderson (1936) (Fig. 7), which has also indicated that this globulin fraction must be insoluble in hot 70% alcohol or be denatured by it. Furthermore their results clearly show that when the globulin is not extracted before the alcohol-soluble fraction, it is always included in the glutelin fraction.

Very similar results to those with potassium sulphate have also been obtained with sodium chloride, the other neutral salt frequently used for the extraction of the salt-soluble proteins. A constant value is reached at a concentration of about 7%. The yield is, however, slightly higher than with

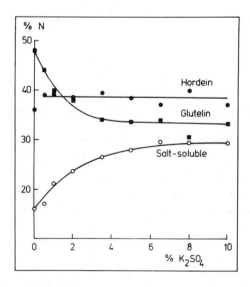

FIG. 5. Percentage of the three nitrogen fractions as a function of the salt concentration of the aqueous extracts (Urion *et al.*, 1944c).

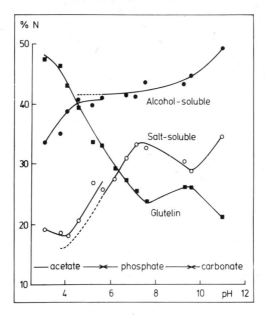

FIG. 6. Percentage of the three nitrogen fractions as a function of the pH of the salt extracts (Urion *et al.*, 1944d).

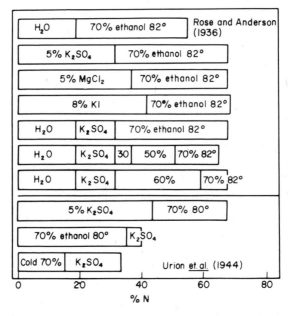

FIG. 7. Influence of the type of solvents (Rose and Anderson, 1936) and of their order (Urion *et al.*, 1944b) on the yield of the water-soluble, salt-soluble and alcohol-soluble fractions.

potassium sulphate, but only by 2%, while that of the hordein fraction remains completely identical. The yield of the hordein fraction also seems to be independent of the type of neutral salt chosen for the extraction of the salt-soluble protein. This conclusion may, however, not apply to all salts. Rose and Anderson (1936) compared the effect of potassium sulphate with that of magnesium chloride and potassium iodide and found (Fig. 7) that the yield of the salt-soluble fraction increased in the order potassium sulphate < magnesium chloride < potassium iodide and that the excess, as compared to potassium sulphate, was in both cases obtained at the expense of the alcohol-soluble fraction. The greater solubility in these two latter salt solutions is perhaps to be correlated with the more structure-disrupting or chaotropic nature of the ions (Hatefi and Handstein, 1969) which is in the order $K^+ <$ Mg^{2+} and $SO_4^{2-} < Cl^- < I^-$, when classified by the lyotropic series of Hofmeister whereby more hydrophobic substances become soluble in water. Very similar results were found in the case of wheat by Gortner et al. (1928), who compared the solubilization of proteins by 21 inorganic salts. They found that the yield of the salt-soluble fraction increased, e.g. from 23 to 64%, when potassium chloride was replaced by potassium iodide.

In order to compare the effect of differing salt concentrations, the temperature must be kept constant, since the yield of the salt-soluble fraction decreases slightly with an increase in temperature, at least from 0 to 30 °C. This decrease causes an increase in both the alcohol-soluble and the glutelin fraction (Urion et al., 1944b).

The pH of the salt solution has also to be adjusted to the most appropriate value, since the yield of the salt-soluble fraction is influenced by this parameter (Fig. 6). The maximum yield is at around pH 7. The decrease in yield from pH 7 to 4 benefits the glutelin fraction, since the yield of the alcohol-soluble fraction remains constant with changing pH. The much lesser solubility of the salt-soluble fraction observed around pH 4 and below is undoubtedly due to an insolubilization by phytin (Kent and Macheboeuf, 1949; Rondelet and Lontie, 1955; Djurtoft, 1961). This component is indeed known to precipitate proteins in acid medium. The data in Fig. 6 also indicate that the hordein fraction shows some solubility in the salt solution at a pH below 4 and that the glutelin fraction only starts to go into solution at a pH above 9·5.

Most of the parameters which affect the extraction of the salt-soluble fraction also apply in the case of the alcohol-soluble proteins and the glutelins. Thus for the extraction of the prolamins the type of alcohol, its percentage, and the extraction temperature should be especially considered. This can, for example, be concluded from the results of Landry et al. (1972), who found that the order of efficiency at room temperature is 60% ethanol < 55% isopropanol < 50% n-propanol and from those of Rose and Anderson (1936) who showed (Fig. 7) that at 82 °C a 30% solution of ethanol extracted much less than a 70% solution. Furthermore, concerning the effect of temperature

Bishop, as long ago as 1928, stated that in order to reach a maximum yield the alcohol solution should be used hot, at least when the salt-soluble proteins were extracted first. To these three parameters must perhaps also be added that of the number of successive steps used to extract the prolamins. Thus Rose and Anderson (1936) found that successive treatments with increasing concentrations of hot alcohol did not extract as much protein as one treatment with hot 70% alcohol (Fig. 7). These treatments, however, yielded successive hordein fractions with a decreasing amide–nitrogen and an increasing arginine–nitrogen content, the least soluble one approaching the composition of the glutelins.

G. ORDER OF THE SOLVENTS

If one wishes to achieve the usual yields, the order in which the solvents are used for extraction by Osborne (1895), has also to be considered. Bishop (1928) was convinced of this, mainly from the comparison of the yield found by him for hordein with that obtained by other workers who preferred to extract the meal directly with alcohol, but also from the fact that small molecular weight nitrogen components, which are normally extracted together with the salt-soluble proteins, were found to be soluble in alcohol. He therefore referred to the work of Brown (1903), who used such a medium as a method for the separation of amides, amino acids and unclassified nitrogen compounds from barley. This assumption of Bishop (1928) was further proved by Urion et al. (1944b) who observed (Fig. 7), that when alcohol was used first, it dissolved more nitrogenous substances than when used after the extraction with potassium sulphate. Furthermore, the percentage extracted afterwards by the salt solution was much smaller than expected from the increase in yield of the alcohol-soluble fraction. This must apparently be due to some denaturation of the salt-soluble proteins, since the yield of this fraction was markedly higher when the previous alcoholic extractions were performed at 20 °C instead of at 70 °C.

Thus, although it is clear that the order used by Osborne (1895) has to be respected, at least when one wants to study the yield of the extracts or the separation of the salt-soluble protein fraction, many workers especially interested in the isolation of the prolamin fraction still preferred to apply the alcoholic extraction directly to the barley flour, following the procedure of Osborne and Clapp (1907). Also Waldschmidt-Leitz and coworkers, although first following the classical extraction scheme of Osborne (Waldschmidt-Leitz and Brutschek, 1955) finally adopted this method when they discovered Waldschmidt-Leitz and Kloos, 1959) that the preparations were relatively richer in the two least mobile of the five components obtained on electrophoresis at pH 2. These two components, referred to as δ- and ε-hordein, are those which contain the highest proportion of glutamic acid. The authors therefore considered the possibility of an association with the salt-soluble proteins during salt extraction, although it is known (Bishop, 1928) that pure hordein is only slightly soluble in water and in salt solution. We presumed that the salt extractions may also have rendered some of the prolamins insoluble in the alcoholic solvent, possibly as a result of an oxidation. That some alcohol-soluble proteins are probably indeed found in the salt-soluble fraction can perhaps be concluded from the work of Grabar and Daussant (1963), who found that an antiserum prepared against the salt-soluble proteins also gave some reaction with the hordein fraction. However, the reaction instead of being due to some hordein in the salt-soluble fraction may also, as mentioned by these authors, be attributed to the presence of some salt-soluble proteins in the hordein fraction.

III. Solubilization and Subfractionation of the Glutelin Fraction

A. USE OF ALKALINE SOLVENTS

The term glutelin, according to the classification of the American Society of Biological Chemists (1908), refers to "simple proteins insoluble in all neutral solvents, but readily soluble in very dilute acids and alkalis". Osborne (1895) found that in the case of barley only part of the nitrogen fraction, left after the extraction of the salt-soluble and alcohol-soluble proteins, could be solubilized by aqueous potassium carbonate, but he did not succeed in recovering the residue. He therefore always considered the whole of the unextracted material as the glutelin fraction, as pointed out earlier. Osborne (1919) reported further that in place of aqueous potassium carbonate other alkaline carbonate or bicarbonate solutions (0·5–1%) or very dilute solutions (0·1–0·2%) of sodium or potassium hydroxide could also be used. Acid solutions could also be employed, but since experiments indicated that seed proteins were more easily altered by acids than by small quantities of alkali, he suggested that they should preferably be avoided.

Larmour (1927), using a 0·2 M NaOH was at first unsuccessful in purifying the glutelin fraction, since the residue became extremely viscous. However, on addition of an equal volume of 95% ethanol a white fibrous coagulum of gumlike nature precipitated, leaving a clear supernatant from which the solubilized glutelin fraction could be precipitated by acidification. He called this fraction "hordenin" in analogy with "glutenin", proposed by Osborne and Harris (1903) for the glutelin fraction of wheat. This fraction was found to contain much less nitrogen as ammonia and much more basic nitrogen than the prolamin fraction.

Csonka and Jones (1929) used 0·2 M NaOH in 50% alcohol and applied it directly to the untreated barley meal. The precipitate obtained on acidification to pH 6·8 was further purified and solubilized in aqueous 0·2% NaOH. Two protein fractions were then isolated from this solution by fractional precipitation with ammonium sulphate: the α-glutelin which separated at 1–3%

saturation and the β-glutelin, in rather small amount, which precipitated at 18% saturation. Both showed a very similar amino acid composition, the α-glutelin having an isoelectric point at pH 6·4.

Rose and Anderson (1936) compared the solubilizing power of alcoholic and of aqueous solutions of 0·05–0·2 N NaOH and of hydrochloric and acetic acids after the extraction of the salt-soluble and the alcohol-soluble proteins. They reported that the alcoholic solutions were more efficient and easier to use, as there was no swelling of the starch. A complete extraction of the glutelin fractions could be obtained with NaOH in 70% alcohol at 82 °C, while only one third was solubilized at room temperature. At 82 °C 0·1 N HCl had only the efficiency of NaOH at room temperature, 0·1 N HOAc solubilized only very little of the glutelin fraction. The authors considered the possibility of subfractionating the glutelins by performing the extractions at a set of increasing temperatures and of NaOH concentrations in 70% alcohol. They reported, moreover, that under the influence of the acid or base, and more particularly at 82 °C, amide nitrogen was hydrolysed from the protein so that it was difficult to isolate fractions of definite composition.

Scriban (1951), fearing amide hydrolysis, never performed an alkaline extraction hot. Biserte and Scriban (1950) used 0·2 N KOH and precipitated the glutelin fraction from the extracts with magnesium sulphate at half saturation as previously done by Bishop (1928). Scriban (1951), mainly following the method of Csonka and Jones (1929), confirmed by paper chromatography the large similarity in amino acid composition of the α- and β-glutelin fractions and found that both contained cystine. A very similar α-glutelin fraction, although with a higher glutamic acid content, was also obtained when the method was applied to barley from which the salt-soluble and alcohol-soluble proteins had been removed. Scriban (1951) found that the method of Csonka and Jones (1929) did not solubilize the glutelin fraction completely. He called the unextracted part "residual nitrogen" since it showed a very atypical amino acid composition. He considered it as either denatured protein or protein of unknown structure from the cellular cytoplasm.

Folkes and Yemm (1956) applied the method of Csonka and Jones (1929) to barley from which the salt-soluble proteins had been removed. They found it possible, during the purification step at pH 6·8 to completely separate the hordenin from the hordein by adjusting the alcohol concentration to 55%. In order to achieve this separation the hordenin was not exposed for more than 1–2 h to the alkaline extraction medium, since otherwise changes occurred which prevented its precipitation at pH 6·8. The amino acid composition of the hordenin fraction was found to differ from the hordein and from the albumins and globulins. The nitrogen left unextracted, on the other hand, had an amide-nitrogen content as low as that of the salt-soluble proteins. Folkes and Yemm (1956), like Scriban (1951), therefore attributed it to some denatured proteins but also considered the possibility that it could correspond to proteins enclosed in unbroken cells.

B. ADDITION OF REDUCING AGENTS

Lontie *et al.* (1953) suggested that, besides the hydrolysis of amide groups, destruction of arginine, dismutation of the disulphide bridges and even hydrolysis of peptide bonds could take place in an alkaline medium. They therefore searched for a less drastic solvent and, following the earlier work of Foster *et al.* (1950) and Bishop (1939), examined the influence of reducing agents on the extraction procedure. They found that it was possible, after the extraction of the salt-soluble proteins and then of the prolamins with 60% 2-propanol at 60 °C, to solubilize another important part of the nitrogen fraction by adding 0·2% sodium metabisulphite to the alcoholic solvent at 60 °C (Fig. 8). They attributed the effect to reduction of the disulphide bridges since the same results were obtained with 1% sodium thioglycollate and also with 1% 2-mercaptoethanol or 1% thioglycollic acid (Lontie and Voets, 1959). The solubilized fraction had an amino acid composition very similar to that of the prolamins (Fig. 9) (Lontie and Voets, 1959) and identical to the fraction isolated in the same way by Waldschmidt-Leitz and Mindemann (1957) and called by them "glutenin".

Like the prolamins, the fraction solubilized in the presence of a reducing agent was thus high in proline, very high in glutamic acid and in amide nitrogen and low in the basic amino acids. It should also, therefore, be

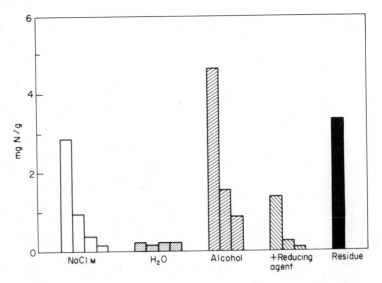

Fig. 8. Nitrogen content/g of successive extracts of meal with 1 M NaCl buffered at pH 7·8, with water, with 60% (v/v) 2-propanol at 60 °C with and without 0·2% $Na_2S_2O_5$ and of the residue. Results taken from Lontie and Voets (1959) and obtained on meal with a nitrogen content of 1·82%, prepared from dehusked Kenia barley and treated with acetone 75% (v/v) at −15 °C.

considered a prolamin rather than a glutelin fraction. It appeared mono-disperse in the ultracentrifuge with a molecular weight of about 46 000 (Waldschmidt-Leitz *et al.*, 1961), instead of 27 500 as found for the classical prolamin fraction (Quensel and Svedberg, 1938). Moreover it separated on electrophoresis into five components, all of which were different from the electrophoretic components obtained from the usual hordein fraction.

The nitrogenous fraction left unextracted after treatment with a reducing agent and called "residue" or "residual fraction" by Lontie and Voets (1959), differed completely in amino acid composition from a prolamin (Fig. 9).

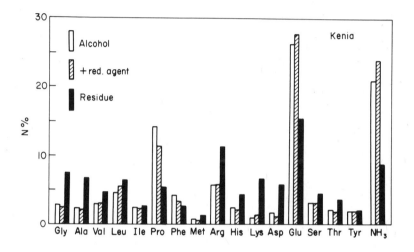

FIG. 9. Amino acid composition, expressed as per cent of the total nitrogen found by ion-exchange chromatography, of the alcoholic fractions of Fig. 8 obtained at 60 °C, in the absence and in the presence of 0.2% $Na_2S_2O_5$, and of the residue (Lontie and Voets, 1959).

Rather, it resembled the salt-soluble proteins and was quite similar to the hordenin fraction of Folkes and Yemm (1956) which was found to be almost free of prolamins. Lontie and Voets (1959) preferred not to use the term "hordenin" in order to avoid any confusion with the "hordenine" alkaloid of the barley rootlets. This residual fraction could be solubilized fairly well by using 0.05 N NaOH in 60% 2-propanol at 60 °C or even better by means of sodium borohydride at a concentration of 0.3–0.5% under the same conditions. Borohydride is also preferable because it is less alkaline, with a pH of about 10.5–11.0. The soluble part was then referred to as the "residual fraction" and the nitrogen left unextracted after that treatment as the "residue".

C. INFLUENCE OF TEMPERATURE

In order to obtain even more typical prolamin and glutelin fractions, alcoholic extractions were carried out in the absence and in the presence of the reducing agents, successively at 20 °C and at 60 °C (Fig. 10). On comparing the amino acid composition of the dialysed and lyophilized fractions with

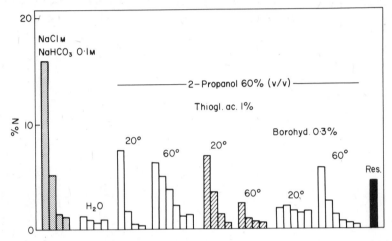

FIG. 10. Yield of the successive extracts, expressed as per cent of the total nitrogen, obtained from barley meal prepared as described on Fig. 8.

each other (Fig. 11) and with those of Fig. 9, it appeared that the method is efficient. Indeed the composition of the prolamin fraction extracted at 20 °C was even more typical (higher proline content) as was the residual fraction solubilized at 60 °C, after preliminary extractions at 20 °C (lower glutamic acid and proline contents, higher content of the basic amino acids). The fraction solubilized at 20 °C in the presence of thioglycollic acid was much more similar to the prolamin fraction of Fig. 9 extracted directly at 60 °C by the alcoholic solvent alone, while the fraction later solubilized at 60 °C was much less prolamin rich. It resembled more the fraction extracted at 20 °C with sodium borohydride which thus still appeared to contain some prolamins. The residual fraction extracted at 60 °C with the alkaline reducing agent, on the other hand, had, as already stated, a definitely more characteristic glutelin composition.

The amount of prolamin-like fraction solubilized with the help of thioglycollic acid depended markedly on the quantity already extracted by the alcoholic solvent alone, since the sum of both fractions was found to be very constant from one experiment to another, at least when the salt extractions and subsequent treatments with water had been performed under the same conditions and when the duration of each extraction was 1 h. This can best

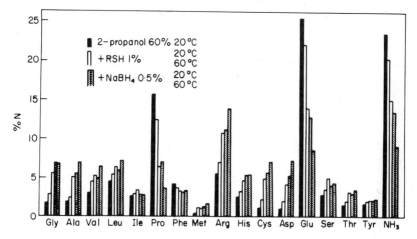

FIG. 11. Amino acid composition, expressed as in Fig. 9, of alcoholic extracts similar to those of Fig. 10, obtained at 20 °C in the absence and at 20 and 60 °C in the presence of 1% thioglycollic acid and of 0·5% sodium borohydride.

be deduced from the comparison of the upper and lower parts of Fig. 12, where the alcoholic extractions (4 of 1 h each per solvent and temperature) were performed either successively at 20 and at 60 °C or simultaneously at 60 °C. In both experiments the same total yield was reached for the whole of the alcoholic extractions without and with thioglycollic acid, even when less nitrogen had been extracted by the alcohol alone.

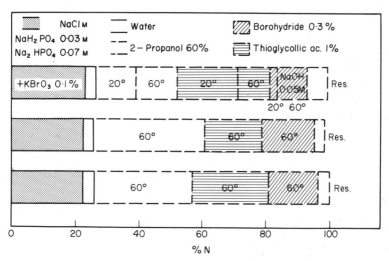

FIG. 12. Influence of the temperature (upper and lower part) and of the duration of the extractions (upper and lower part 60 min, middle part 35 min) on the yield of the prolamin fraction solubilized by 60% 2-propanol successively in the absence and in the presence of 1% thioglycollic acid. Experiments performed on barley meal prepared as described on Fig. 8.

Reducing the duration of the extractions from 60 to 35 min (Fig. 12, middle part) increased the yield of the alcohol-soluble prolamins but did not allow complete extraction of the prolamin-like fraction solubilized in the presence of thioglycollic acid. When the extraction of the globulin fraction was omitted, the yield of the nitrogen extracted by alcohol alone was not affected (Fig. 13; see also Fig. 5). However, the yield of the fraction solubilized in the

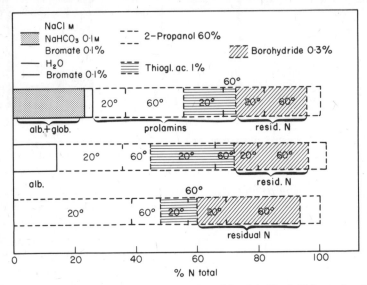

FIG. 13. Influence of the omission of the extraction of the globulins (middle part) and of the albumins and globulins (lower part) on the yield of the subsequent extractions compared with the usual scheme (upper parts). Experiments performed on barley meal prepared as described on Fig. 8.

presence of thioglycollic acid increased a corresponding amount, indicating that the globulins must apparently be soluble in that medium. This was further confirmed by solubility measurements performed on a purified globulin fraction.

When the extraction of the albumin fraction was omitted too (Fig. 13), total yield of nitrogen extracted by alcohol without and with thioglycollic acid increased only very little, as compared with the experiment where only the extraction of the globulin had been suppressed. A marked increase was, however, obtained in the residual fraction solubilized at 60 °C. In contrast with the globulins, the albumin fraction thus appears to be insoluble in both prolamin solvents with the exception, however, of a small part probably corresponding to the low molecular weight components normally extracted with it by the aqueous solutions. Thus, when using the extraction scheme indicated in Fig. 10, the prolamin fraction obtained in the presence of the acid reducing agent may be contaminated by unextracted globulins and the residual fraction extracted in the presence of sodium borohydride by unextracted albumins.

Comparison can be made of the total nitrogen extracted by alcohol without and with thioglycollic acid, when both the albumin and globulin fractions are (Fig. 13, upper part) or are not (Fig. 13, lower part) extracted first. The assumption can also be made that the globulin, but not the albumin fraction, is soluble in 60% 2-propanol in the presence of thioglycollic acid. When this is done and extraction is applied directly to the meal, then the alcoholic solution alone seems to solubilize the whole of the prolamin fraction, without the need of thioglycollic acid, even at 20 °C. From these results and from the constancy of the total amount of nitrogen extracted by alcohol with and without thioglycollic acid, when the salt-soluble proteins have been removed first (Fig. 12), one may conclude that the nitrogen fractions extracted by alcohol, both in the presence and absence of thioglycollic acid, must, in the absence of any interference of the globulins, correspond to one and the same fraction, part of which becomes insoluble during the prior salt and water extractions, probably as a result of oxidation. That oxidation occurs can also be deduced from the work of Ewart (1968), who used starch gel electrophoresis in 8 M urea and lactate buffer, successively in the presence and in the absence of 2-mercaptoethanol, to study the glutelin fraction solubilized by 0·1 N acetic acid after pre-extraction of the salt-soluble and classical alcohol-soluble fractions. He found that in the presence of the reducing agent some bands appeared in the region of the alcohol-soluble proteins, at the expense of some slow moving components, but that the mobility of the albumin- and globulin-like components, which were also found in these extracts, remained unaffected. He thus concluded that some of the components of the glutelin fraction, as defined by Osborne (1895), which apparently did not correspond to salt-soluble proteins, are largely cross-linked by disulphide bonds. However, he also considered the possibility of other types of cross-links. Ewart (1972) found that thiol groups indeed became titratable with phenylmercury acetate when the glutelin fraction extracted with 0·1 N HOAc was reduced with $Na_2S_2O_5$ in the presence of guanidinium chloride.

Thus the higher yields obtained when the alcoholic extractions are performed directly on the meal, as observed by Bishop (1928) and by Urion et al. (1944b) (cf. Fig. 7), may apparently be attributed to solubilization of more prolamin. It seems wise to follow this procedure (Osborne and Clapp, 1907; Waldschmidt-Leitz and Kloos, 1959), when one is especially interested in the study of the prolamins.

With regard to the "residual fraction", solubilized in the presence of sodium borohydride, i.e. the non-prolamin part of the glutelin fraction according to Osborne (1895), it is still not clear whether it corresponds to some denatured salt-soluble proteins, to proteins enclosed in unbroken cells or in some sub-cellular structures, or to a definite membrane fraction insoluble because bound or imbedded in an insoluble matrix.

That subcellular bodies and bound proteins are indeed to be found in barley can be deduced from the electron micrographs of Jones (1969) and from the freeze-etching study of Buttrose (1971), who observed that besides the spherosomes and the aleurone grains containing crystalloids and globoids, the aleurone cells also contained structures similar to microbodies and rough endoplasmic reticulum characterized by the presence of numerous polyribosomes. Furthermore, some material apparently corresponding to proteins was found to adhere to several of the large starch granules (Pomeranz and Sachs, 1972; Munck, 1972) and latent β-amylase to be bound to the hordein-containing protein bodies (Tronier and Ory, 1970). Also many other hydrolase activities were found to be associated with protein bodies (Ory and Henningsen, 1969) such as phytase, which appears to be linked mainly with the fine structure of the appendages attached to the membrane of some of them (Tronier et al., 1971).

D. OTHER PROCEDURES

In order to solve the problem of the origin of the residual fraction, some further experiments remain to be done. For example, the whole of the extraction scheme of Fig. 10 ought to be applied to meal without any intact cells or to endosperm-rich fractions, as prepared, e.g. by Normand et al. (1965), by Novacek et al. (1966) or by Pomeranz et al. (1971) or even to some subcellular bodies such as the hordein-rich "protein bodies" as isolated by Ory and Henningsen (1969).

Extractions in the presence of cellulase preparations were found to increase the yield of the water-soluble nitrogen fraction of barley (Préaux et al., 1973), of several cereals (Hirano et al., 1963), of coconuts (Chandrasekaran and King, 1967), and of mung beans (Hang et al., 1970a) or peas (Hang et al., 1970b). Another possible origin of the residual fraction is, e.g. insolubilization that could have occurred during the extractions as a result of interaction with the non-nitrogenous components (e.g. polyphenols and phytin) of the barley grain. In order to reduce the possible interaction with the polyphenols, extractions are always performed on meal prepared from dehusked barley (Rondelet and Lontie, 1953) and pre-extracted with acetone 75% (v/v) at $-15\,^{\circ}C$ (Rondelet and Lontie, 1955). In order to avoid an interaction with phytin, salt extractions are performed at a pH value above the isoelectric point of most of the proteins, at least at pH 6·6 (Préaux et al., 1963) or even above (Rondelet and Lontie, 1955; Préaux et al., 1963).

Instead of using NaBH$_4$ or NaOH for the extraction of the residual fraction remaining after alcoholic extractions in the presence of a reducing agent, excellent solubilization can also be obtained (Ivanko, 1971) in borate buffer of pH 10 by the successive addition of 0·6% mercaptoethanol and 0·5% sodium dodecyl sulphate (SDS), which are known to release nucleic acids and lipids from proteins. A 0·5% solution of SDS containing 0·6% mercaptoethanol, which is capable of completely solubilizing the proteins of membranes, was also found to be very efficient and, except for a small proportion, even to solubilize the whole of the nitrogen of untreated meal (Landry et al., 1972).

To summarize, it may be said that the glutelin fraction, as defined by Osborne (1895), still contains some prolamins, but these can be removed very efficiently with the classical prolamin solvent, to which an acid or neutral reducing agent has been added. The nitrogen fraction left unextracted after such treatment has an amino acid composition very similar to that of the salt-soluble proteins. It can be solubilized fairly well, and under less alkaline conditions than before, with the help of sodium borohydride or with sodium dodecyl sulphate (Ivanko, 1971). The reason why part of the prolamins becomes oxidized during the aqueous extractions still remains unclear and further experiments remain to be done to determine the precise origin of the true residual salt-soluble fraction.

ACKNOWLEDGEMENTS

We wish to express our gratitude to the Institut pour l'Encouragement de la Recherche Scientifique dans l'Industrie et l'Agriculture and to the Centre Technique et Scientifique de la Brasserie, de la Malterie et des Industries Connexes for the grants awarded to our research projects. We are also grateful to Miss A. Monshouwer for her skilful assistance.

REFERENCES

American Society of Biological Chemists and American Physiological Society: recommendations on protein nomenclature (1908). *J. biol. Chem.* **4**, Proc. 48–51.
Biserte, G. and Scriban, R. (1950). *Bull. Soc. Chim. Biol.* **32**, 959–968.
Bishop, L. R. (1928). *J. Inst. Brew.* **34**, 101–118.
Bishop, L. R. (1929). *J. Inst. Brew.* **35**, 316–322.
Bishop, L. R. (1930). *J. Inst. Brew.* **36**, 336–346.
Bishop, L. R. (1939). "The Proteins of Barley Grains with Special Reference to Hordein". M.Sc. Thesis, University of Birmingham.
Bishop, L. R. (1961). *J. Inst. Brew.* **67**, 244–248.
Brown, H. T. (1903–6). *Trans. Guin. Res. Lab.* (taken from Bishop, 1928).
Buttrose, M. S. (1971). *Planta* **96**, 13–26.
Chandrasekaran, A. and King, K. W. (1967). *Agric. Fd Chem.* **15**, 305–309.
Csonka, F. A. and Jones, D. B. (1929). *J. biol. Chem.* **82**, 17–21.
Djurtoft, R. (1961). Salt-Soluble Proteins of Barley (*Hordeum distichum*). *Dansk Videnskabs Forlag* A/S København.
Enari, T.-M. and Mikola, J. (1961). European Brew. Conv., Vienna pp. 62–68.
Enari, T.-M., Nummi, M., Mikola, J. and Mäkinen, V. (1962). *Finska Kemists. Medd.* **71**, 44–52.
Ewart, J. A. D. (1968). *J. Sci. Fd Agric.* **19**, 241–245.
Ewart, J. A. D. (1972). *J. Sci. Fd Agric.* **23**, 567–579.
Fink, H. and Kunisch, G. (1937). *Wochschr. Brauerei* **54**, 209–212.
Folkes, B. F. and Yemm, E. W. (1956). *Biochem. J.* **62**, 4–11.
Foster, J. F., Yang, J. T. and Yui, N. H. (1950). *Cereal Chem.* **27**, 477–487.
Gortner, R. A., Hoffman, W. F. and Sinclair, W. B. (1928). *Kolloid Z.* **44**, 97–108.
Grabar, P. and Daussant, J. (1963). European Brew. Conv., Brussels, p. 43–58.

Hang, Y. D., Wilkens, W. F., Hill, A. S., Steinkraus, K. H. and Hackler, L. R. (1970a). *Agric. Fd Chem.* **18**, 9–11.

Hang, Y. D., Wilkens, W. F., Hill, A. S., Steinkraus, K. H. and Hackler, L. R. (1970b). *Agric Fd Chem.* **18**, 1083–1085.

Hatefi, Y. and Hanstein, W. G. (1969). *Proc. natn. Acad. Sci.* **62**, 1129–1136.

Hermbstädt (1831). *J. Chem. Techn. Ökonom Chim.* **12**, 1–53.

Hirano, K., Urashima, Y. and Kuroda, A. (1963). *Hakko Kogaku Zasshi* **41**, 288–292.

Hofman-Bang, G. (1930). *J. Inst. Brew.* **36**, 381–388.

Hofman-Bang, G. (1931). *J. Inst. Brew.* **37**, 72–80.

Ivanko, S. (1971). *Biol. Pl.* **13**, 155–164.

Jones, R. L. (1969). *Planta* **85**, 359–375.

Kent, N. and Macheboeuf, M. (1949). European Brew. Conv., Luzern pp. 1–6.

Kleinschmitt, A. (1907). *Z. physiol. Chem.* **54**, 110–118.

Landry, J., Moureaux, T. and Huet, J.-C. (1972). *Bios* (Paris) **7–8**, 281–292.

Larmour, R. K. (1927). *J. agric. Res.* **35**, 1091–1120.

Lontie, R., Rondelet, J. and Dulcino, J. (1953). European Brew. Conv., Nice, pp. 33–38.

Lontie, R. and Voets, T. (1959). European Brew. Conv., Rome, pp. 27–36.

Munck, L. (1972). *Hereditas* **72**, 1–128.

Normand, F. L., Hogan, J. T. and Deobald, H. J. (1965). *Cereal Chem.* **42**, 359–367.

Novacek, E. J., Petersen, C. F. and Slinkard, A. E. (1966). *Cereal Chem.* **43**, 384–399.

Ory, R. L. and Henningsen, K. W. (1969). *Pl. Physiol.*, *Lancaster*, **44**, 1488–1498.

Osborne, T. B. (1895). *J. Am. chem. Soc.* **17**, 539–567.

Osborne, T. B. (1908). *Science, N.Y.* **28**, 417–427.

Osborne, T. B. and Harris, I. F. (1903). *J. Am. chem. Soc.* **22**, 323–353.

Osborne, T. B. and Clapp, S. M. (1907). *Am. J. Physiol.* **19**, 117–124.

Osborne, T. B. (1919). "The Vegetable Proteins". Longmans, Green and Co., London.

Pomeranz, Y., Ke, H. and Ward, A. D. (1971). *Cereal Chem.* **48**, 47–58.

Pomeranz, Y. and Sachs, I. B. (1972). *Cereal Chem.* **49**, 1–4.

Préaux, G., Van Gijsel, A., Robin, C., Awouters, F. and Lontie, R. (1963). Europear Brew. Conv., Brussels, pp. 59–77.

Préaux, G., Boni, L., Galer, J., Hofman, M. and Lontie, R. (1973). FEBS Specia Meeting Dublin, Abstract 95.

Proust (1817). *Ann. Chim. Phys.* **5**, 337–350.

Quensel, O. and Svedberg, T. (1938). *C. r. trav. Lab. Carlsberg, Sér. chim.* **22**, 441–448.

Rondelet, J. and Lontie, R. (1953). European Brew. Conv., Nice, pp. 22–32.

Rondelet, J. and Lontie, R. (1955). European Brew. Conv., Baden-Baden, pp. 90–97

Rose, R. C. and Anderson, J. A. (1936). *Can. J. Res.* **14C**, 109–116.

Scriban, R. (1951). "Les protides de l'orge, du malt et du moût". Lille, Imprimeri Morel et Corduant.

Tronier, B. and Ory, R. L. (1970). *Cereal Chem.* **47**, 464–471.

Tronier, B., Ory, R. L. and Henningsen, K. W. (1971). *Phytochemistry* **10**, 1207 1211.

Urion, E. and Golovtchenko, V. (1940). *Bull. Soc. Chim. biol.* **22**, 203–213.

Urion, E., Lejeune, G. and Golovtchenko, V. (1944a). *Bull. Soc. Chim. biol.* **2(** 221–227.

Urion, E., Lejeune, G. and Golovtchenko, V. (1944b). *Bull. Soc. Chim. biol.* **2(** 303–310.

Urion, E., Lejeune, G. and Collin, M. (1944c). *Bull. Soc. Chim. biol.* **26**, 310–31(

Urion, E., Lejeune, G. and Thiec, J. (1944d). *Bull. Soc. Chim. biol.* **26**, 316–323.
Waldschmidt-Leitz, E. and Brutschek, H. (1955). *Brauwissenschaft* **8**, 278–283.
Waldschmidt-Leitz, E. and Kloos, G. (1959). *Z. physiol. Chem.* **314**, 218–223.
Waldschmidt-Leitz, E. and Mindemann, R. (1957). *Hoppe-Seyler's Z. physiol. Chem.* **308**, 257–262.
Waldschmidt-Leitz, E., Mindemann, R. and Keller, L. (1961). *Hoppe-Seyler's Z. Physiol. Chem.* **323**, 93–97.

CHAPTER 5

Mechanism of Protein Synthesis in Higher Plants

ORIO CIFERRI

Institute of Plant Physiology, University of Pavia, Italy

I. INTRODUCTION

Although quite a few of the first reports on protein synthesis in *in vitro* systems dealt with higher plants (for reviews see Ts'o, 1962; Boulter, 1965; Mans, 1967; Cocking, 1968; Allende, 1969), studies on the mechanism of protein synthesis in these organisms have not progressed at the same pace as those on bacteria and animal tissues. One of the reasons was that at the beginning there was considerable controversy about the efficiency of the protein synthesizing systems isolated from higher plants. When this controversy was resolved, it became apparent that the plant systems were not as active as originally claimed. On the contrary, the incorporation of amino acids in such systems was substantially lower than that obtained in the systems prepared from prokaryotes and animal tissues. In addition, the preparation of cell-free extracts and the fractionation of the different components from plants was more difficult than from other organisms, possibly because of the presence of higher levels of nucleases and of the difficulties inherent in the breakage of the cell wall. Furthermore another difficulty, albeit realized *a posteriori*, was due to the presence in photosynthetic eukaryotes of different synthesizing systems in the cytoplasm, in the chloroplast and in the mitochondrion. The last point makes it more difficult to separate and characterize the components responsible for protein synthesis in plants. Indeed, the unambiguous characterization of each system depends on preparing protein synthesizing systems from cytoplasm, chloroplasts and mitochondria with little or no cross-contamination. This implies that the organelles must be isolated in substantial quantities, intact and uncontaminated by other cell components. The last requirement is particularly vexing, since chloroplast preparations are frequently contaminated by mitochondria, and vice versa. The availability of chloroplast-free plants, either naturally occurring forms or artificially induced mutants, has provided material for obtaining preparations of plant mitochondria uncontaminated by chloroplasts. The reverse situation, that is a plant endowed with chloroplasts but devoid of mitochondria, does not exist or, at least, has never been found. In addition, while it is possible to discriminate between protein synthesis in the cytoplasm and in organelles, for instance by using selective inhibitors, it is much more difficult to do so in the case of mitochondria and chloroplasts since the characteristics of their protein synthesizing machinery are very similar (see below). Despite these limitations, enough is known about the mechanism of protein synthesis in the cytoplasm of eukaryotic plants to indicate that the mechanism is basically the same as that operating in analogous systems developed from prokaryotes or from animal tissues or cells.

It must be added that protein synthesis covers a wide range of topics, including the transcription of nuclear DNA, the synthesis and assembly of ribosomes, the synthesis of precursors such as nucleotides and amino acids, and the influence of the physiological conditions, including the action of hormones, on the synthesis of proteins. Here we shall deal only with the mechanism for protein synthesis in the cytoplasm of plants, mainly as deduced by studies with *in vitro* systems. In addition, some of the interrelations among the processes of protein synthesis in the cytoplasm, in the chloroplasts and in the mitochondria will be briefly outlined.

II. Ribosomes

The existence of at least two classes of ribosomes in plant cells was established as early as 1962 by Lyttleton in the case of spinach (Lyttleton, 1962). Those with a higher sedimentation coefficient, the so called "80S ribosomes" ("eukaryotic ribosomes"), were found to be localized in the cytoplasm while the ribosomes with a sedimentation coefficient of 70S such as that of bacterial ribosome ("70S" or "prokaryotic ribosomes") were present in the chloroplasts. While there may be wide differences in the sedimentation coefficients of the ribosomes isolated from organelles, especially from mitochondria, the ribosomes present in the cytoplasm of all eukaryotes so far examined have always been found to be of the 80S type. This difference in the basic structure of the ribosome is reflected in a number of functional differences. Thus all the evidence so far gathered indicates that "eukaryotic" ribosomes respond in *in vitro* assays to elongation, and possibly initiation and termination factors isolated from the cytoplasm of eukaryotes and not to the analogous factors isolated from prokaryotes. Conversely "prokaryotic" ribosomes, whether from bacteria, blue-green algae, chloroplasts or mitochondria, respond to the factors isolated from prokaryotes or organelles but not to those from the cytoplasm of eukaryotes (Ciferri, 1972). This basic difference in the structure and the function of the two types of ribosomes is further reflected in the sensitivity to some selective inhibitors of protein synthesis. For example, an antibiotic such as chloramphenicol inhibits protein synthesis *in vitro* and *in vivo* in prokaryotes and cell organelles but not in the cytoplasm of eukaryotes. On the other hand, cycloheximide blocks protein synthesis in the cytoplasm of almost all eukaryotic cells so far tested but is ineffective in the case of prokaryotes or of cell organelles. Furthermore, attempts to prepare "hybrid" ribosomes employing isolated ribosomal subunits from different organisms has been successful only when the subunits came from ribosomes of the same type (e.g. from bacteria or from the cytoplasm of eukaryotes) but not when attempts were made to cross the borderline prokaryotes/eukaryotes (Lederberg and Mitchison, 1962). As it will be discussed later, in at least one case the partial reconstitution of a hybrid ribosome between the ribosomal subunits from a bacterium and those from chloroplasts has been successful. It would certainly be very interesting to explore the possibility of preparing other hybrid ribosomes of the prokaryotic type, employing different subunits from bacteria, blue-green algae, chloroplasts and mitochondria. This would be especially interesting in the case of the ribosomes from the latter organelles since their sedimentation coefficients vary considerably (from the 55–60S of the mitochondrial ribosomes from animals, to the 70–75S of those from fungi and the 78–80S of those from higher plants).

III. Mechanism of Protein Synthesis

The basic mechanism for the synthesis of proteins in the cytoplasm of eukaryotic organisms is summarized schematically in Table I. Briefly, three different sequences of reactions occur in this process, namely the initiation of the peptide chain, the elongation of that chain and the termination of its synthesis. The initiation (see top of Table I) and termination (see bottom of Table I) reactions occur once for every peptide chain synthesized, while the elongation step (middle portion of Table I) is repeated as many times as there are mRNA codewords for specifying amino acids. It may be added that ribosomes, translation factors, mRNA, tRNAs and aminoacyl-tRNA synthetases act catalytically in such reactions, since the total process only consumes amino acids, ATP and GTP.

TABLE I

Cytoplasmic protein synthesis in eukaryotes

INITIATION

Initiator met-tRNA binds to smaller ribosomal subunit. mRNA and larger ribosomal subunit added to initiation complex.
Initiation factors required, GTP required and hydrolysed.

ELONGATION

Aminoacyl$_1$-tRNA brought to ribosome by *elongation factor-1* corresponding to second codon on mRNA. Peptide bond synthesized on ribosome. Met-aminoacyl$_1$-tRNA translocated on ribosome by *elongation factor-2*. GTP required and hydrolysed.
Aminoacyl$_2$-tRNA brought by EF-1 corresponding to third codon on mRNA, etc. . . .

TERMINATION

Corresponding to terminating codon(s) on mRNA, completed peptidyl-tRNA released from ribosome-mRNA complex. *Release factors* required, GTP required and hydrolysed. Ribosomes dissociate into subunits.

A. INITIATION

It is well established that peptide chains are synthesized in prokaryotes starting with methionine formylated in the *N*-terminal position. In these organisms, a special methionyl-tRNA functions as initiator of the synthesis of the peptide chain after a preliminary formylation catalysed by a transformylase that is probably absent from the cytoplasm of eukaryotes. In contrast in the cytoplasm of higher plants, peptide chains are initiated with nonformylated methionine as previously found for the analogous systems isolated from animal tissues. Two methionyl-tRNAs can be isolated from the cytoplasm of wheat embryos (Marcus *et al.*, 1970; Leis and Keller, 1970a, b; Tarrago *et al.*, 1970; Ghosh *et al.*, 1971) and from different beans (Yarwood *et al.*, 1971; Guillemaut *et al.*, 1973a). Neither of the two forms is formylated by the transformylase isolated from the same plant, that is probably of chloroplast or mitochondrial origin. One of the two met-tRNAs, designated as either met-tRNA$_1$ or met-tRNA$_2$, transfers unformylated methionine exclusively, or almost so, to the N-terminal position of the peptide chain. In addition, this tRNA binds specifically at low Mg^{2+} concentration to the 40S ribosomal subunit in the presence of the initiator codon AUG or to the ribosomes in the presence of a natural (viral) mRNA. The other species, met-tRNA$_m$ or met-tRNA$_2$, transfers methionine to the internal positions of the peptide chain and its activity is not affected by an inhibitor of the initiation reactions such as aurintricarboxylic acid (Marcus et al., 1970a, b). Thus, as reported for the cytoplasmic systems from animals, the first tRNA appears to function as initiator tRNA while the second is that responsible for the insertion of methionine in the internal positions of the peptide chain. Some authors have reported that in plants, in contrast with the animal and bacterial systems, the initiation factors appear to present in the high speed (post-ribosomal) supernatant rather than on the ribosomes (Marcus, 1970a, b). However, more recently, initiation factors have been isolated also from the salt washings of cytoplasmic ribosomes from pea (Wells and Beevers, 1973). The universality of the mechanism for peptide chain initiation (and elongation) in prokaryotes and eukaryotes is elegantly demonstrated in experiments involving the *in vitro* translation of the RNA from Satellite Necrosis Tobacco Virus (SNTV) (Klein et al., 1972; Klein and Clark, 1973). In these experiments the terminal peptide of the coat protein synthesized by the system prepared from *Escherichia coli* is fmet-ala-lys . . . while that synthesized by a system from wheat germ is met-ala-lys. . . . In addition, more recently it has been demonstrated that a system from wheat germ translates with good efficiency and high fidelity the mRNA for rabbit haemoglobin (Roberts and Paterson, 1973).

B. ELONGATION

Elongation of the peptide chain in the plant cytoplasmic systems requires the concomitant presence of two protein factors. One, designated as EF-1, is responsible for the binding, in the presence of GTP, of the aminoacyl-tRNA to the ribosomes. The second factor, EF-2, is responsible for the translocation of the peptidyl-tRNA from the ribosomal acceptor site to the donor site. Such elongation factors have been isolated from a variety of plants and resolved into the two complementary fractions in the case of the preparations obtained from wheat germ (Legocki and Marcus, 1970) and broad bean (Yarwood et al., 1971). From wheat germ, Legocki (Golinska and Legocki, 1973; Twardowski and Legocki, 1973) has purified an apparently homogeneous protein with a molecular weight of 70 000 that shows all the characteristics of EF-2. From this material, EF-1 was also considerably purified but it still appears to be heterogeneous, possibly because it contains different oligomers of the same protein with a molecular weight of approximately 60 000. The problem of the existence of multiple forms of EF-1 has frequently arisen during the isolation of such factor(s) from eukaryotes (Schneir and Moldave, 1968; McKeehan and Hardesty, 1969; Lanzani et al., 1974) and it would be of a great interest to ascertain which form is the active one *in vivo* and what role the possible monomer–oligomer transition plays in the regulation of the protein synthetic activity.

The reactions for peptide chain elongation in plant cytoplasm thus appear to be very similar to those observed in the case of the systems from animals. This is in good agreement with early observations on the interchangeability of the factors between plants and animals (Parisi *et al.*, 1967) that have been subsequently confirmed by using separated elongation factors (Yarwood *et al.*, 1971). One characteristic feature of the elongation factors from prokaryotes is the possibility of further fractionating EF-T (functionally equivalent to the EF-1 of eukaryotes) into two complementary factors, $EF-T_u$ and $EF-T_s$. The former factor is required for the actual binding of aminoacyl-tRNA to the ribosomes while the function of the latter is possibly that of regenerating an active form of $EF-T_u$. Up to now, there is no evidence for the existence of a third elongation factor in the cytoplasm of eukaryotes nor for different forms or aggregates of EF-1 having different activities, so that it must be surmised that one single protein, EF-1, catalyses the reactions that in bacterial systems are carried on by $EF-T_u$ and $EF-T_s$. Furthermore, it is possible that in the cytoplasm of photosynthetic eukaryotes the active form of the elongation factors is a complex composed of both EF-1 and EF-2. Legocki (see above) has presented evidence that may indicate the presence of a stable complex of the two elongation factors in imbibed wheat germs while only the separated factors would be present in the dry, quiescent seed. Such results indicate that during the early stage of germination the elongation factors could pass from an inactive form or state to an active one characterized by the formation of a multienzyme complex. Such a transition is somewhat analogous to that previously reported for the formation of polysomes during water imbibition of seeds (Marcus and Feeley, 1965).

C. TERMINATION

Nothing is known about peptide chain termination in the cytoplasm of plants. The apparent universality of the genetic code that has been confirmed in part also in the case of the cytoplasmic ribosomes from wheat (Basilio *et al.*, 1966) makes it very likely that the codons UAG, UAA and UGA, demonstrated to act as chain terminators in *in vitro* systems from bacteria and animal tissues, do so also in the case of plants. To our knowledge, no data have been reported demonstrating the presence in plants of release factors analogous to those responsible for peptide chain termination in bacteria or mammals.

D. RIBOSOME CYCLE

Since the initiation of peptide chain synthesis takes place on the small ribosomal subunit, it is obvious that at a certain stage of protein synthesis dissociation of the ribosomal monomers into two subunits must take place. The mechanism proposed for *Escherichia coli* envisages that once a polypeptide chain is completely synthesized and the ribosome reaches the end of the mRNA, it dissociates immediately into its subunits. By density labelling techniques, Kaempfer (1969) has demonstrated the formation in the cytoplasm of *Candida krusei* of ribosomes with a hybrid density, after transferring a culture growing in a medium containing heavy isotopes to a medium containing light isotopes. Such ribosomes with hybrid density could not have been formed except by random association of preformed (heavy) subunits and newly synthesized (light) subunits. No other evidence for a ribosome cycle in the cytoplasm of eukaryotes has been presented. On the contrary, studies on the ribosomal profiles of plant tissues of different age or in different physiological conditions seem to indicate that a decrease in the polyribosomes is accompanied by an increase in the monosome region and a very slight increase in the areas of ribosomal subunits (Lin and Key, 1967; Lin and Key, 1971; Payne *et al.*, 1971; Lin *et al.*, 1973). Such findings do not, however, exclude the existence of a ribosome cycle in plants since the monosomes may be those arising from the breakage of polysomes during the fairly harsh conditions required for the preparation of cell-free extracts.

IV. tRNA and Aminoacyl-tRNA Synthetase

The limited literature available on the occurrence and characteristics of tRNAs and aminoacyl-tRNA synthetases in the cytoplasm of higher plants has been recently reviewed (Lea and Norris, 1972). The situation is somewhat confused since higher plants contain separate sets of tRNAs for cytoplasmic, chloroplastic, and mitochondrial protein synthesis. Thus, if one assumes that mitochondria from higher plants may contain approximately 20 distinct mitochondrial tRNAs as demonstrated in the case of the mitochondria from yeast (Reijnders and Borst, 1972) and possibly the same number, if not more, may be present in the chloroplasts, the number of different tRNA species present in a single plant cell may be around 100. Also the number of the synthetases is higher than 20. The characterization of these macromolecules is further complicated by the fact that their specificity may vary considerably. Thus, for instance, there are cytoplasmic synthetases that aminoacylate only cytoplasmic tRNAs and not organellar tRNAs, or synthetases from the cytoplasm that charge both cytoplasmic and organellar tRNAs. The same differences in specificity have been shown to occur in the case of the few synthetases of organellar origin so far investigated (see below). In addition, different tissues or organs from the same plant may contain different complements of synthetases and tRNAs for the same amino acid, as demonstrated, for instance, for leucine in the case of soyabean (Anderson and Cherry, 1969; Kanabus and Cherry, 1971). Therefore although diligent efforts have been made to isolate specific tRNAs and synthetases by employing non-green tissues or chloroplast-free mutants or else by isolating tRNAs and enzymes from preparations of presumably intact cell organelles uncontaminated by other cell constituents, present data indicating a lack of specificity must at the moment be viewed with caution. Certainly when more is known about the distribution and the specificity of the different tRNAs and synthetases in different cell compartments and tissues and in the same tissue at different stages of development, it will be possible to assess the role of these macromolecules in the development of the cell and in the differentiation of the plant. An additional problem is that of the possible function of the cytokinins present in many tRNA hydrolysates. Cytokinins have been found frequently in the tRNAs from plants (Skoog and Armstrong, 1970), but their presence in tRNAs from non-photosynthetic organisms such as bacteria, yeast and even animal tissues seems to argue against a special role for them in higher plants. Indeed, it may be simply that some of the precursors of certain modified bases present in tRNA happen to have cytokinin activity. Alternatively, bases with cytokinin activity may represent breakdown products of the turnover of RNAs, although their levels in certain tissues is so high that their formation cannot be accounted for simply on the basis of the hydrolysis of tRNA.

At the moment, we can only conclude that there is ample evidence that many tRNAs contain bases endowed with cytokinin activity. The facts that the concentrations of tRNA is highest in tissues in rapid growth, that leaves and germinating embryos, for instance, contain relatively more tRNA than roots or quiescent seeds and that conspicuous changes are evident in the content and characteristics of the tRNAs and/or the synthetases during development of the plant cell, indicate a possible role of tRNA in the control of protein synthesis and in cell differentiation. The effect of cytokinins on plant growth and morphogenesis may be the expression of changes in the relative rates of protein synthesis brought about by hormonal control of the synthesis and the function of particular tRNA species. The contention that tRNAs play an important role in the control of protein synthesis is partly substantiated by the intriguing aspects of the specific aminoacylation of the terminal 3′ end of the RNA from certain plant viruses. Chapeville and coworkers (Litvak et al., 1970; Pinck et al., 1970; Yot et al., 1970) have shown that the 3′ end of the RNA of turnip-yellow mosaic virus (TYMV) possesses a tRNA-like structure that is recognized as a substrate by the tRNA nucleotidyl transferase from E. coli. After addition of the terminal adenosine, the viral RNA is aminoacylated specifically with valine by the synthetase from E. coli. On the other hand, tobacco mosaic virus (TMV) RNA is specifically aminoacylated with histidine by a yeast synthetase and, to a lesser extent, by the synthetase from wheat embryos (Litvak et al., 1973). Aminoacylation of TMV RNA does not require a preliminary addition of adenosine to the 3′ end since the RNA already has a 3′ terminal sequence -pCpCpA, as all other tRNAs. Even more interesting is the finding that the aminoacylated viral RNAs and not the unacylated ones interact with the wheat embryo elongation factor EF-1 and GTP to give a complex that is similar to the EF-1 . . .

GTP ... aminoacyl-tRNA complex formed during protein synthesis. It is not known if aminoacylation is a prerequisite for viral RNA infectivity nor whether the possible interaction of the aminoacylated viruses with plant EF-1 proves that aminoacylation precedes the replication or the expression of the viral genome. However, by analogy with the involvement of *E. coli* elongation factor T in the enzyme responsible for the replication of the RNA of phage QB (Blumenthal *et al.*, 1972), one may speculate that interaction of host EF-1 with viral RNA may be required for the replication of the plant virus RNA or for its translation by the protein synthetic machinery of the plant. The role played by tRNAs or tRNA-like sequences in the viruses is quite puzzling, as demonstrated also by the fact that certain animal viruses carry in their particles free tRNA molecules capable of accepting up to 15 different amino acids (Bonar *et al.*, 1967; Travnicek, 1969).

V. Products of Cytoplasmic Protein Synthesis

Up to now, there has been no report of the synthesis of a single well-characterized protein by a cell-free system prepared from plant material. Certainly this is in part due to the fact that nothing corresponding to the haemoglobin/reticulocyte relationship in animals exists in plants. Nevertheless, by accurately choosing the right type of material and the right stage of development, it may be possible to isolate relatively large amounts of mRNAs specific for one, or a few, proteins or prepare cell-free systems which synthesize fairly high amounts of a single, or a few, well-characterized proteins. For instance, during the final stages of development of many seeds, almost all the proteins synthesized are storage globulins (Danielson, 1956) that are probably synthesized on a special class of ribosomes (see below). In addition, one could use for such studies the proteins or the enzymes that are induced either by light, such as many photosynthetic enzymes or nitrate reductase (Zucker, 1972), or during development of the plant, e.g. α-amylase in germinating seeds. The synthesis of the latter enzyme, which appears to be formed *de novo* in fairly large quantities during germination of cereal seeds, is stimulated specifically by gibberellins either endogenously synthesized or exogenously supplied (Briggs, 1973). Indeed in naturally embryo-less wheat grains, synthesis of the enzyme occurs only if gibberellin is supplied (Khan *et al.*, 1973). In addition, even from plants it is probably possible to isolate classes of ribosomes that synthesize only certain proteins. Boulter and co-workers (Payne and Boulter, 1969; Briarty *et al.*, 1969; Payne *et al.*, 1971) have demonstrated the presence of free and membrane-bound ribosomes in *Vicia faba* seeds and concluded that, in this plant, storage globulins are synthesized mainly or exclusively on the membrane-bound ribosomes rather than on those that are free. In fact, the relative amounts of the two classes of ribosomes vary with the stage of development of the seed. During the phase of rapid division prior to the synthesis of storage protein, the number of free ribosomes is greater than those that are membrane-bound. Shortly before the synthesis of the storage proteins begins, the number of membrane-bound ribosomes increases drastically. These ribosomes appear to be synthesized *de novo* rather than arising from attachment to the membranes of preexisting free ribosomes. On the contrary, when the seed has reached maturity, an increase in the amount of free ribosomes becomes evident due to the detachment from the membranes of the membrane-bound ribosomes.

VI. PROTEIN SYNTHESIS IN MITOCHONDRIA AND CHLOROPLASTS AND THEIR RELATION TO CYTOPLASMIC PROTEIN SYNTHESIS

It is now clear that chloroplasts and mitochondria, in addition to their own DNA and RNA, contain the complete machinery for DNA replication, transcription and translation. Among points that remain to be clarified is one regarding the contribution of these organelles to the synthesis of organellar proteins and the degree of autonomy of such syntheses. How much do these two types of organelle depend for their activity on the products of the nucleo-cytoplasm? Since other contributors to this volume are dealing with the structure and the biogenesis of chloroplasts and mitochondria from higher plants (see Chapters 6 and 7), we shall limit ourselves to a brief outline of the differences and the similarities of the protein synthesizing systems from the cytoplasm, the mitochondrion and the chloroplast of plants. Before doing so, it must be mentioned that very few data are available in the case of mitochondria from higher plants and it is necessary to assume that what has been found in the case of mitochondria from nonphotosynthetic plants, notably yeast and *Neurospora crassa*, is true for those from higher plants.

A. ORGANELLAR RIBOSOMES

Perhaps the most striking difference between the systems for protein synthesis in the cytoplasm and in either type of organelle is the finding, already mentioned, that the ribosomes from chloroplasts and mitochondria are much more similar to those present in prokaryotes than to those present in the cytoplasm of eukaryotes (Borst and Grivell, 1971; Kroon *et al.*, 1972) (Table II). As in the case of ribosomes from prokaryotes, the sedimentation coefficient of chloroplast ribosomes is approximately 70S and that of the ribosomal RNAs is 23, 16 and 5S. Values in the same range have been reported for the mitochondrial ribosomes from fungi (Küntzel and Noll, 1967; Grivell *et al.*, 1971) and somewhat higher values apply to those from photosynthetic plants (Leaver and Harmey, 1972). On the other hand, mitochondrial ribosomes

TABLE II

Sedimentation coefficient of ribosomes and ribosomal RNAs from representative organisms and organelles

	Cytoplasm	Mitochondrion	Chloroplast	Organism
EUKARYOTES Fungi	80 [26[a], 17[a], 5·8[b], 5[b]]	73 [23[a], 16[a], ?]		a. *Neurospora crassa* (Küntzel and Noll, 1967). b. *Saccharomyces cerevisiae* (Udem et al., 1971).
Higher plants	80 [25[c], 18[c], 5·8[d], 5[e]]	78 [24[f], 18[f], 5[f]]	70 [23[c], 16[c], 5[e]]	c. *Phaseolus vulgaris* (Loening and Ingle, 1967). d. *Pisum sativum* (Sy and McCarty, 1970). e. *Vicia faba* (Dyer and Leech, 1968). f. *Brassica rapa* (Leaver and Harmey, 1972; Leaver, 1973).
Animals	80 [28[g], 18[g], 5·8[d], 5[h]]	55 [17[i], 13[i], ?]		d. Rat liver (Sy and McCarty, 1970). g. Rat liver (Loening, 1968). h. Rat liver (Galibert et al., 1965). i. Rat liver (Aaij and Borst, 1970).
PROKARYOTES	70 [23[l] 16[l], 5[l]]			l. *E. coli* (Kurland, 1960).

from animal cells have sedimentation coefficients around 55–60S (O'Brien and Kalf, 1967; Attardi and Ojala, 1971; Brega and Vesco, 1971) and, as in the case of those from *N. crassa* mitochondria (Lizardi and Luck, 1971), lack a 5S RNA (Attardi and Attardi, 1971). A fourth RNA, the so called 5·8S or 7S RNA (Monier, 1972) that may be isolated on denaturation of the ribosomal RNAs from eukaryotic ribosomes (Pene *et al.*, 1968; Sy and McCarty, 1970; King and Gould, 1970) seems to be absent from all ribosomes of the prokaryotic type (Payne and Dyer, 1972). Regardless of differences in sedimentation coefficient, the ribosomes from organelles, whatever their origin, appear to be of the prokaryotic type on the basis of at least the following criteria:

(1) Requirement of high Mg^{2+} concentration for stability (i.e. 10 mM Mg^{2+} as compared to 1 mM Mg^{2+} required in the case of eukaryotic ribosomes).

(2) Peptide chain initiates with *N*-formyl-methionyl-tRNA and not with methionyl-tRNA. Ribosomes from organelles are active in *in vitro* assays in the presence of initiation factors from prokaryotes.

(3) Peptide chain elongation *in vitro* takes place in the presence of elongation factors from organelles or from prokaryotes but not in the presence of elongation factors from the cytoplasm of eukaryotes.

(4) Protein synthesis is inhibited *in vivo* and *in vitro* by antibiotics inhibiting bacterial ribosomes (e.g. chloramphenicol or lincomycin) and not by those inhibiting the activity of eukaryotic ribosomes (e.g. cycloheximide or anisomycin).

In addition, at least in one case, hybrid ribosomes that still retain activity in an *in vitro* assay have been prepared by mixing the appropriate ribosomal subunits from *E. coli* and from chloroplasts of *Euglena gracilis* (Lee and Evans, 1971).

The site of codification of the components of the organellar ribosomes has been established for both chloroplasts and mitochondria. Hybridization experiments have clearly demonstrated that the ribosomal RNAs of chloroplasts and mitochondria are coded in the organellar DNA and not in the nuclear DNA (Tewari, 1971; Borst, 1972). The genetic information for such RNAs is transcribed most probably in the organelle itself by a specific RNA polymerase that appears to be coded in the nuclear DNA (references in Table III). On the other hand, all the evidence so far available indicates that most, or perhaps all, the proteins of the chloroplast and the mitochondrial ribosomes are coded in the nuclear DNA and translated in the cytoplasm (references in Table III). Thus, the biosynthesis of the organellar ribosomes requires the cooperative gene action of the nucleus and of the organelle itself.

B. ORGANELLAR tRNA AND AMINOACYL-tRNA SYNTHETASE

As early as 1967, Barnett and Brown showed the presence in *Neurospora crassa* mitochondria of tRNAs and synthetases that were different from their cytoplasmic counterparts. These results were soon confirmed in the case of other organisms, including higher plants, and extended to the chloroplasts (Barnett *et al.*, 1969; Burkard *et al.*, 1970; Leis and Keller, 1970b; Merrick and Dure III, 1971; Guderian *et al.*, 1972; Parthier *et al.*, 1972). In a few cases it has been shown that the synthesis of the chloroplast-specific tRNA's and/or aminoacyl-tRNA synthetases is induced on exposure of cells or plants to light. Thus, if cells of *E. gracilis* are exposed to light, new species of tRNA appear in the chloroplast and new synthetases are formed. The latter seem to aminoacylate only the light-induced tRNA's (Barnett *et al.*, 1969; Reger *et al.*, 1970). A similar stimulation by light of the synthesis of plastid-specific tRNA's has been observed in bean (Williams and Williams, 1970; Burkard *et al.*, 1972) and cotton (Merrick and Dure III, 1971).

It does appear, then, that chloroplasts and mitochondria contain tRNAs and aminoacyl-tRNA synthetases distinct from those present in the cytoplasm and which also differ in the two types of organelle. Indeed, Guillemaut *et al.* (1973a) have provided evidence showing that chloroplasts and mitochondria from bean contain distinct tRNAs for N-formyl-methionine and different tRNAs and synthetases for leucine have been separated in the case of mito-chondria and chloroplasts from tobacco leaves (Guderian *et al.*, 1972). However, up to now, it seems that the number of the organelle-specific tRNAs is far smaller than that of the amino acid codewords and possibly even smaller than that of the amino acids present in proteins so that one has to assume that cytoplasmic tRNAs and synthetases are also utilized for organellar protein synthesis.

Some, perhaps all, of the organelle-specific tRNA appear to be coded in the organellar DNA. Specific hybridization of mitochondrial DNA with mitochondrial tRNAs has been reported in the cases of yeast (Casey *et al.*, 1972; Cohen *et al.*, 1972), Hela cells (Aloni and Attardi, 1971), rat liver (Nass and Buck, 1970) and *Xenopus laevis* (Dawid, 1972).

Less is known concerning the coding site of the tRNAs present in the chloroplasts. Total tobacco chloroplast tRNA hybridizes with chloroplast DNA but it is not known whether it hybridizes also with nuclear DNA (Tewari and Wildman, 1970). Leucyl-tRNA from chloroplasts of bean leaves appears to hybridize with both nuclear and chloroplastic DNA although these experiments were performed using the mixture of isoaccepting tRNAs[leu] present in the chloroplasts, two of which seem identical to the cytoplasmic tRNA[leu] species (Guillemaut *et al.*, 1973b). The possibility that some of the chloroplast-specific tRNAs are coded in the nuclear DNA and others in the organellar DNA cannot be discarded at the present. The former may be the species that are aminoacylated by the cytoplasmic enzymes and the latter only by the

TABLE III

Genetic localization and sites of transcription and translation of the macromolecules involved in chloroplast (*) and mitochondrial (†) protein synthesis

	Coded in		Translated in		Organism
	Nuclear DNA	Organellar DNA	Cytoplasm	Organelle	
RNA polymerase	†[1] *[a]		†[2,3] *[b]		1. Yeast (Tsai et al., 1971; Wintersberger, 1970). 2. Neurospora crassa (Barath and Kuntzel, 1972a; Barath and Kuntzel, 1972b). 3. Tetrahymena pyriformis (Millis and Suyama, 1972). a. Chlamydomonas reinhardii (Surzycki et al. 1970). b. Pisum sativum (Ellis and Hartley, 1971).
Ribosomal proteins	*[c,d]		†[4–6] *[e,f]		4. Neurospora crassa (Kuntzel, 1969a; Lizardi and Luck, 1972; Neupert et al., 1969b). 5. Yeast (Schmitt, 1971). 6. Tetrahymena pyriformis (Millis and Suyama, 1972). c. Acetabularia mediterranea (Kloppstech and Schweiger, 1973). d. Nicotiana tabacum (Bourque and Wildman, 1973). e. Acetabularia mediterranea (Kloppstech and Schweiger, 1974). f. Chlamydomonas reinhardii (Margulies, 1971).

Ribosomal RNAs	*g ?		†7–12 *g,h,i,l
tRNAS (some)	*m ?		†13–16 *m
Aminoacyl-tRNA synthetases (some)	†17 *n	*n,o	
Elongation factors	†17,18	†17–19	

7. HeLa cells (Aloni and Attardi, 1971).
8. N. crassa (Schafer and Kuntzel, 1972; Wood and Luck, 1969).
9. Yeast (Fukuhara, 1967; Wintersberger and Viehhauser, 1968).
10. T. pyriformis (Suyama, 1967).
11. Rat liver (Aaij et al., 1970).
12. Xenopus laevis (Dawid, 1972).
g. Various plants (Ingle et al., 1970).
h. N. tabacum (Tewari and Wildman, 1968).
i. Euglena gracilis (Scott and Smillie, 1967).
l. E. gracilis (Stutz and Rawson, 1970).

13. Yeast (Casey et al., 1972; Halbreich and Rabinowitz, 1971; Wintersberger and Viehhauser, 1968).
14. Rat liver (Nass and Buck, 1970).
15. X. laevis (Dawid, 1972).
16. HeLa cells (Aloni and Attardi, 1971).
m. Phaseolus vulgaris (Guillemaut et al., 1973b).

17. N. crassa (Gross et al., 1968).
n. E. gracilis (Reger et al., 1970).
o. E. gracilis (Parthier, 1973).

17. N. crassa (Barath and Küntzel, 1972b).
18. Yeast (Parisi and Cella, 1971; Scragg, 1971; Richter, 1971).
19. Chlorella vulgaris (Ciferri et al., 1974).

chloroplast enzymes. The only report on the site of codification of a chloroplast-specific synthetase indicates that the enzyme is coded in the nuclear
DNA (Reger *et al.*, 1970) as has been found also in the case of a leucyl-tRNA
synthetase from *N. crassa* mitochondria (Gross *et al.*, 1968).

C. ORGANELLAR TRANSLATION FACTORS

It has already been mentioned that mitochondrial and chloroplast ribosomes respond in *in vitro* assays to initiation and elongation factors that are
different from those present in the cytoplasm and possibly even different in
the same organism for each type of organelle (Ciferri and Tiboni, 1973). In
addition, organellar ribosomes have been shown to catalyse the reactions for
peptide chain initiation and elongation in the presence of the corresponding
factors obtained from prokaryotes but not in the presence of those active on
the ribosomes from the cytoplasm of eukaryotes (Kuntzel, 1969b; Sala *et al.*,
1970; Sala and Kuntzel, 1970; Richter and Lipmann, 1970; Grivell *et al.*,
1971; Grivell and Groot, 1972; Swanson, 1973). In the case of two fungi and
one alga, it appears that the genetic information for the mitochondrial
elongation factors is coded in the nuclear DNA and translated on the cytoplasmic ribosomes since such mitochondrion-specific factors are present in
cells in which mitochondrial protein synthesis has been inhibited or even in
yeast cells that do not contain mitochondrial DNA (Parisi and Cella, 1971;
Scragg, 1971; Richter, 1971; Barath and Kuntzel, 1972b; Ciferri *et al.*,
1974). Nothing is known about the site of genetic coding of the translation
factors involved in chloroplast protein synthesis. However, in contrast to
what happens in the mitochondria, the synthesis or activity of translation
factors is controlled by the chloroplast itself, since in *Chlorella vulgaris* the
chloroplast-specific factors are undetectable in an apochlorotic mutant or in
the wild-type strain grown in the dark under heterotrophic conditions
(Ciferri and Tiboni, 1973).

In conclusion, it appears that mitochondria and chloroplasts are capable of
performing a semi-autonomous protein synthesis. Such syntheses are carried
out utilizing machineries distinct and different from those operating in the
cytoplasm. However, with the exception of the RNA of the organellar ribosomes and some of the tRNAs, all the protein components of such systems so
far studied appear to be coded in the nuclear DNA and translated in the cytoplasm (Table III). The dependency of the protein synthetic activity of chloroplasts and mitochondria on the nucleocytoplasm is therefore absolute.

VII. PRODUCTS OF ORGANELLAR PROTEIN SYNTHESIS

Since all the proteins involved in chloroplast and mitochondrial protein
synthesis so far studied appear to be products of the nucleocytoplasm, the
question now arises concerning the nature of the proteins synthesized in these
organelles.

The direct characterization of the products of organellar protein synthesis is at the moment rather unsatisfactory, because the most straightforward approach, that is the incubation of isolated organelles with radioactive amino acids and the analysis of the radioactive protein(s) synthesized, has in general failed to show direct amino acid incorporation into any readily identifiable mitochondrial or chloroplast protein. In these experiments, most of the radioactivity is associated with highly insoluble proteins, presumably associated with the membrane fraction. Another widely used approach relies on the administration to intact cells or tissues of selective inhibitors of cytoplasmic or organellar protein synthesis (e.g. cycloheximide or chloramphenicol) to detect which protein is synthesized and hence becomes labelled when one of the protein synthesizing systems is inhibited (Boulter et al., 1972; Bradbeer, 1973; Ellis et al., in press). This approach is especially useful when the antibiotics are administered during development of the organelles such as when dark-grown cells or plants are exposed to light or when anaerobically-grown cells are exposed to oxygen.

The use of selective inhibitors in vivo allows one to estimate the relative contribution of the organellar system to the total protein synthetic activity. In the case of mitochondria approximately 5–15% of the total mitochondrial protein is synthesized in the mitochondrion itself (Neupert et al., 1969a; Sebald et al., 1969; Beattie, 1970; Schweyen and Kaudewitz, 1970). The figure is much higher in the case of chloroplasts since, for instance, in Chlamydomonas reinhardii approximately 50% of the protein appears to be contributed by the organelle itself (Hoober et al., 1969). The tentative conclusions that may be drawn from studies utilizing the two approaches outlined above is that chloroplasts synthesize autonomously some hydrophobic proteins that constitute the inner membrane of the organelle including the chloroplast lamellae. In addition, some photosynthetic cytochromes as well as the large subunit of ribulose-1,5-diphosphate carboxylase are synthesized in the chloroplast (Table IV) while the small subunit of the enzyme appears to be synthesized in the cytoplasm (Kawashima, 1970; Criddle et al., 1970; Ellis and Hartley, 1971; Kawashima and Wildman, 1972; Ellis et al., in press). The fact that ribulose-1,5-diphosphate carboxylase (Fraction I protein) is by far the most abundant of all chloroplast proteins (Kawashima and Wildman, 1970) may explain why chloroplast ribosomes, although they synthesize very few proteins, may contribute approximately half of the total protein present in the organelle.

Besides membrane proteins, mitochondria appear to synthesize autonomously the three major polypeptides of cytochrome oxidase, cytochrome b and some coupling factors (e.g. the peptides associated with the F_1-ATPase complex) (Schatz and Mason, 1974). Thus the contribution of the protein synthetic activity of organelles seems to be limited to some, probably not all, of the proteins of the organellar inner membrane, including some coupling factors, a few cytochromes, and certain subunits of a few organellar enzymes.

TABLE IV

Possible products of chloroplast protein synthesis

	Organism
Proteins of the inner membrane including thylakoid-constituting proteins	*Vicia faba* (Machold, 1971). *Phaseolus vulgaris* (Bradbeer, 1973). *Pisum sativum* (Ellis *et al.*, in press). *Chlamydomonas reinhardii* (Hoober, 1970). *Euglena gracilis* (Harris *et al.*, 1973).
Larger subunit of ribulose-1,5-diphosphate carboxylase	*P. vulgaris* (Bradbeer, 1973). *Hordeum vulgare* (Criddle *et al.*, 1970). *Nicotiana tabacum* (Kawashima, 1970). *Euglena gracilis* (Smillie *et al.*, 1971; Schiff, 1971). *Chl. reinhardii* (Armstrong *et al.*, 1971). *P. sativum* (Ellis and Hartley, 1971; Blair and Ellis, 1973).
Cytochrome f	*P. vulgaris* (Bradbeer, 1973).
Cytochromes of the b type	*P. vulgaris* (Bradbeer, 1973). *E. gracilis* (Smillie *et al.*, 1971; Schiff, 1971). *Chl. reinhardii* (Armstrong *et al.*, 1971).

Unless one assumes that quite a number of proteins endowed with a regulatory function, such as the replication factor postulated for yeast mitochondria (Williamson, *et al.*, 1971), are present and made in the organelles from the information contained in the organellar DNA, one reaches two disturbing conclusions. First, in the mitochondrial DNA, and more so in the chloroplast DNA, there is apparently much more "genetic space" than is taken up by the organellar ribosomal RNAs, the organellar-specific tRNAs and the organelle-coded proteins. Second, the presence of three complete protein synthesizing systems, each comprising approximately 150–200 different macromolecules, in the cytoplasm, in the chloroplast and in the mitochondrion seems hardly justified if one considers that almost all organellar proteins, are made in the cytoplasm, presumably from the information encoded in the nuclear DNA.

From our present limited knowledge, the situation seems to be irrational and uneconomic for the cell but, when we know more, it will no doubt appear to be justified and advantageous.

ACKNOWLEDGEMENTS

The author thanks Drs R. J. Ellis, C. J. Leaver, A. B. Legocki, B. Parthier and J. H. Weil for access to their manuscripts before publication. The experimental work performed in the author's laboratory was supported by a grant from the Laboratorio di Genetica Biochimica ed Evoluzionistica of the Consiglio Nazionale delle Ricerche.

REFERENCES

Aaij, C. and Borst, P. (1970). *Biochim. biophys. Acta* **217**, 560.

Aaij, C. Saccone, C., Borst, P. and Gadaleta, M. N. (1970). *Biochim. biophys. Acta* **199**, 373.

Allende, J. E. (1969). *In* "Techniques in Protein Synthesis" (P. N. Campbell and J. R. Sargent, eds), Vol. II, p. 55. Academic Press, London and New York.

Aloni, Y. and Attardi, G. (1971). *J. molec. Biol.* **55**, 271.

Anderson, M. B. and Cherry, J. H. (1969). *Proc. natn. Acad. Sci. U.S.A.* **62**, 202.

Armstrong, J. J., Surzycki, S. J., Moll, B. and Levine, R. P. (1971). *Biochemistry* **10**, 692.

Attardi, B. and Attardi, G. (1971). *J. molec. Biol.* **55**, 231.

Attardi, G. and Ojala, D. (1971). *Nature, New Biol.* **229**, 133.

Barath, Z. and Küntzel, H. (1972a). *Nature, New Biol.* **240**, 195.

Barath, Z. and Küntzel, H. (1972b). *Proc. natn. Acad. Sci. U.S.A.* **69**, 1371.

Barnett, W. E. and Brown, D. H. (1967). *Proc. natn. Acad. Sci. U.S.A.* **57**, 452.

Barnett, W. E., Pennington, C. J. Jr. and Fairfield, S. A. (1969). *Proc. natn. Acad. Sci. U.S.A.* **63**, 1261.

Basilio, C., Bravo, M. and Allende, J. E. (1966). *J. biol. Chem.* **241**, 1917.

Beattie, D. S. (1970). *FEBS Letters* **9**, 232.

Blair, G. E. and Ellis, R. J. (1973). *Biochim. biophys. Acta* **319**, 223.

Blumenthal, T., Landers, T. A. and Weber, K. (1972). *Proc. natn. Acad. Sci. U.S.A.* **69**, 1313.

Bonar, R. A., Sverak, L., Bolognesi, D. P., Langlois, A. J., Beard, D. and Beard, J. W. (1967). *Cancer Res.* **27**, 1138.

Borst, P. (1972). *A. Rev. Biochem.* **41**, 333.

Borst, P. and Grivell, L. A. (1971). *FEBS Letters* **13**, 73.

Boulter, D. (1965). *In* "Biosynthetic Pathways in Higher Plants" (J. B. Pridham and T. Swain, eds), p. 101. Academic Press, London and New York.

Boulter, D., Ellis, R. J. and Yarwood, A. (1972). *Biol. Rev.* **47**, 113.

Bourque, D. P. and Wildman, S. G. (1973). *Biochem. biophys. Res. Commun.* **50**, 532.

Bradbeer, J. W. (1973). *In* "Biosynthesis and its Control in Plants" (B. V. Milborrow, ed.), p. 279. Academic Press, London and New York.

Brega, A. and Vesco, C. (1971). *Nature, New Biol.* **229**, 136.

Briarty, L. G., Coult, D. A. and Boulter, D. (1969). *J. exp. Bot.* **20**, 358.

Briggs, D. E. (1973). *In* "Biosynthesis and its Control in Plants" (B. V. Milborrow, ed.), p. 219. Academic Press, London and New York.

Burkard, G., Guillemaut, P. and Weil, J. H. (1970). *Biochim. biophys. Acta* **224**, 184.

132 ORIO CIFERRI

Burkard, G., Vaultier, J. P. and Weil, J. H. (1972). *Phytochemistry* **11**, 1351.
Casey, J., Cohen, M., Rabinowitz, M., Fukuhara, H. and Getz, G. S. (1972). *J. molec. Biol.* **63**, 431.
Ciferri, O. (1972). *Symp. Biol. Hung.* **13**, 263.
Ciferri, O. and Tiboni, O. (1973). *Nature, New Biol.* **245**, 209.
Ciferri, O., Tiboni, O., Lazar, G. and Van Etten, J. L. (1974). *In* "The biogenesis of mitochondria" (A. M. Kroon and C. Saccone, eds), p. 107. Academic Press, London and New York.
Cocking, E. C. (1968). *In* "Plant Cell Organelles" (J. B. Pridham, ed.), p. 198. Academic Press, London and New York.
Cohen, M., Casey, J., Rabinowitz, M. and Getz, G. S. (1972). *J. molec. Biol.* **63**, 441.
Criddle, R. S., Dau, B., Kleinkopf, G. A. and Huffaker, R. C. (1970). *Biochem. biophys. Res. Commun.* **41**, 621.
Danielson, C. E. (1956). *A. Rev. Pl. Physiol.* **7**, 215.
Dawid, I. B. (1972). *J. molec. Biol.* **63**, 201.
Dyer, T. A. and Leech, R. M. (1968). *Biochem. J.* **106**, 689.
Ellis, R. J., Blair, G. E. and Hartley, M. R. (1973). *In* "Nitrogen metabolism in plants" (T. W. Goodwin and R. M. S. Smellie, eds), pp. 137–162. Academic Press, London and New York.
Ellis, R. J. and Hartley, M. R. (1971). *Nature, New Biol.* **233**, 193.
Fukuhara, H. (1967). *Proc. natn. Acad. Sci. U.S.A.* **58**, 1065.
Galibert, F., Larsen, C. J., Lelong, J. C. and Boiron, M. (1965). *Nature, Lond.* **207**, 1039.
Ghosh, K., Grishko, A. and Ghosh, H. P. (1971). *Biochem. biophys. Res. Commun.* **42**, 462.
Golinska, B. and Legocki, A. B. (1973). *Biochim. biophys. Acta* **324**, 156.
Grivell, L. A. and Groot, G. S. P. (1972). *FEBS Letters* **25**, 21.
Grivell, L. A., Reijnders, L. and Borst, P. (1971). *Biochim. biophys. Acta* **247**, 91.
Gross, S. R., McCoy, M. T. and Gilmore, E. B. (1968). *Proc. natn. Acad. Sci. U.S.A.* **61**, 253.
Guderian, R. H., Pulliam, R. L. and Gordon, M. P. (1972). *Biochim. biophys. Acta* **262**, 50.
Guillemaut, P., Burkard, G., Steinmetz, A. and Weil, J. H. (1973a). *Pl. Sci. Lett.* **1**, 141.
Guillemaut, P., Steinmetz, A., Burkard, G. and Weil, J. H. (1973b). *C.R. Soc. Biol.* **167**, 961.
Halbreich, A. and Rabinowitz, M. (1971). *Proc. natn. Acad. Sci. U.S.A.* **68**, 294.
Harris, E. H., Preston, J. F. and Eisenstadt, J. M. (1973). *Biochemistry* **12**, 1227.
Hoober, J. K. (1970). *J. biol. Chem.* **245**, 4327.
Hoober, J. K., Siekevitz, P. and Palade, G. E. (1969). *J. biol. Chem.* **244**, 2621.
Ingle, J., Possingham, J. V., Wells, R., Leaver, C. J. and Loening, U. E. (1970). *In* "Control of Organelle Development" (P. L. Miller, ed.), p. 302. Sympos. Soc. Exper. Biol. XXIV. Cambridge University Press.
Kaempfer, R. (1969). *Nature, Lond.* **222**, 950.
Kanabus, J. and Cherry, J. H. (1971). *Proc. natn. Acad. Sci. U.S.A.* **68**, 873.
Kawashima, N. (1970). *Biochem. biophys. Res. Commun.* **38**, 119.
Kawashima, N. and Wildman, S. G. (1970). *A. Rev. Pl. Physiol.* **21**, 325.
Kawashima, N. and Wildman, S. G. (1972). *Biochim. biophys. Acta* **262**, 42.
Khan, A. A., Verbeek, R., Waters, E. C. Jr. and Van Onckelen, H. A. (1973). *Pl. Physiol. Lancaster* **51**, 641.
King, H. W. S. and Gould, H. (1970). *J. molec. Biol.* **51**, 687.

Klein, W. H. and Clark, G. M. Jr. (1973). *Biochemistry* **12**, 1528.

Klein, W. H., Nolan, C., Lazar, J. M. and Clark, J. M. Jr. (1972). *Biochemistry* **11**, 2009.

Kloppstech, K. and Scheweiger, H. G. (1973). *Expl. Cell. Res.* **80**, 69.

Kloppstech, K. and Scheweiger, H. G. (1974). *Pl. Sci. Lett.* **2**, 101.

Kroon, A. M., Agsteribbe, E. and De Vries, H. (1972). *In* "The Mechanism of Protein Synthesis and its Regulation" (L. Bosch, ed.), p. 539, North-Holland, Amsterdam.

Küntzel, H. (1969a). *Nature, Lond.* **222**, 142.

Küntzel, H. (1969b). *FEBS Letters* **4**, 140.

Küntzel, H. and Noll, H. (1967). *Nature, Lond.* **215**, 1340.

Kurland, C. G. (1960). *J. molec. Biol.* **2**, 83.

Lanzani, G. A., Bollini, R. and Soffientini, A. N. (1974). *Biochim. biophys. Acta* **335**, 275.

Lea, P. J. and Norris, R. D. (1972). *Phytochemistry* **11**, 2897.

Leaver, C. J. (1973). *Biochem. J.* **135**, 237.

Leaver, C. J. and Harmey, M. A. (1972). *Biochem. J.* **129**, 37p.

Lederberg, S. and Mitchison, J. M. (1962). *Biochim. biophys. Acta* **55**, 104.

Lee, S. G. and Evans, W. R. (1971). *Science, N.Y.* **173**, 241.

Legocki, A. B. and Marcus, A. (1970). *J. biol. Chem.* **245**, 2814.

Leis, J. P. and Keller, E. B. (1970a). *Biochem. biophys. Res. Commun.* **40**, 416.

Leis, J. P. and Keller, E. B. (1970b). *Proc. natn. Acad. Sci. U.S.A.* **67**, 1593.

Lin, C. Y. and Key, J. L. (1967). *J. molec. Biol.* **26**, 237.

Lin, C. Y. and Key, J. L. (1971). *Pl. Physiol., Lancaster* **48**, 547.

Lin, C. Y., Travis, R. L., Chia, L. S. Y. and Key, J. L. (1973). *Phytochemistry* **12**, 515.

Litvak, S., Carrè, D. S. and Chapeville, F. (1970). *FEBS Letters* **11**, 316.

Litvak, S., Tarrago, A., Tarrago-Litvak, L. and Allende, J. E. (1973). *Nature, New Biol.* **241**, 88.

Lizardi, P. M. and Luck, D. J. L. (1971). *Nature, New Biol.* **229**, 140.

Lizardi, P. M. and Luck, D. J. L. (1972). *J. Cell Biol.* **54**, 56.

Loening, U. E. (1968). *J. molec. Biol.* **38**, 355.

Loening, U. E. and Ingle, J. (1967). *Nature, Lond.* **215**, 363.

Lyttleton, J. W. (1962). *Expl Cell. Res.* **26**, 312.

Machold, O. (1971). *Biochim. biophys. Acta* **238**, 324.

Mans, R. J. (1967). *A. Rev. Pl. Physiol.* **18**, 127.

Marcus, A. (1970a). *J. biol. Chem.* **245**, 955.

Marcus, A. (1970b). *J. biol. Chem.* **245**, 962.

Marcus, A., Bewley, J. D. and Weeks, D. P. (1970a). *Science, N.Y.* **167**, 1735.

Marcus, A. and Feeley, J. (1965). *J. biol. Chem.* **240**, 1675.

Marcus, A., Weeks, D. P., Leis, J. P. and Keller, E. B. (1970b). *Proc. natn. Acad. Sci. U.S.A.* **67**, 1681.

Margulies, M. M. (1971). *Biochem. biophys. Res. Commun.* **44**, 539.

McKeehan, W. L. and Hardesty, B. (1969). *J. biol. Chem.* **244**, 4330.

Merrick, W. C. and Dure III, L. S. (1971). *Proc. natn. Acad. Sci. U.S.A.* **68**, 641.

Millis, A. J. T. and Suyama, Y. (1972). *J. biol. Chem.* **247**, 4063.

Monier, R. (1972). *In* "The Mechanism of Protein Synthesis and its Regulation" (L. Bosh, ed.), p. 353. North-Holland, Amsterdam.

Nass, M. M. K. and Buck, C. A. (1970). *J. molec. Biol.* **54**, 187.

Neupert, W., Sebald, W., Schwab, A. J., Massinger, P. and Bücher, Th. (1969a). *Europ. J. Biochem.* **10**, 589.

134 ORIO CIFERRI

Neupert, W., Sebald, W., Schwab, A. J., Pfaller, A. and Bücher, Th. (1969b). *Europ. J. Biochem.* **10**, 585.

O'Brien, T. W. and Kalf, G. F. (1967). *J. biol. Chem.* **242**, 2172.

Parisi, B. and Cella, R. (1971). *FEBS Letters* **14**, 209.

Parisi, B., Milanesi, G., Van Etten, J. L., Perani, A. and Ciferri, O. (1967). *J. molec. Biol.* **28**, 295.

Parthier, B. (1973). *FEBS Letters* **38**, 70.

Parthier, B., Krauspe, R. and Samtleben, S. (1972). *Biochim. biophys. Acta* **277**, 335.

Payne, P. I. and Boulter, D. (1969). *Planta* **84**, 263.

Payne, P. I. and Dyer, T. A. (1972). *Nature, New Biol.* **235**, 145.

Payne, E. S., Brownrigg, A., Yarwood, A. and Boulter, D. (1971). *Phytochemistry* **10**, 2299.

Pene, J. J., Knight, E. Jr. and Darnell, J. E. Jr. (1968). *J. molec. Biol.* **33**, 609.

Pinck, M., Yot, P., Chapeville, F. and Duranton, H. M. (1970). *Nature, London* **226**, 954.

Reger, B. J., Fairfield, S. A., Epler, J. L. and Barnett, W. E. (1970). *Proc. natn. Acad. Sci. U.S.A.* **67**, 1207.

Reijnders, L. and Borst, P. (1972). *Biochem. biophys. Res. Commun.* **47**, 126.

Richter, D. (1971). *Biochemistry* **10**, 4422.

Ritcher, D. and Lipmann, F. (1970). *Biochemistry* **9**, 5065.

Roberts, B. E. and Paterson, B. M. (1973). *Proc. natn. Acad. Sci. U.S.A.* **70**, 2330.

Sala, F. and Küntzel, H. (1970). *Europ. J. Biochem.* **15**, 280.

Sala, F., Sensi, S. and Parisi, B. (1970). *FEBS Letters* **10**, 89.

Schäfer, K. P. and Küntzel, H. (1972). *Biochem. biophys. Res. Commun.* **46**, 1312.

Schatz, G. and Mason, T. L. (1974). *A. Rev. Biochem.* **43**, 51.

Schiff, J. A. (1971). *In* "Autonomy and Biogenesis of Mitochondria and Chloroplasts" (N. K. Boardman, A. W. Linnane and R. M. Smillie, eds), p. 98. North-Holland, Amsterdam.

Schmitt, H. (1971). *FEBS Letters* **15**, 186.

Schneir, M. and Moldave, K. (1968). *Biochim biophys. Acta* **166**, 58.

Schweyen, R. and Kaudewitz, F. (1970). *Biochem. biophys. Res. Commun.* **38**, 728.

Scott, S. N. and Smillie, R. M. (1967). *Biochem. biophys. Res. Commun.* **28**, 598.

Scragg, A. H. (1971). *FEBS Letters* **17**, 111.

Sebald, W., Hofstötter, Th., Hacker, D. and Bücher, Th. (1969). *FEBS Letters* **2**, 177.

Skoog, F. and Armstrong, D. J. (1970). *A. Rev. Pl. Physiol.* **21**, 359.

Smillie, R. M., Bishop, D. G., Gibbons, G. C., Graham, D., Grieve, A. M., Raison, J. K. and Reger, B. J. (1971). *In* "Autonomy and Biogenesis of Mitochondria and Chloroplasts" (N. K. Boardman, A. W. Linnane and R. M. Smillie, eds), p. 422. North-Holland, Amsterdam.

Stutz, E. and Rawson, J. R. (1970). *Biochim. biophys. Acta.* **209**, 16.

Surzycki, S. J., Goodenough, U. W., Levine, R. P. and Armstrong, J. J. (1970). *In* "Control of Organelle Development" (P. L. Miller, ed.), p. 13. Symp. Soc. Exp. Biol. XXIV. Cambridge University Press.

Suyama, Y. (1967). *Biochemistry* **6**, 2829.

Swanson, F. R. (1973). *Biochemistry* **12**, 2142.

Sy, J. and McCarty, K. S. (1970). *Biochim. biophys. Acta* **199**, 86.

Tarrago, A., Monasterio, O. and Allende, J. E. (1970). *Biochem. biophys. Res. Commun.* **41**, 765.

Tewari, K. K. (1971). *A. Rev. Pl. Physiol.* **22**, 141.

Tewari, K. K. and Wildman, S. G. (1968). *Proc. natn. Acad. Sci. U.S.A.* **59**, 569.
Tewari, K. K. and Wildman, S. G. (1970). *In* "Control of Organelle Development" (P. L. Miller, ed.), p. 147. Symp. Soc. Exp. Biol. XXIV. Cambridge University Press.
Travnicek, M. (1969). *Biochim. biophys. Acta* **182**, 427.
Tsai, M., Michaelis, G. and Criddle, R. S. (1971). *Proc. natn. Acad. Sci. U.S.A.* **68**, 473.
Ts'o, P. O. P. (1962). *A. Rev. Pl. Physiol.* **13**, 45.
Twardowski, T. and Legocki, A. B. (1973). *Biochim. biophys. Acta* **324**, 171.
Udem, S. A., Kaufman, K. and Warner, J. R. (1971). *J. Bact.* **105**, 101.
Wells, G. N. and Beevers, L. (1973). *Pl. Sci. Lett.* **1**, 281.
Williams, G. R. and Williams, A. S. (1970). *Biochem. biophys. Res. Commun.* **39**, 858.
Williamson, D. H., Maroudas, N. G. and Wilkie, D. (1971). *Molec. gen. Genet.* **111**, 209.
Wintersberger, E. (1970). *Biochem. biophys. Res. Commun.* **40**, 1179.
Wintersberger, E. and Viehhauser, G. (1968). *Nature, Lond.* **220**, 699.
Wood, D. D. and Luck, D. J. L. (1969). *J. molec. Biol.* **41**, 211.
Yarwood, A., Boulter, D. and Yarwood, J. N. (1971). *Biochem. biophys. Res. Commun.* **44**, 353.
Yarwood, A., Payne, E. S., Yarwood, J. N. and Boulter, D. (1971). *Phytochemistry* **10**, 2305.
Yot, P., Pinck, M., Haenni, A., Duranton, H. M. and Chapeville, F. (1970). *Proc. natn. Acad. Sci. U.S.A.* **67**, 1345.
Zucker, M. (1972). *A. Rev. Pl. Physiol.* **23**, 133.

CHAPTER 6

The Biogenesis of Plant Mitochondria

C. J. LEAVER

Department of Botany, University of Edinburgh, Scotland

I. INTRODUCTION

Reliable information on the metabolic activities and biogenesis of plant mitochondria has lagged behind similar studies on ascomycete and animal mitochondria. This is largely due to technical difficulties in extracting suitable quantities of intact and uncontaminated mitochondria from plant tissues. It is to be hoped that with improved techniques our understanding of the biogenesis of plant mitochondria will soon be on a par with our knowledge of similar activities in chloroplasts.

Several excellent reviews on the autonomy and biogenesis of ascomycete and animal mitochondria have appeared in recent years (Ashwell and Work, 1970; Beattie, 1971; Borst, 1972; Dawid, 1972) and the reader in referred to these for a broader and more comprehensive appreciation of the subject.

The aim of this article is first of all to discuss briefly the limitations of mitochondrial autonomy and their replication, and then review our present knowledge of the composition and function of the protein synthetic machinery of mitochondria. This will be followed by a consideration of those areas of research which might lead to a better understanding of the mode of mitochondrial replication and biogenesis particularly with reference to the way in which this is linked to their main metabolic function of respiration and geared to the development of the plant cell.

II. REPLICATION OF MITOCHONDRIA AND LIMITATIONS OF MITOCHONDRIAL AUTONOMY

A. THE ORIGIN OF MITOCHONDRIA

The exact evolutionary and cellular origins of mitochondria is still a matter of considerable controversy. The idea that mitochondria may have evolved from endosymbiotic bacteria (see Ashwell and Work, 1970) has in recent years aroused some interest because of the discovery of mitochondria specific DNA, and an associated protein-synthesizing system which was found to have several functional similarities with the bacterial system. An alternative suggestion is that mitochondria evolved by evolutionary divergence within the cell of a dual system of genetic control, the mitochondria being formed by the encapsulation within a membrane of a specific portion of nuclear genetic material at some time in the past (Mahler and Perlman, 1971). Obviously both theories are untestable and evidence in support of either theory will be difficult to get.

Our knowledge as to the continuity and replication of mitochondria within the cell is slightly more advanced than speculation on their evolutionary origin but is still very incomplete. Three main theories are generally proposed (see Beattie, 1971; Baxter, 1971 and Öpik, 1968, for reviews):

1. Formation from other membraneous structures (cell organelles).
2. Formation by the growth and division of pre-existing mitochondria.
3. *De novo* formation from submicroscopic precursors.

A variety of observations from electron micrographs of cell sections, without supporting biochemical evidence, have over the years been advanced in support of the suggestion that mitochondria may be formed from pre-existing cell membranes, including those of the plasmalemma (Robertson, 1964), nuclear membrane (Bell and Mühlethaler, 1964) and the endoplasmic reticulum. Preliminary work on membrane structure, synthesis and enzymic properties (see Baxter, 1971) has indicated several points of similarity between the outer (but not the inner) mitochondrial membrane and the endoplasmic reticulum which may indicate a common biosynthetic relationship, but at this time the evidence for suggesting the generation of the former by the latter is weak.

The biochemical work of Luck (1963 and 1965) with *Neurospora crassa*, taken together with electron microscopical evidence for mitochondrial division by fission in a variety of plants and other organisms, is consistent with the interpretation that mitochondria grow by accretion of new materials, inserted randomly into the existing membrane structure and that the mitochondrial population increases by simple division. Luck's conclusions were based on experiments in which he followed the incorporation of radioactive choline (Luck, 1963) and lecithin (Luck, 1965) into constituent phospholipids of the mitochondria of a choline requiring mutant of *N. crassa*. He found by autoradiography that choline and lecithin was introduced continuously and randomly into all existing mitochondria, observations which are not consistent with the hypothesis that mitochondria form from pre-existing non-mitochondrial membraneous precursors or *de novo*, in which case the labelling pattern would have been non random.

There is little convincing evidence for the *de novo* synthesis of mitochondria, which would presumably require a very rapid co-ordinated integration of preformed subunits into a functional organelle.

As has been pointed out by several reviewers in this field, with the limited evidence available it is quite possible that different methods of mitochondrial replication may take place in different tissues and at different stages of development.

B. MITOCHONDRIAL DNA

It is now generally accepted that mitochondria contain their own DNA (mtDNA) (see Borst, 1972, for review), and a considerable body of information concerning the characterization of a range of mtDNAs has become available and is in part summarized in Table I. The buoyant densities of mtDNAs differs from that of the homologous nuclear DNAs in the organisms so far studied. In a range of higher plants the density is remarkably constant at around $1 \cdot 706 - 1 \cdot 707$ g/cm^3 (Suyama and Bonner, 1966; Wells and Ingle, 1970; Leaver and Harmey, 1972), in contrast to the nuclear DNAs which vary in density between $1 \cdot 691$ and $1 \cdot 702$ g/cm^3 (Fig. 1).

As can be seen from Table I, with the exception of *Tetrahymena*, mtDNAs exist in the form of closed circles, which vary in contour length from approximately 5 μm in animals, to 20–26 μm in the ascomycetes and at least 30 μm in higher plants. Borst (1972) and other workers have analysed the genetic complexity of mtDNA by quantitative renaturation studies of the heat-denatured DNA. Renaturation studies on mtDNA as compared to nuclear DNA suggest the absence of large scale heterogeneity, while the rates of renaturation indicate that the potential genetic information available in mtDNA is roughly equivalent to its genome size.

The genetic complexity of animal mtDNA expressed as the molecular weight of the genome is of the order of $1 \cdot 0 \times 10^7$ (see Table I) the total information content of which corresponds to about 15 000 base pairs. (For double-helical DNA, 2×10^6 daltons = 1 μm). The small size of this genome to-

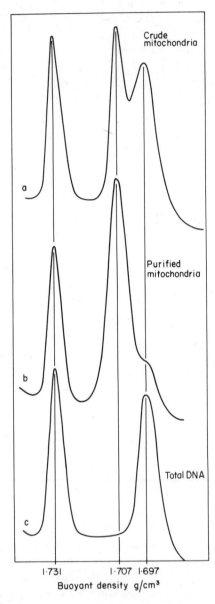

FIG. 1. Microdensitometer tracings of u.v. photographs of mitochondrial and total cellular DNA species from turnip, obtained after analytical CsCl-density-gradient centrifugation. The DNA samples were adjusted to a density of 1·720 g/cm³ and centrifuged at 44 000 rev./min for 20 h at 25 °C. The marker *Micrococcus lysodeikticus* DNA had a density of 1·731 g/cm³. (a) "Crude" mitochondrial DNA; (b) "purified" mitochondrial DNA; (c) total cellular DNA, the major band being the nuclear DNA, 1·697 g/cm³.

TABLE I

Some characteristics of mitochondrial DNA from a number of organisms

Source of mitochondria	Confor-mation	Size (µm)	Genetic complexity based on quanti-tative renaturation experiments	Buoyant density in CsCl (g/cm³)		References
				mt-DNA	Nuclear DNA	
ANIMAL TISSUES	Circular	4·6–5·9	$1·0 \times 10^7$	1·686–1·711	1·694–1·707	Borst and Flavell (1972)
PROTOZOA						
Tetrahymena pyriformis	Linear	15	$3·0–4·0 \times 10^7$	1·684	1·688	Suyama and Miura (1968) Flavell and Jones (1970)
FUNGI						
Neurospora crassa						
Strain Em 5256	Linear	26	$6·6 \times 10^7$	1·701	1·712	Luck and Reich (1964)
Strain 5256	Circular	19	—	—	—	Agsterribbe et al. (1972)
Saccharomyces cerevisiae	Circular	25	5×10^7	1·679	1·693	Hollenberg et al. (1970) Tewari et al. (1966)
HIGHER PLANTS						
Phaseolus vulgaris	Linear	19·5	$1·0 \times 10^8$	1·707	1·693	Wolstenholme and Gross (1968) Suyama and Bonner (1966)
Lactuca sp.	Linear	—	$1·4 \times 10^8$	1·706	1·706	Wells and Birnstiel (1969)
Pisum sativum	Circular	30	$7·4 \times 10^7$	1·706	1·698	Kolodner and Tewari (1972)

gether with many other kinds of evidence (Borst, 1970, 1972) has made it clear that animal mtDNA can code for only a small fraction of the total mitochondrial components. Hybridization studies with mtRNA and homologous mtDNA from *Xenopus* (Dawid, 1970) and HeLa cells (Aloni and Attardi, 1971) would suggest that mitochondrial rRNAs and at least some of the tRNAs are coded for by mtDNA, thereby accounting for some 25–30% of the genome, leaving approximately 11 000 base pairs which would be sufficient to code for only about 20 small proteins. If we can extend to higher plants the observation that animal and ascomycete mtDNA contain only one cistron for each rRNA and *ca* 15 cistrons for tRNAs, then less than 10% of the mtDNA will be involved in coding for the stable RNA molecules in plant mitochondria. Assuming higher plant mtDNA has a potential genetic information equivalent to its genome size, which is at least 6–7 times the size of the mitochondrial genome of animal tissues, then it could code for as many as 200 proteins. It is therefore advisable to be cautious in the indiscriminate extrapolation of results obtained with animal mtDNA to higher plant mtDNA, because about 90% of the genes present in the latter are absent in the former, and the possibility exists that higher plant mitochondria may have a greater degree of autonomy within the cell than mitochondria of animal tissues.

C. MITOCHONDRIAL DNA REPLICATION AND TRANSCRIPTION

Isolated mitochondria will incorporate radioactively labelled deoxyribonucleotide triphosphates into a product which has been identified as mtDNA (Wintersberger, 1966), and in recent years several groups have purified and partially characterized a specific mitochondrial DNA polymerase, e.g. in rat liver (Meyer and Simpson, 1968), in yeast (Iwashima and Rabinowitz, 1969; Wintersberger and Wintersberger, 1970).

The synthesis of mtRNA by isolated mitochondria has been reported by several workers (see Borst, 1972) although the product has been poorly characterized. Recently Kuriyama and Luck (1973) have demonstrated the synthesis of a high molecular weight rRNA which is subsequently processed into the two stable mt-rRNAs in mitochondria from *Neurospora crassa*.

Using the same organism Küntzel and Schäfer (1971) have purified a mitochondrial RNA polymerase with a strong preference for *Neurospora* mtDNA as template. To date, however, there is no direct proof that mtDNA is transcribed into a messenger RNA product which is subsequently translated into an identifiable protein.

III. MITOCHONDRIAL PROTEIN SYNTHESIZING MACHINERY

It is now generally agreed that mitochondria contain specific ribosomes, tRNAs and enzymes which function in mitochondrial protein synthesis. For a complete summary of the subject the reader is referred to reviews by Borst (1972) and Dawid (1972).

A. MITOCHONDRIAL RIBOSOMES AND RIBOSOMAL RNA

Table II gives the sedimentation characteristics of mitochondrial ribosomes and their component ribosomal RNAs from a variety of organisms. It can be seen that the progressive decrease in information content of mtDNA from higher plants to animals appears to be associated with or involve the ribosomal cistrons, which have been reduced in animal mitochondria to an unprecedented small size.

The mitochondrial ribosome is generally smaller than the homologous cytoplasmic ribosome and sediments more slowly. Animal mitochondria contain so-called "miniribosomes" which sediment between 55 S and 60 S (with subunits of 40 S and 30 S), while a 70–74 S mt-ribosome (with subunits of 50 S and 40 S), has been isolated from several fungi. The available information on higher plant mt-ribosomes is somewhat sparse; Wilson et al. (1968)

TABLE II

Sedimentation characteristics (S values) of mitochondrial ribosomes and their RNA components compared with *Escherichia coli*

Organism	Mito-chondrial ribosomes	Mitochondrial ribosomal RNAs	References
ANIMALS			
Xenopus laevis	60	21/13	Dawid (1970)
HeLa cells	60	16/12	Attardi et al. (1970)
Rat (liver)	55	16/13	O'Brien and Kalf (1967) Ashwell and Work (1970) Groot et al. (1970)
FUNGI			
Neurospora crassa	73	23/16	Kuntzel and Noll (1967)
Aspergillus nidulans	67	23·5/15·5	Edelman et al. (1970)
Saccharomyces cerevisiae	74 ± 1	21–22/14–15	Stegeman et al. (1970) Grivell et al. (1971)
PROTOZOA			
Tetrahymena	"80"	21/14	Chi and Suyama (1970)
Euglena gracilis	"71"	21/16	Avadhani and Buetow (1972)
HIGHER PLANTS			
Brassica rapa L.	"78"	24/18·5	Leaver and Harmey (1972)
Zea mays L.	"78"	24–25/18–19	Pring (1973)

have described corn mt-ribosomes which sediment at 66 S, whilst Leaver and Harmey (1972) from a range of plant mitochondria and Pring (1973) from corn mitochondria, have described a mt-ribosome sedimenting at approximately 77–78 S. Although mt-ribosomes differ in size and physical properties from the bacterial 70 S ribosome, they are functionally similar with respect to their sensitivity to a range of antibiotics and also in their ability to interchange factors involved in peptide chain initiation and elongation (Boulter *et al.*, 1972). A number of unusual physical and chemical properties exhibited by

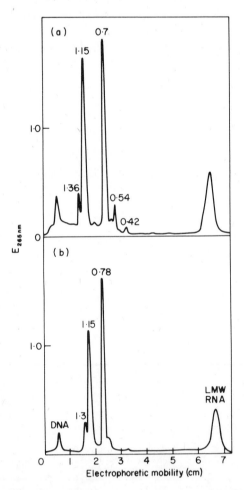

FIG. 2. Polyacrylamide-gel fractionation of nucleic acids from purified mitochondria of (a) turnip, (b) mung bean. Nucleic acids were prepared by the detergent-phenol procedure and fractionated on 2·4% (w/v) polyacrylamide gels for 2 h at 50 V (5 mA/8 cm gel) and then scanned at 265 nm in a Joyce–Loebl Chromoscan (see Leaver and Ingle, 1971). RNA components are referred to as their molecular weight in millions. LMW RNA = low molecular weight RNA.

mt-rRNAs from various organisms has in the past led to difficulties in characterizing the precise size of these molecules.

The low mol % guanine plus cytosine (GC) content (from as low as 26% in yeast to as high as 44% in some animals) of mt-rRNA in contrast to most bacterial and cytoplasmic rRNAs, may account for the observations that fungal mt-rRNA (Edelman et al., 1970, 1971) melts at a lower temperature than cytoplasmic rRNA and also shows anomalous behaviour on gel electrophoresis. Several workers using animal (Groot et al., 1970) and ascomycete (Edelman et al., 1971) mt-rRNAs have shown that the electrophoretic mobility of mt-rRNAs relative to homologous cytoplasmic rRNA and E. coli rRNA, is strongly dependent on ionic strength and temperature. As a consequence of the variable dependence of secondary structure of mt-rRNAs on physical and chemical environmental conditions, their molecular weights cannot always be reliably inferred from their sedimentation behaviour ·or electrophoretic mobility. However, bearing this in mind (see Table II), animal mt-rRNAs sediment at 12–13 S and 16–17 S relative to E. coli rRNA (taken as 16 S and 23 S) and have calculated molecular weights of 0·3–0·36, and 0·35–0·56 × 10⁶ daltons. Fungal mt-rRNAs are larger and sediment at 14–16 S and 21–24 S corresponding to molecular weights of approximately 0·63–0·72 × 10⁶ daltons and 1·23–1·28 × 10⁶ daltons.

Apart from a preliminary report by Baxter and Bishop (1968) that soyabean mitochondria contain two high-molecular-weight RNA species, intermediate in size between the two high-molecular-weight cytoplasmic rRNAs, little was known of the mt-rRNAs of higher plants until comparatively recently. Using highly purified mitochondria from a range of higher plants, Leaver and Harmey (1972) have shown the presence of two major high-molecular-weight rRNA components in the mt-ribosome with apparent molecular weights in the range 1·12–1·18 × 10⁶ and 0·69–0·78 × 10⁶ daltons as estimated by gel electrophoresis (Fig. 2). In addition, Pring (1973) has reported that corn mitochondria contain rRNAs with estimated molecular weights of 0·74–0·75 and 1·26 × 10⁶ daltons. Table III lists the major high-molecular-weight RNAs isolated from a range of plant mitochondria, together with values for homologous cytoplasmic rRNAs isolated under similar conditions.

The two cytoplasmic rRNAs as separated by gel electrophoresis (Fig. 3a) occur in a ratio of approximately 1·9:1, while the ratio of the mt-rRNAs is less than 1·64:1 (Fig. 3b), the value expected from a ribosome containing equimolar proportions of the two rRNAs. Furthermore, in all these fractionations additional lower molecular weight components were present. The inclusion of RNAase inhibitors in the extraction medium did not alter the proportions of the heavy to light mt-rRNAs. If, however, the extraction and fractionation was carried out in the presence of Mg^{2+} ions (Fig. 4a) or at a temperature below 5 °C (Leaver, 1973) the two high-molecular-weight mt-rRNAs are nicely separated with a ratio of 1·7:1. When a similar RNA sample is fractionated in the normal EDTA-containing electrophoresis buffer

146 C. J. LEAVER

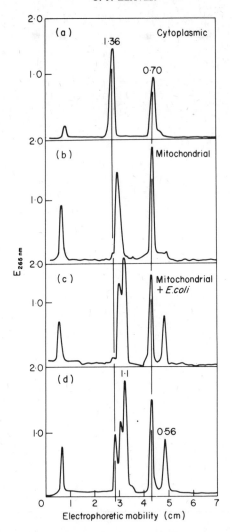

FIG. 3. Polyacrylamide gel electrophoresis of (a) turnip cytoplasmic nucleic acids; (b) turnip mitochondrial nucleic acids; (c) mitochondrial nucleic acids plus *E. coli* nucleic acids; (d) mitochondrial, cytoplasmic and *E. coli* nucleic acids. Preparations were fractionated by electrophoresis on 2·4% (w/v) polyacrylamide gels for 3·5 h. Low-molecular-weight RNAs have run off the gels.

at room temperature (Fig. 4b) the proportion of $1·15 \times 10^6$ rRNA decreases relative to the $0·7 \times 10^6$ rRNA (to 1·0:1) and there are increases in the amounts of several minor lower molecular weight components.

Turnip mt-rRNAs sediment at 24S and 18·5S relative to standards of cyto-plasmic rRNAs (25 S and 18 S) and *E. coli* rRNAs (23 S and 16 S) on sucrose-

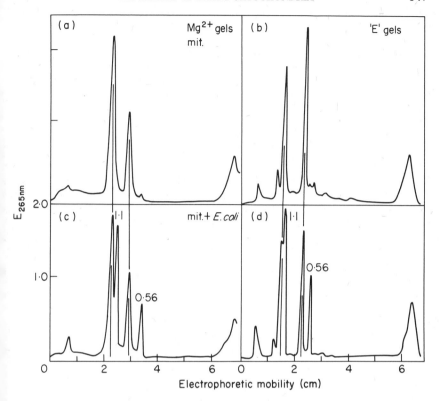

FIG. 4. The effect of Mg^{2+} and EDTA on the preparation and fractionation of turnip mitochondrial RNA. Nucleic acids were prepared from turnip mitochondria in the presence of 10 mM $MgCl_2$ and fractionated (a) in a Mg^{2+}-containing buffer (see Leaver and Ingle, 1971); (b) in a Mg^{2+}-containing buffer + *E. coli* rRNA ($1·1 \times 10^6$ and $0·56 \times 10^6$ daltons); (c) in the normal EDTA-containing buffer; (d) in the normal EDTA-containing buffer + *E. coli* rRNA. The Mg^{2+}-containing gels were subjected to electrophoresis for 3 h at 3 mA/8 cm gel and the normal EDTA-containing gels for 2 h at 6 mA/8 cm gel.

density-gradient centrifugation and furthermore the rRNA components are present in the ratio 1·7:1 (Leaver and Harmey, 1973).

B. MITOCHONDRIAL LOW MOLECULAR WEIGHT RNA

Several studies have shown that animal and ascomycete mitochondria contain transfer RNAs (so-called 4 S RNA with a molecular weight of *ca* 25 000) which are distinct from cytoplasmic tRNAs of the same cells and which hybridize to homologous mt-DNA (Barnett and Brown, 1967; and see Borst, 1972). In addition mitochondria also contain aminoacyl-tRNA synthetases which show some specificity with respect to the mt-tRNA which they will acylate (Barnett *et al.*, 1967).

TABLE III

The molecular weights of mitochondrial and cytoplasmic ribosomal RNAs from a range of higher plants

Tissue	Mitochondrial		Cytoplasmic	
	Heavy	Light	Heavy	Light
Turnip root	1·15	0·70	1·36	0·70
Mung bean hypocotyl	1·15	0·78	1·30	0·70
Potato tuber	1·12	0·70	1·30	0·70
Cauliflower inflorescence	1·18	0·69	1·30	0·69
Pea stem	1·15	0·75	1·30	0·69

The molecular weights were determined by polyacrylamide gel electrophoresis using *Escherichia coli* rRNA of mol. wt $1·10 \times 10^6$ and $0·56 \times 10^6$ as a standard. Mol. wts of homologous chloroplast rRNAs were in the range $1·06–1·12 \times 10^6$ and $0·56–0·57 \times 10^6$.

In higher plants we have shown the presence of mitochondria-specific 4S RNA (Figs 2 and 5a) (Leaver and Harmey, 1972) although an ability to accept amino acids was not demonstrated. Meng and Vanderhoef (1972) have demonstrated the presence of a single mitochondrial tyrosyl tRNA, similar but not identical to *E. coli* tRNA^{Tyr}, and distinct from the three cytoplasmic species in soyabean. Guillemaut *et al.* (1973) have extracted tRNA and demonstrated the presence of a methionyl-tRNA which can be formylated and of a transformylase, in "pure and intact" bean (*Phaseolus vulgaris*) mitochondria. These data suggest that in common with mitochondria from other organisms, and with chloroplasts and bacteria, the initiation of protein synthesis in plant mitochondria requires a formylated methionyl-tRNA. This is in contrast with the process in the cytoplasm of eukaryotic cells where a non-formylated tRNA is required.

Ribosomes from bacteria, chloroplasts and the cytoplasm of eukaryotic cells contain a 5S rRNA (*ca* 40 000 molecular weight) in addition to the two high-molecular weight rRNA species. This RNA is part of the large ribosomal subunit and has no known function. We have isolated such a 5S rRNA component from total mitochondrial nucleic acid preparations (Fig. 5a) and shown its presence in approximately equimolar proportions with mt-rRNA in purified mt-ribosomes (Leaver and Harmey, 1972). These results are in contrast with reports which suggest the absence of an RNA with the electrophoretic mobility of 5S RNA in *Neurospora* (Lizardi and Luck, 1971) yeast or animal mitochondria (see Borst and Grivell, 1971).

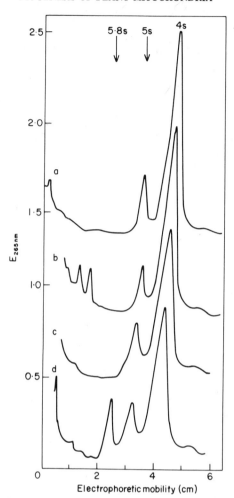

FIG. 5. Polyacrylamide gel-electrophoresis of the low-molecular weight RNA components. (a) Purified turnip mitochondria, (b) as (a) but heated at 70 °C for 10 min and cooled rapidly by immersion in liquid N_2 before fractionation, (c) total turnip nucleic acid, (d) as (c) but heated to 70 °C before fractionation. The RNA was fractionated on 7·5% (w/v) polyacrylamide gels by electrophoresis for 3 h at 9 V/cm of gel (Payne and Dyer, 1972).

IV PROTEIN SYNTHESIS BY ISOLATED MITOCHONDRIA

The earliest report of *in vitro* amino acid incorporation into protein by mitochondria appeared in the late 1950s. The extensive literature on the subject, which has accumulated since then, has recently been comprehensively reviewed and the reader is referred to Beattie (1971), Borst (1972), Dawid (1972), Mason *et al.* (1972) and Kroon and Arendzen (1972). Earlier disagreements concerning the requirements for the incorporation process are now being resolved and precise, reproducible conditions are being established.

Two important requisites of any study of protein synthesis are that the mitochondria should be free of cytoplasmic contamination and that bacterial contamination should be minimized because of their possible contribution to the observed amino acid incorporation. A need for a source of energy to support amino acid incorporation is now generally accepted and recent reports (Beattie and Ibrahim, 1973) suggest that identical rates of incorporation are obtained when either an ATP and ATP-generating system are supplied or ATP is generated by respiratory chain-linked phosphorylations using ADP and a substrate such as succinate.

An important consideration which has been overlooked in some cases is that amino acid incorporation into the protein of mitochondria involves not only the formation of the peptide bond but also the active transport of amino acids across the mitochondrial membrane and other energy-dependent processes which under normal conditions are closely coupled with ATP synthesis by oxidative phosphorylation (Wheldon and Lehninger, 1966). It should thus be borne in mind that any limitations imposed by experimental conditions, such as ionic or osmotic composition of the medium, on any of these associated processes may drastically affect protein synthesis itself.

In most respects the mitochondrial incorporation process resembles protein synthesis in the typical bacterial system, conditions varying slightly depending on the source of mitochondria and their physiological state (see Beattie, 1971, for further details). Studies on protein synthesis by higher plant mitochondria are limited to those by Moore *et al.* (1971) who characterized a mitochondrial amino acid incorporating system from etiolated soyabean hypocotyls, a short report by Goswami *et al.* (1973) using *Vigna sinensis* mitochondria and our recent work with mitochondria from etiolated mung bean hypocotyls.

In our own studies, mitochondria were prepared from 5-day-old etiolated mung bean (*Phaseolus aureus*) hypocotyls grown under sterile conditions. All mitochondrial preparations were routinely checked for purity and intactness by electron microscopy, oxygen electrode polarography and spectrophotometry. Bacterial counts of each reaction mixture indicated that despite the sterile procedure some contamination occurred; however, the values never exceeded 10^4 bacteria/incubation mixture and were normally far less. In addition the incorporation, as will be shown, is energy dependent, has a plateau-shaped progress curve, is specific for certain oxidizable substrates, is negligible in the presence of acetate, requires added Mg^{2+} ions, adenine nucleotides and phosphate. It is sensitive to osmotic shock, lysis by low concentrations of detergent, requires coupled mitochondria, shows no correlation with bacterial counts and follows mitochondria precisely on sucrose gradients, all of which suggest that bacteria do not contribute significantly to the measured incorporation (Roodyn and Grivell, 1969).

Incorporation is dependent upon the supply of an oxidizable substrate ADP, and P_i, and "coupled" mitochondria in a simple buffered medium (Table IV). When some of the commoner TCA cycle intermediates were used as substrate, the efficiency of the substrates in promoting amino acid incorporation reflected the P/O values and suggests that the incorporation rates were due to the energy yield of these substrates (Fig. 6). Incorporation is linear for the first 20–30 min and then gradually plateaus between 30 and 50 min (Fig. 6). It is noteworthy that under the experimental conditions, when between 1 and 2 mg mitochondrial protein were included in an incubation volume of 0·5 ml, the onset of anaerobiosis would be quite rapid, occurring during the first few minutes of incubation and as the mixtures were shaken but not stirred the available oxygen due to back diffusion would not be significant in view of QO_2N values of 426 µl O_2 mg protein/h. The formation of ATP therefore proably occurs quite rapidly during the first few minutes of incubation, and thereafter supports a diminishing rate of amino acid incorporation until the supply of ATP is exhausted. Mitochondria reisolated from incubation mixtures (maintained at 25 °C) at intervals, retained the ability to oxidize substrate but became progressively uncoupled so that the ATP generating mechanisms could no longer function, thus contributing another factor causing a decrease in incorporation.

The optimal concentration of ADP was found to be 1 µmol/ml, higher or lower levels producing an inhibition of incorporation. Using a value of 240

TABLE IV

Requirements for the incorporation of ^{14}C-amino acids into the protein of isolated mung bean mitochondria

System	^{14}C-Amino acid incorporated (cpm/mg protein/40 min)	% of complete system
Complete	5320	100
Minus malate	265	5
Minus ADP	1279	24
Minus phosphate	586	11
Minus malate + succinate (10 mM)	3780	71
Minus malate + acetate (20 mM)	957	18

Tris pH 7·2	12·5 mM	$2·8 \times 10^4$ Bacteria/inc. mix.	
KH_2PO_4	5 mM	1·13 mg mt Protein/inc. mix.	
$MgCl_2$	15 mM		
Malate	20 mM		
KCl	12·5 mM		
ADP	1·0 mM		
Mannitol	200 mM		

0·5 µCi Amino acid C^{14} (U) mix. 54 mCi/matom C.

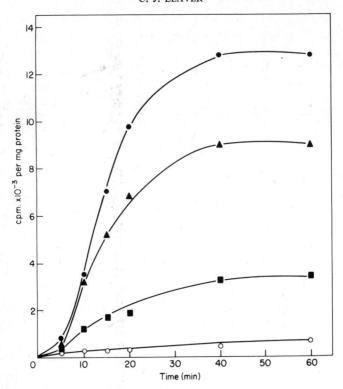

FIG. 6. Time course of ^{14}C-amino acid incorporation into the protein of isolated mung bean mitochondria. Incubation conditions as in Table IV except that various oxidizable substrates were included: (—○—) minus substrate; (—●—) + 20 mM malate; (—▲—) + 10 mM succinate; (—■—) + 1 mM NADH. (1·13 mg protein/incubation mixture. 5 × 10^3 bacteria/ incubation mixture).

TABLE V

The effect of inhibitors of oxidative phosphorylation on the incorporation of ^{14}C-amino acids into the protein of isolated mung bean mitochondria

Additions	^{14}C-Amino acid incorporated (cpm/mg protein/40 min)	% of control	% inhibition of respiration
None	12 830	100	—
KCN (0·2 mM)	3857	30	85
DNP (0·1 mM)	769	6	—
Antimycin (0·4 μg)	2440	19	85
Triton (0·05%)	23	0	92

20 mM Malate as substrate.

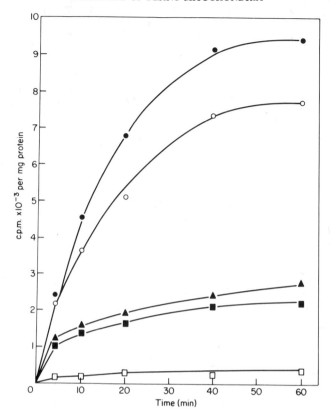

FIG. 7. The effect of different energy sources on the time course of ^{14}C-amino acid incorporation into protein of isolated mung bean mitochondria. Incubation conditions as in Table IV except that different sources of ATP were included: (—□—) minus substrate; (—■—) + 20 mM malate; (—●—) + 20 mM malate + 1 mM ADP; (—▲—) + 20 mM malate + 1 mM ATP + phosphoenol pyruvate–pyruvate kinase; (—○—) + 20 mM malate + 1 mM ATP. (2·68 mg protein/incubation mixture; 8 × 10³ bacteria/incubation mixture).

μM for the concentration of dissolved oxygen, the maximum amount of ATP generated due to oxidation of malate with a P/O ratio of 3, approximates to that obtained as the optimal value, thus emphasizing the close coupling of the amino acid incorporation to the endogenous phosphorylating systems. When ATP is substituted for ADP in the reaction mixture, the final level of incorporation is some 84% of the control value (Fig. 7) while the provision of a phosphoenol pyruvate–pyruvate kinase energy generating system results in a level of incorporation only 28% of the control. These results are consistent with those reported by McCoy and Doeg (1972) who found that the addition of phosphoenol pyruvate inhibited *in vitro* amino acid incorporation by rat liver mitochondria, presumably by stimulation of the adenine translocase system thus reducing the ATP levels available for incorporation.

Inhibitors and uncouplers of respiration inhibit amino acid incorporation, their effect apparently being directly related to their effect on respiration (Table V). The widespread use of differential inhibitors to distinguish between organellar and cytoplasmic protein synthesis has led to a suggestion that not all the commonly used inhibitors of protein synthesis are specific for the process (see Wilson and Moore, 1973). Chloramphenicol, assumed to be a specific inhibitor of organellar and bacterial protein synthesis, has an effect at relatively high concentrations on respiration. Cycloheximide, an inhibitor of cytoplasmic ribosome function, has been shown to stimulate oxygen uptake in plant tissue (Ellis and MacDonald, 1970), possibly as a result of uncoupling oxidative phosphorylation.

To avoid attributing inhibitor effects on respiration to specific inhibition of amino acid incorporation by our tightly coupled mitochondria, parallel measurements of the effects of several inhibitors on both amino acid incorporation and respiration as monitored in the oxygen electrode, were made. Cycloheximide, at concentrations which markedly inhibit cytoplasmic protein synthesis, had little effect on amino acid incorporation or respiration (Table VI), while lincomycin, previously shown to inhibit protein synthesis in bacteria and chloroplasts (see Ellis, 1970), had a marginal effect on amino acid incorporation and slightly inhibited respiration at 100 µg/ml (Table VI). The insensitivity of intact mitochondria to lincomycin could be due to the fact that the mitochondrial membranes under normal conditions are impermeable to the drug, as has been found in rat liver mitochondria by Kroon and De Vries (1971).

In common with other *in vitro* mitochondrial systems, the active D-*threo* isomer of chloramphenicol (Table VI) markedly inhibited amino acid incorporation over a range of 50–500 µg/ml, while only slightly inhibiting respiration at the highest concentration. The L-*threo* isomer was a much less effective inhibitor of incorporation and had a greater effect on respiration, an observation confirmed by the work of Wilson and Moore (1973). The inhibition of amino acid incorporation by L-*threo*-chloramphenicol has also been found by Baxter (1972) in soyabean mitochondria and could be attributable to an effect on respiration.

Isolation and fractionation of ribosomes from mitochondria which had been allowed to incorporate ^{14}C-amino acids for 45 min under standard conditions (Fig. 8), confirmed the association of hot-trichloracetic acid-precipitable radioactivity ("nascent protein") with the "77–78 S" mitochondrial ribosomes.

The work summarized above suggests that isolated plant mitochondria incorporate amino acids into protein by a mechanism with functional similarities to the process in bacteria. To date, however, there are no reports which identify the products of protein synthesis by higher plant mitochondria although some limited progress has recently been made with mitochondria from other organisms.

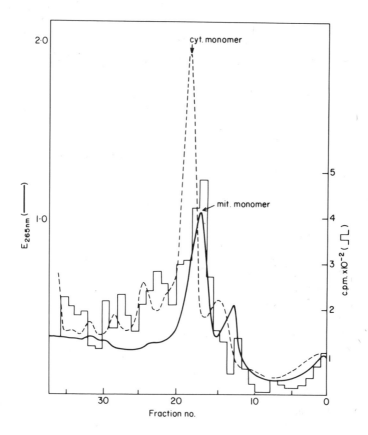

FIG. 8. Sucrose-density-gradient fractionation of mitochondrial ribosomes extracted from *in vitro* ^{14}C-amino acid labelled mung bean mitochondria. Mitochondria were allowed to incorporate ^{14}C-amino acids for 45 min under conditions described in Table IV, with malate as substrate. Labelled mitochondria were recovered by centrifugation, lysed in 10 mM-Tris-HCl buffer, pH 8·5, 50 mM-KCl and 10 mM-MgCl$_2$ (TKM) containing 4% (v/v) Triton X-100 and centrifuged at 10 000 *g* for 10 min. The resulting supernatant was layered over a "cushion" of 1 M-sucrose in TKM buffer and the mt-ribosomes pelleted by centrifugation (105 000 g for 5 h in the Spinco type 40 rotor). Mt-ribosomes were resuspended in TKM and layered on 10–35% (w/v) exponential sucrose-density-gradients in the same buffer and centrifugation was for 5·5 h at 24 000 rev./min in the Spinco SW 25·1 rotor at a chamber temperature of 0 °C. Fractionation and estimation of radioactivity allowed resolution of the (——) mitochondrial (78 S) ribosomes from the marker (————) ^{32}P-labelled cytoplasmic (80 S) ribosomes and identification of hot trichloracetic acid-precipitable radioactivity (⎍⎍) in ^{14}C-amino acid labelled nascent protein.

TABLE VI

The effect of inhibitors of protein synthesis on ^{14}C-amino acid incorporation into protein by isolated mung bean mitochondria

Additions	μg/ml	^{14}C-Amino acid incorp. (cpm/mg protein/45 min)	Incorporation of control	% inhibition of respiration
Control	—	7445	100	0
L-threo-	25	6030	81	0
Chloram-	50	5658	76	0
phenicol	125	4467	60	0
	250	3425	47	20
D-threo-	50	2382	32	0
Chloram-	100	2084	28	0
phenicol	250	1861	25	0
	500	1638	22	5
Lincomycin	10	6580	89	0
	100	6020	81	15
Cycloheximide	10	7480	101	0
	100	6360	86	5

Bacteria $1·01 \times 10^4$/incubation mixture. $2·16$ mg mt Protein/incubation mixture.
Incubation mixture: Tris pH 7·2   12·5 mM   KCl 12·5 mM
                     KH$_2$PO$_4$ 5·0 mM   ADP 1·0 mM
                     MgCl$_2$ 15·0 mM   Mannitol 200·0 mM
                     Malate   20·0 mM   0·5 μc AA C^{14} (U)mix54mCi/ matom C.

V. THE PRODUCTS OF MITOCHONDRIAL PROTEIN SYNTHESIS

The isolation and characterization of proteins synthesized by mitochondria is of considerable interest as they are obviously essential for the development of normal mitochondria and may play an important regulatory role in mitochondrial biogenesis.

Recent advances in our knowledge of the products of the mitochondrial protein-synthesizing machinery have come from both *in vivo* and *in vitro* studies. These systems have been exploited by being used in conjunction with specific inhibitors of either mitochondrial or cytoplasmic protein synthesis; the radioactively labelled products of synthesis being analysed by SDS-polyacrylamide gel electrophoresis which allows for fractionation, on a size basis, of the relatively insoluble mitochondrial translation products.

Studies of protein synthesis in isolated mitochondria to date suggest that if mitochondria are allowed to incorporate radioactive amino acids *in vitro*, the majority, if not the only, proteins synthesized are localized in the membrane fraction of the mitochondria (see Beattie, 1971). Preparation of an inner mitochondrial membrane fraction by either digitonin treatment (Schnaitman *et al.*, 1967) or by osmotic shock and sonication (Sottocasa *et al.*, 1967) indicates that the majority of the incorporated radioactive amino acids were in the insoluble lipoprotein or so called structural protein. Almost no radioactivity was present in the outer membrane fractions or the soluble proteins of the matrix (Beattie *et al.*, 1967).

Polyacrylamide gel electrophoresis of the labelled "structural" proteins purified by a variety of techniques indicate that several (between one and ten), but by no means all, of the protein bands contain radioactivity (Beattie, 1970; see also Borst, 1972). The precise identity of this (these) protein(s) is still in doubt although it is being suggested that the labelled proteins of the inner mitochondrial membrane may be part of the cytochrome oxidase and the ATPase enzyme complexes (see Mason *et al.*, 1972).

Unlike experiments with isolated mitochondria, *in vivo* labelling procedures are not subject to problems of energy supply, proper incubation media or isolation artifacts, and also allow for more efficient labelling of the mitochondrial translation product. They have normally involved the use of specific inhibitors which allow a distinction to be made both quantitatively and qualitatively between the respective contributions of the cytoplasmic and mitochondrial protein-synthesizing systems to the production of mitochondrial proteins and the biogenesis of active mitochondria. Low concentrations of chloramphenicol, erythromycin and similar drugs inhibit mitochondrial protein synthesis without having an effect on cytoplasmic protein synthesis. In contrast, the cytoplasmic system is inhibited by cycloheximide leaving the mitochondrial system unaffected.

Inhibition of *in vivo* cytoplasmic protein synthesis in rats, followed by administration of radioactive amino acids and isolation of the labelled mitochondrial protein indicate that labelling of the proteins was inhibited by some 85% (Beattie, 1970) and furthermore that the incorporation was mainly into insoluble membrane proteins similar to those labelled *in vitro*. Similar results have been obtained with *Neurospora* by Schweyen and Kaudewitz (1970) and Neupert and Ludwig (1971). They found that cycloheximide inhibited labelling of the outer mitochondrial membrane indicating that most, if not all, of the outer membrane is synthesized by the cytoplasmic system. Parallel experiments using chloramphenicol blocked synthesis of the inner membrane proteins.

It is now generally accepted that the RNA of mitochondrial ribosomes is a transcription product of the mitochondrial genome. The intracellular site of synthesis of the structural proteins of the mitochondrial ribosome, which have been shown to be distinct from their cytoplasmic counterparts (Küntzel, 1969; Vasconcelos and Bogorad, 1971) has, in contrast, been in doubt for several years. However, a recent comprehensive study of the intracellular sites of synthesis of the mitochondrial ribosomal proteins in *Neurospora crassa*

158   C. J. LEAVER

produces convincing evidence that all 53 structural proteins are synthesized
by cytoplasmic ribosomes (Lizardi and Luck, 1972).

Results of *in vivo* and *in vitro* studies in general support each other (Ibrahim *et al.*, 1973) and have led to estimates of between 8 and 15% (see Borst,
1972) for the minimal contribution of mitochondrial ribosomes to the synthesis of mitochondrial proteins. These proteins have been tentatively identified
as structural proteins of the inner membrane and may be part of the cytochrome oxidase and ATPase enzyme systems.

VI. BIOGENESIS OF MITOCHONDRIA

A. BIOGENESIS OF MITOCHONDRIA IN YEAST

From the earlier discussion it is obvious that mtDNA has insufficient coding
capacity to specify all the mtRNA species and mitochondrial proteins which
are found in a functional mitochondrion. Furthermore, inhibitor and other
studies (see Beattie, 1971; Borst, 1972) suggest that the majority of mitochondrial proteins, specifically those of the outer membrane, the matrix and
a large proportion of the proteins of the inner membrane, are coded for by
nuclear DNA, synthesized on cytoplasmic ribosomes and subsequently

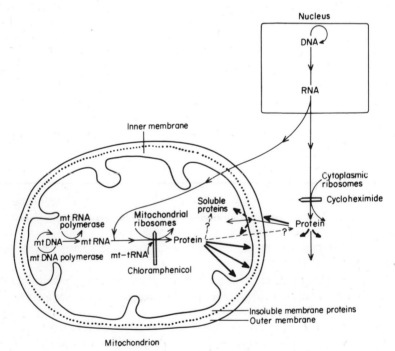

FIG. 9. Possible interrelationships between the cytoplasmic and mitochondrial genetic and
protein synthesizing systems.

transferred into the mitochondrial structure by an, as yet, unknown mechanism (see Fig. 9). The fact that the remaining 5–10% of the mitochondrial proteins are made on mt-ribosomes does not prove that the sequence of these proteins is coded for by mtDNA. In fact it is possible that in addition to proteins, a proportion of the messenger RNA translated by mt-ribosomes is of nuclear origin and is also imported by some mechanism into the mitochondrion (see Fig. 9). Dawid (1972) has even gone so far as to suggest that animal mt-DNA codes only for mitochondrial rRNA and tRNA, supporting his hypothesis by quoting the work of Swanson (1971) who has shown that isolated *Xenopus* mitochondria incorporate synthetic polynucleotides (polyuridylate) which combine with mt-ribosomes and stimulate incorporation of phenylalanine into trichloracetic acid-insoluble material. Borst (1972) on the other hand, using a variety of indirect arguments, suggests that the few proteins so far identified as products of the mitochondrial protein-synthesizing system do not exceed the coding capacity of mtDNA, and that messenger RNA import into mitochondria is quantitatively unimportant if it occurs at all. Although the majority of experimental evidence to date suggests that most mitochondrial proteins are synthesized on the cycloheximide-sensitive, cytoplasmic ribosomes, the successful integration and assembly of these proteins and enzymes into the functional mitochondrial unit requires the co-ordinated synthesis of proteins by the chloramphenicol-sensitive, mitochondrial protein-synthesizing system (see Beattie, 1971). Such a closely integrated system is almost certainly controlled by a feedback mechanism, the precise nature of which is still unknown.

One example of an excellent system which has been exploited in the study of the interdependence of the two protein-synthesizing systems during the assembly of functional mitochondria, which may have parallels in higher plants, is that of yeast cells which are undergoing respiratory adaptations. The early work of Linnane et al. (see Linnane et al., 1972) and more recently Criddle and Schatz (1969) has shown that anaerobically grown yeast or yeast growing on a fermentable carbon source (e.g. glucose), contain very simple mitochondrial structures called promitochondria. These promitochondria lack cytochrome oxidase and other respiratory chain intermediates, as well as having little if any visible ultrastructure. However, they do contain mtDNA (Schatz and Saltzgaber, 1969) and possess the ability for protein synthesis.

When anaerobically grown cells are aerated or when glucose in the medium is consumed, derepression of the mitochondrial enzyme systems occurs with a parallel increase in internal structure and the development of a normal active respiratory chain. The application of chloramphenicol or cycloheximide to cells undergoing such a transition have shown that during short-term experiments, both the mitochondrial and cytoplasmic protein-synthesizing systems can function independently of each other. Longer term experiments have suggested that the successful replication of DNA and the biogenesis and assembly of functional mitochondria requires a highly complex set of interrelationships between the two systems for protein synthesis (see Beattie, 1971). Although these differential inhibitors are powerful tools for studying the interplay of these two systems in mitochondrial biogenesis, we should, as has been suggested by Dawid (1972), be careful of over-interpretation of the results of such studies.

B. BIOGENESIS OF MITOCHONDRIA IN HIGHER PLANTS

A recent study by Öpik (1973) has examined a comparable situation to the aerobic adaptation of yeast and the accompanying mitochondrial development, in higher plants. She has made an attempt to correlate respiratory rate, cytochrome oxidase activity and mitochondrial structure in coleoptiles of rice (*Oryza sativa* L.) germinated under aerobic and anaerobic conditions. The rice coleoptile apparently emerges and elongates considerably even under complete anaerobiosis, although the respiration rate is considerably lower and the cytochrome oxidase activity, on a per cell basis, is *ca* 25% of the aerobically grown controls. In striking contrast to the situation in yeast, however, the internal mitochondrial structure in terms of cristae development as viewed by electron microscopy is similar in both aerobic and anaerobically grown coleoptiles. Öpik concludes, from the fact that mitochondria from anaerobically grown rice coleoptiles still have a high density of cristae in spite of a very low cytochrome oxidase activity, that mitochondrial membranes can be formed even when some membrane component is almost lacking, cytochrome oxidase being an integral part of the inner membrane and cristae. Her observation that cytochrome oxidase activity rapidly increases when anaerobically grown coleoptiles are transferred to air, suggests that the enzyme can be incorporated into membranes formed some time previously. Obviously the anaerobically grown rice coleoptiles could prove a valuable system for the study of the process of integration of certain respiratory enzymes into the mitochondrial membrane structure.

Another system which will obviously play an increasingly important role in the study of mitochondrial biogenesis in plants is in the cotyledons of maturing and germinating seeds. During the early development of the seed in many members of the Leguminosae, the cotyledons show a high metabolic activity. During the final phase of development, as the seed matures, water is lost, metabolic activity decreases, and it has been observed that the fine structure of the mitochondria becomes disorganized (Bain and Mercer, 1966a). Accompanying this change Kollöffel (1970) has demonstrated in peas that there is a gradual decrease in the phosphorylation efficiency and respiratory control ratios of isolated mitochondria, possibly as a result of the loss of the biochemical integrity of the electron transport chain, as evidenced by a selective and marked decrease in the succinate and malate oxidase systems, and a relatively small decrease of succinate and malate dehydrogenase activity.

During germination and rehydration of the legume seeds, the biochemical integrity of the mitochondria from the cotyledons as judged by their phosphorylation efficiency and respiratory control increases rapidly during the first 20 h. At the same time the activity of the succinate and malate oxidase systems increases markedly, whereas the corresponding dehydrogenase activities increases only very little (Kollöffel, 1970). Accompanying these biochemical changes Öpik (1965) and Bain and Mercer (1966b) showed by electron microscopy a rebuilding of the ultrastructure of mitochondria. Nawa and Asahi (1971) found a shift to a higher density of mitochondria fractionated on sucrose gradients, which was associated with an increase in mitochondrial protein and phospholipid content. Breidenbach et al. (1965, 1967) showed increases in cytochrome, protein and DNA contents of purified mitochondrial fractions prepared from peanut cotyledons by gradient centrifugation during germination over a period of 10 days. Only further study will resolve whether these developmental changes in mitochondrial activity, structure and composition in germinating seeds are due to an increase in the respiratory capacity of the mitochondria themselves and/or an increase in the number of mitochondria.

The active biogenesis of mitochondria in the absence of cell division has also been reported to occur during the ageing of tissue slices from several roots and tubers (e.g. potato and sweet potato). The early work of Lee and Chasson (1966) and Asahi and his colleagues (Asahi et al., 1966, 1969) and more recent reports by Sakano et al. (1971a, b) have shown that the mitochondrial population almost doubles in 24 h. This increase in mitochondria is accompanied by increases in respiratory activity, the activities of several respiratory enzymes including cytochrome oxidase and succinate dehydrogenase (Sakano and Asahi, 1971a) and the lipid and acid-insoluble nitrogen content of isolated mitochondrial fractions (Sakano et al., 1968; Nakano and Asahi, 1970).

In a further report (Sakano and Asahi, 1969), mitochondrial biogenesis in sweet potato was shown to be dependent on the activation of protein synthesis which took place within 8 h after slicing, and was accompanied by the production of a new population of mitochondria which were less dense and more intensively labelled with radioactive leucine in vivo than the pre-existing mitochondria. The authors concluded (Sakano and Asahi, 1971a, b) that the various mitochondrial proteins are synthesized at different stages during the biogenesis of the mitochondria. The synthesis of soluble and "heme" proteins, together with a proportion of the structural proteins, is apparently required before the synthesis of a class of structural proteins (which were bile-salt insoluble) which accompanies the development of the biological function of the mitochondria. Experiments involving the use of differential inhibitors suggested that the early synthesis of the soluble, structural and "heme" proteins was dependent upon both cytoplasmic and mitochondrial protein synthesis, while the synthesis of the bile-salt insoluble, structural protein, occurred on cytoplasmic ribosomes.

The developmental changes in mitochondrial activity and structure described above should provide excellent systems for the study of the biosynthesis and assembly of functional mitochondria.

VII. CONCLUSIONS AND PERSPECTIVES

It has now been established that, like their animal and fungal counterparts, plant mitochondria contain distinct genetic and protein-synthesizing systems (see Fig. 9), but they are not autonomous organelles.

If a comparison with the work on animal and fungal mitochondria is valid then mtDNA codes for organelle ribosomal RNA and tRNA molecules, the remaining genes presumably code for protein or have some control function. Probably less than 10% of the total mitochondrial protein is synthesized on organelle ribosomes, and this relatively insoluble protein is localized in the inner mitochondrial membrane. The majority of mitochondrial proteins are coded for by nuclear DNA, synthesized in the cytoplasm and subsequently integrated into the mitochondrial structure. This process is known to require the close co-ordination of mitochondrial and cytoplasmic protein-synthesizing systems, but to date we know little of the mechanisms involved in the transfer of such proteins (or for that matter RNA) across the mitochondrial membrane. The complex interrelationships and controls which must be involved during mitochondrial biogenesis in higher plants and fungi may differ from those found in animal mitochondria, which contain less than one sixth as much DNA and hence information content. This also raises the question as to why higher plant mitochondria contain so much more genetic information to perform the same main functions as animal mitochondria, those of respiration and generation of energy. Identification of the products of protein synthesis by plant mitochondria should go some way towards answering this question as well as clarifying the relative contributions of the nuclear and mitochondrial genomes to the biogenesis of the organelle.

A study of the co-ordination of synthesis and assembly of the various mitochondrial proteins during the developmental changes in mitochondrial activity which are known to occur in higher plants may advance our knowledge of the cellular origins of mitochondria and the factors controlling their biogenesis and turnover.

REFERENCES

Agsterribbe, E., Kroon, A. M. and Van Bruggen, E. F. J. (1972). *Biochim. biophys. Acta* **269**, 299–303.

Aloni, Y. and Attardi, G. (1971). *J. molec. Biol.* **55**, 251–270.

Asahi, T and Majima, R. (1969). *Pl. Cell Physiol.* **10**, 317–323.

Asahi, T., Honda, Y. and Uritani, I. (1966). *Archs Biochem. Biophys.* **113**, 498–499.

Ashwell, M. and Work, T. S. (1970). *A. Rev. Biochem.* **39**, 251–90.

Attardi, G., Aloni, Y., Attardi, B., Ojala, D., Pica-Mattoccia, L., Robberson, D. L. and Storrie, B. (1970). *Cold Spring Harbor Symp. Quant. Biol.* **35**, 599–619.
Avadhani, N. G. and Buetow, D. E. (1972). *Biochem. J.* **128**, 353–365.
Bain, J. M. and Mercer, F. V. (1966a). *Aust. J. biol. Sci.* **19**, 49–67.
Bain, J. M. and Mercer, F. V. (1966b). *Aust. J. biol. Sci.* **19**, 69–84.
Barnett, W. E. and Brown, D. H. (1967). *Proc. natn. Acad. Sci. U.S.A.* **57**, 452–458.
Barnett, W. E., Brown, D. H. and Epler, J. L. (1967). *Proc. natn. Acad. Sci. U.S.A.* **57**, 1175–1181.
Baxter, R. and Bishop, D. H. L. (1968). *Biochem. J.* **109**, 13p–14p.
Baxter, R. (1971). "Results and Problems in Cell Differentiation". Vol. 2 (J. Reinert and H. Ursprung, eds). Springer-Verlag, Berlin, 46–64.
Baxter, R. (1972). Personal communication.
Beattie, D. S. (1970). *FEBS Letters* **9**, 232–234.
Beattie, D. S. (1971). *Sub-Cell. Biochem.* **1**, 1–23.
Beattie, D. S., Basford, R. E. and Koritz, S. B. (1967). *Biochemistry* **6**, 3099–3106.
Beattie, D. S. and Ibrahim, N. G. (1973). *Biochemistry* **12**, 176–180.
Bell, P. R. and Mühlethaler, K. (1964). *J. Cell Biol.* **20**, 235–248.
Borst, P. (1970). *In* "Control of Organelle Development". (P. L. Miller, ed.), pp. 201–225. Cambridge University Press.
Borst, P. (1972). *A. Rev. Biochem.* **41**, 333–376.
Borst, P. and Flavell, R. A. (1972). *In* FEBS. Proceedings of the Eighth Meeting, Amsterdam. "Mitochondria", pp. 1–19. North-Holland/American Elsevier.
Borst, P. and Grivell, L. A. (1971). *FEBS Letters* **13**, 73–88.
Boulter, D. (1970). *A. Rev. Pl. Physiol.* **21**, 91–114.
Boulter, D., Ellis, R. J. and Yarwood, A. (1972). *Biol. Rev. Cambridge phil. Soc.* **47**, 113–175.
Breidenbach, R. W., Castelfranco, P. and Peterson, C. (1966). *Pl. Physiol., Lancaster* **41**, 803–809.
Breidenbach, R. W. and Criddle, R. S. (1967). *Pl. Physiol., Lancaster* **42**, 1035–1041.
Chi, J. C. H. and Suyama, Y. (1970). *J. molec. Biol.* **53**, 531–556.
Criddle, R. S. and Schatz, G. (1969). *Biochemistry* **8**, 322–334.
Dawid, I. B. (1970). *In* "Control of Organelle Development" (P. L. Miller, ed.), pp. 227–246. Cambridge University Press.
Dawid, I. B. (1972). *In* FEBS. Proceedings of the Eighth Meeting, Amsterdam. "Mitchondria", pp. 35–51. North-Holland/American Elsevier.
Edelman, M., Verma, I. M. and Littauer, U. Z. (1970). *J. molec. Biol.* **49**, 67–83.
Edelman, M., Verma, I. M., Herzog, R., Galun, E. and Littauer, U. Z. (1971). *Eur. J. Biochem.* **19**, 372–378.
Ellis, R. J. (1970). *Planta* **91**, 329–335.
Ellis, R. J. and MacDonald, I. R. (1970). *Pl. Physiol., Lancaster* **46**, 227–232.
Flavell, R. A. and Jones, I. G. (1970). *Biochem. J.* **116**, 811–817.
Goswami, B. B. Chakrabarti, S., Dube, D. K. and Roy, S. C. (1973). *Biochem. J.* **134**, 815–816.
Grivell, L. A., Reijnders, L. and Borst, P. (1971). *Eur. J. Biochem.* **19**, 64–72.
Groot, P. H. E., Aaij, C. and Borst, P. (1970). *Biochem. Biophys. Res. Commun.* **41**, 1321–1326.
Guillemaut, P., Burkard, G., Steinmetz, A. and Weil, J. H. (1973). *Pl. Sci. Lett.* **1**, 141–149.
Hollenberg, C. P., Borst, P. and Van Bruggen, E. F. J. (1970). *Biochim. biophys. Acta* **209**, 1–15.
Iwashima, A. and Rabinowitz, M. (1969). *Biochim. biophys. Acta* **228**, 492–502.
Ibrahim, N. G., Burke, J. P. and Beattie, D. S. (1973). *FEBS Letters* **29**, 73–76.

Kollöffel, C. (1970). *Planta* 91, 321–328.

Kolodner, R. and Tewari, K. K. (1972). *Proc. natn. Acad. Sci. U.S.A.* 69, 1830–1834.

Kroon, A. M. and De Vries, H. (1971). *In* "Autonomy and Biogenesis of Mitochondria and Chloroplasts" (N. K. Boardman, A. W. Linnane and R. M. Smillie, eds.), pp. 318–327. North-Holland, Amsterdam.

Kroon, A. M. and Arendzen, A. J. (1972). *In* FEBS. Proceedings of the Eighth Meeting, Amsterdam. "Mitochondria", pp. 71–83. North-Holland/American Elsevier.

Küntzel, H. (1969). *Nature, Lond.* 222, 142–146.

Küntzel, H. and Noll, H. (1967). *Nature, Lond.* 215, 1340–1345.

Küntzel, H. and Schäfer, K. P. (1971). *Nature New Biol.* 231, 265–269.

Kuriyama, Y. and Luck, D. J. L. (1973). *J. molec. Biol.* 73, 425–437.

Leaver, C. J. (1973). *Biochem. J.* 135, 237–240.

Leaver, C. J. and Ingle, J. (1971). *Biochem. J.* 123, 235–243.

Leaver, C. J. and Harmey, M. A. (1972). Unpublished observations.

Leaver, C. J. and Harmey, M. M. (1972). *Biochem. J.* 129, 37–38p.

Leaver, C. J. and Harmey, M. A. (1973). *Biochem. Soc. Symp.* 38, 175–193.

Lee, S. G. and Chasson, R. M. (1966). *Physiologia Pl.* 19, 199–206.

Linnane, A. W., Haslam, J. M., Lukins, H. B. and Nagley, P. (1972). *A. Rev. Microbiol.* 26, 163–198.

Lizardi, P. M. and Luck, D. J. L. (1971). *Nature New Biol.* 229, 140–142.

Lizardi, P. M. and Luck, D. J. L. (1972). *J. Cell Biol.* 54, 56–74.

Luck, D. J. L. (1963). *J. Cell Biol.* 16, 483–499.

Luck, D. J. L. (1965). *J. Cell Biol.* 24, 461–470.

Luck, D. J. L. and Reich, E. (1964). *Proc. natn. Acad. Sci. U.S.A.* 52, 931–938.

Mahler, H. K. and Perlman, P. S. (1971). *Biochemistry* 10, 2979–2990.

Mason, T., Ebner, E., Poyton, R. O., Saltzgaber, J., Wharton, D. C., Mennocci, L. and Schatz, G. (1972). *In* FEBS. Proceedings of the Eighth Meeting, Amsterdam. "Mitchondria", pp. 53–69. North-Holland/American Elsevier.

McCoy, G. D. and Doeg, K. A. (1972). *Biochem. biophys. Res. Commun.* 46, 1411–1417.

Meng, R. L. and Vanderhoef, L. N. (1972). *Pl. Physiol., Lancaster* 50, 298–302.

Meyer, R. R. and Simpson, M. V. (1968). *Proc. natn. Acad. Sci. U.S.A.* 61, 130–137.

Moore, A. L., Borck, K. and Baxter, R. (1971). *Planta* 97, 299–309.

Nakano, M. and Asahi, T. (1970). *Pl. Cell Physiol.* 11, 499–502.

Nawa, Y. and Asahi, T. (1971). *Pl. Physiol., Lancaster* 48, 671–674.

Neupert, N. and Ludwig, G. D. (1971). *Eur. J. Biochem.* 19, 523–532.

O'Brien, T. W. and Kalf, G. F. (1967). *J. biol. Chem.* 247, 2172–2179.

Öpik, H. (1965). *J. exp. Bot.* 16, 667–682.

Öpik, H. (1968). *In* "Plant Cell Organelles" (J. B. Pridham, ed.), pp. 47–88. Academic Press, London and New York.

Öpik, H. (1973). *J. Cell Sci.* 12, 725–739.

Payne, P. I. and Dyer, T. A. (1972). *Nature New Biol.* 235, 145–147.

Pring, D. R. (1974). *Pl. Physiol., Lancaster* 53, 677–683.

Robertson, J. D. (1964). *In* "Cellular Membranes in Development" (E. Racker, ed.), pp. 1–82. Academic Press, New York and London.

Roodyn, D. B. and Grivell, L. A. (1969). *FEBS Symposium* 17, 161–177.

Sakano, K., Asahi, T. and Uritani, I. (1968). *Pl. Cell Physiol.* 9, 49–60.

Sakano, K. and Asahi, T. (1969). *Agric. biol. Chem.* 33, 1433–1439.

Sakano, K. and Asahi, T. (1971a). *Pl. Cell Physiol.* 12, 417–426.

Sakano, K. and Asahi, T. (1971b). *Pl. Cell Physiol.* 12, 427–436.

Schatz, G. and Saltzgaber, J. (1969). *Biochim. biophys. Acta* 180, 186.

Schaintman, C., Erwin, V. and Greenwalt, J. (1967). *J. Cell Biol.*, **32**, 719–735.
Schweyen, R. and Kaudewitz, F. (1970). *Biochem. biophys. Res. Commun.* **38**, 728–735.
Sottocasa, G., Kuylenstierna, B. and Ernster, L. (1967). *J. Cell Biol.* **32**, 415–438.
Stegeman, N. J., Cooper, C. S. and Avers, C. J. (1970). *Biochem. biophys. Res. Commun.* **39**, 69–76.
Suyama, Y. and Bonner, W. D. (1966). *Pl. Physiol.*, *Lancaster*, **41**, 383–388.
Suyama, Y. and Miura, K. (1968). *Proc. natn. Acad. Sci. U.S.A.* **60**, 235–242.
Swanson, F. R. (1971). *Nature, Lond.* **231**, 31–34.
Tewari, K. K., Vötsch, W., Mahler, H. R. and Mackler, B. (1966). *J. molec. Biol.* **20**, 453–481.
Vasconcelos, A. C. L. and Bogorad, L. (1971). *Biochim. biophys. Acta* **228**, 492–502.
Wells, R. and Birnstiel, M. L. (1969). *Biochem. J.* **112**, 777–786.
Wells, R. and Ingle, J. (1970). *Pl. Physiol.*, *Lancaster* **46**, 178–179.
Wheldon, L. W. and Lehninger, A. L. (1966). *Biochemistry* **5**, 3533–3545.
Wilson, R. H., Hanson, J. B. and Mollenhauer, H. H. (1968). *Pl. Physiol.*, *Lancaster* **43**, 1874–1877.
Wilson, S. B. and Moore, A. L. (1973). *Biochim biophys. Acta* **292**, 603–610.
Wintersberger, E. (1966). *Biochem. Biophys. Res. Commun.* **25**, 1–7.
Wintersberger, U. and Wintersberger, E. (1970). *Eur. J. Biochem.* **13**, 20–27.
Wolstenholme, D. R. and Gross, N. J. (1968). *Proc. natn. Acad. Sci. U.S.A.* **61**, 245–252.

CHAPTER 7

The Biogenesis of Chloroplasts

B. PARTHIER, R. KRAUSPE, D. MUNSCHE AND R. WOLLGIEHN

*Institute of Plant Biochemistry, Research Centre for Molecular Biology and
Medicine, Academy of Sciences, 401 Halle/E., German Democratic
Republic*

1. INTRODUCTION

A. STRUCTURE AND FUNCTION OF CHLOROPLASTS

Chloroplasts are unique organelles in the cells of green plants. They harbour
the machinery for the transformation of light energy into chemical energy

used for the reduction of CO_2 to carbohydrates. Thus chloroplasts are not only useful for the plant cells in which they reside, but are also essential to life because they produce basic organic metabolites and energy. Since chloroplasts contain up to 60% of the total protein of green cells, they are regarded as sites of intensive protein synthesis.

Chloroplasts consist of a rigid lamellar system and a mobile soluble phase. In the highly structured and stacked double membranes called thylakoids (Menke, 1961), lipoproteins together with phospholipids, galactolipids, porphyrin derivatives and carotenoids are arranged in a meaningful manner to form flat vesicles with a granular substructure (Mühlethaler, 1971). This substructure of the thylakoid membrane is today best considered as a rigid sheet of bimolecular lipid layers and subunits of structural proteins to which other enzymatically active proteins are attached. The arrangement of the constituents of the lamellae is a prerequisite for the light-driven reactions of photosynthesis, photon capture, electron emission and transport, which are monitored by the pigment–lipoprotein complexes of photosystems I and II. The lamellae can appear as homogeneous membranes, e.g. in the chloroplasts of algae, or grana and intergrana regions can alternate randomly as in higher plants, where the grana regions consist of a larger number of thylakoids.

The thylakoid system is surrounded by the soluble stroma, a compartment in which the dark reactions following photosynthetic H_2O cleavage take place. Beside the process of carbohydrate synthesis a number of other enzymes active in the metabolism of nucleic acids, proteins, lipids, and pigments are localized in the stroma or associated with membranes.

B. TWO ASPECTS OF CHLOROPLAST BIOGENESIS

Biogenesis of organelles includes reproduction and growth. Plastids are self-duplicating bodies, reproducing themselves by fission. There is no evidence to suggest that plastids arise in any other way. They possess their own plastid-specific genetic material, the plastom. Moreover, together with the mitochondria chloroplasts represent a special type of protein synthesizing system which is clearly differentiated from the nucleocytoplasmic system by the prokaryotic features of its components (Table I). The presence of organelle-specific genetic material with all the components necessary for gene expression would indicate, superficially, complete autonomy, for the chloroplast as a "cell within the cell". However, in recent years much evidence has accumulated which demonstrates that this is not the case; indeed, the maintenance of the chloroplast is heavily dependent on the support of the surrounding cytoplasm. Thus, plastids are regarded as semiautonomous organelles although the chloroplast DNA does control a number of organelle functions as shown by geneticists. The continuity of the plastids in phanerogamous plants is guaranteed by the proplastid transmission to the offspring via female

TABLE I

Elements of gene expression in blue-green algae, chloroplasts and nucleo-cytoplasm of eukaryotic plants: comparison of some properties

Element	Property	Blue-green algae	Chloroplast	Nucleo-cytoplasm
DNA	Buoyant density (g/cm³)	1·705 ± 0·010	1·697 ± 0·004	1·715 ± 0·008[a]
	G + C content (%)	48 ± 9	38 ± 2	57 ± 7
	Renaturation	slow; extensive	rapid; complete	slow; incomplete
	Circular conformation	− (+)[b]	+	−
	Presence of histones	−	−	+
	5'-methylcytosine	+	−	+
Ribosomes	S values; also of subunits	70 (50 + 30)	70 (50 + 30)	80 (60 + 40)
rRNA	S values of mature molecules	23 + 16	23 + 16	25 + 18
	Precursor molecules (S values)	small (23′ + 16′)	small (23′ + 16′)	large (>37)
rProteins	Electrophoretic pattern	all are different from each other		
	Immunological properties	all are different from each other		
tRNA	Recognition by cognate chloroplast synthetases			
aa-tRNA synthetases	Acylation of chloroplast tRNAs	+	+	−
	Thermostability	high	high	low
RNA polymerase	Sensitivity to rifampicin	+	+	−
Protein synthesis	Chain initiation by F-met-tRNA$_F$	+	+	−
	Inhibition by chloramphenicol	+	+	−
	Inhibition by cycloheximide	−	−	+

a. Algae. b. Bacteria.

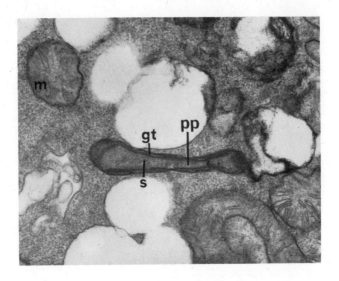

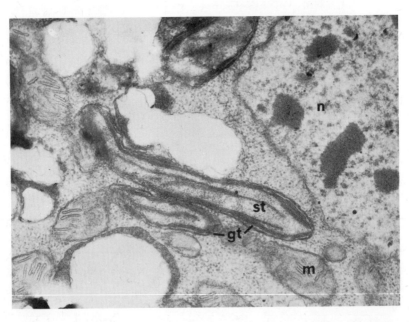

Fig. 1. Electron micrograph of dark-grown and illuminated *Euglena gracilis* showing pro-plastid and chloroplasts. (a) Dark-grown cells in heterotroph medium; (b) same culture illuminated with 1200 lux for 8 h; (c) same culture illuminated with 1200 lux for 60 h; (d) autotroph cells grown at 6000 lux. PP, proplastid; GT, girdle-like thylakoids; ST, straight thylakoids; M, mitochondria; N, nucleus; S, stroma; P, chloroplast.

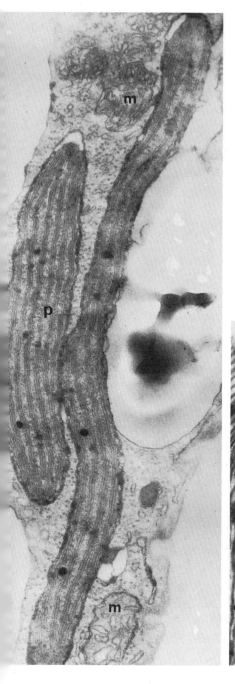

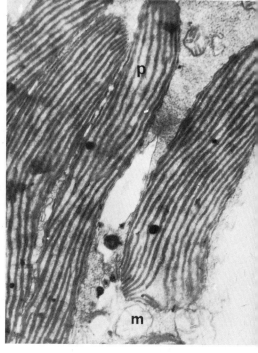

gametes. Biparental inheritance, however, is also possible (Hagemann, 1967; Walles, 1971).

A second aspect of chloroplast biogenesis is characterized by the light-induced transformation of the organelle precursors into mature, photosynthetically active chloroplasts. Such precursor organelles, proplastids, or etioplasts in higher plants, are characterized by the lack of chlorophyll and consequently photosynthesis and by the limited number of thylakoids (Fig. 1a). Illumination causes a dramatic increase of RNA and chloroplast enzyme proteins. The concomitant multiplication of lamellae (Fig. 1b–d) is connected with a transformation of protochlorophyllides to chlorophyllides. The light-triggered primary reactions of chloroplast formation are still unknown, but the observed activation of the genetic material resulting in the massive synthesis of macromolecules seems to be a consequence rather than the cause of the response to light (see section V).

C. EXPERIMENTAL APPROACHES

The main methods of studying chloroplast biogenesis are the genetical and biochemical approaches as already mentioned. Investigators employing sophisticated analytical procedures determine the localization, amount, properties and synthetic rates and sites of the genetic elements. Secondly, using the light-induction of macromolecular synthesis as well as other physiological controls, they attempt to elucidate the system that regulates chloroplast biogenesis.

Three main procedures are used: (i) determination of the synthetic capacity of isolated organelles so that the formation of chloroplast constituents can be estimated without the participation of cytoplasmic components; (ii) determination of the synthetic capacity *in vivo* in the presence of inhibitors of macromolecule synthesis specific for either prokaryotic or eukaryotic systems, thus suppressing the activity of one of the two genetic or synthetic systems; and (iii) the use of mutants. Each of these methods may not yield convincing evidence, and the conclusions drawn can be faulty because of the pitfalls of the methods, e.g. *in vitro* artifact formation, metabolic side-effects of the inhibitors and incorrect localization of the mutation. However, the combined results from all three approaches have led to considerable advances in our knowledge during the last decade.

In recent years, the biogenesis of chloroplasts has frequently been reviewed (Boardman *et al.*, 1971; Boulter *et al.*, 1972; Goodwin, 1966; Hagemann, 1967; Kirk, 1970, 1971; Kirk and Tillney-Bassett, 1967; Parthier, 1970; Parthier and Wollgiehn, 1966; Schiff and Zeldin, 1968; Schnepf and Brown, 1971; Woodcock and Bogorad, 1971). In this present account, therefore, attention is mainly focused on our own studies on chloroplast RNA and protein synthesis.

II. Plastid DNA—Basis of Genetic Information

That DNA occurs in chloroplasts (cDNA) was demonstrated with certainty more than 10 years ago, although it was predicted much earlier from genetic studies (Hagemann, 1967; Walles, 1971). Meanwhile, ample evidence for its existence, properties and synthesis has been obtained from cytochemistry, electron microscopy, radioautography, and analytical and biochemical work (Granick and Gibor, 1967; Iwamura, 1966; Kirk, 1971a, b; Tewari, 1971; Woodcock and Bogorad, 1971).

A. LOCALIZATION AND PROPERTIES

In one of the pioneer studies in this field (Ris and Plaut, 1962), fibrils of 25 Å thickness were observed in chloroplasts of *Chlamydomonas* and could be proved to represent DNA. Electron microscopic studies using different techniques of chloroplast isolation and breakage revealed circular doublestranded DNA molecules in the chloroplasts of algae (Green and Burton, 1970; Manning and Richards, 1972; Manning *et al.*, 1971; Nass and Ben-Shaul, 1972; Rochaix, 1972). In *Euglena* the covalently closed DNA circles appeared supercoiled; their mean length was determined to be 44 μm, corresponding to a mol. wt of $8·3 \times 10^7$ daltons (Manning *et al.*, 1971; Nass and Ben-Shaul, 1972).

According to several reports, circular DNA cannot be detected in osmotically shocked chloroplasts of higher plant cells (Mache and Waygood, 1970; Odintsova *et al.*, 1970; Woodcock and Fernández-Morán, 1968), and the situation is the same in blue-green algae (Kung *et al.*, 1972). However, fragmentation during the preparation cannot be excluded, since circular cDNA molecules of 37–42 μm contour length have recently been demonstrated in pea chloroplast lysates (Kolodner and Tewari, 1972). The calculated mol. wt of 9×10^7 daltons agrees with that of the cDNA in *Euglena* (Manning *et al.*, 1971). In spinach chloroplasts, the linear DNA molecules show contour lengths between 30 and 150 μm and are rarely longer (Woodcock and Fernández-Morán, 1968). The genetic material can occur in double-stranded and tertiary helicoidal conformations and appears to be attached to internal chloroplast membranes (Mache and Waygood, 1970; Woodcock and Fernández-Morán, 1968).

The content and properties of chloroplast DNAs from algae and from higher plants are compared in Table II. The amount of DNA has been estimated as $10^{-15}–10^{-14}$ g per organelle, corresponding to mol. wts between 5×10^8 and 5×10^9 daltons. This is *ca.* 0·01% of the total DNA content of the nucleus (Kung *et al.*, 1972). The buoyant densities of all cDNAs are very similar (1·697 ± 0·001) and consequently the gross GC-contents at 37·5 ± 1%. Thus cDNA is more constant in base composition than nuclear DNAs

TABLE II

Comparison of some properties of chloroplast DNAs from green algae,
higher plants, and total DNA of blue-green algae

Property	Organism			
	Euglena	*Chlamy-domonas*	Higher plant	*Anacystis nidulans*
Amount per chloroplast (g × 10^{-15})	10	8·6	2–6	30
Circular conformation	yes	yes	yes	no
Presence of 5′-methylcytosine	no	no	no	yes
Renaturation velocity	rapid	rapid	rapid	slow
Buoyant density (g/cm³)	1·697	1·695	1·698	1·715
G + C content (%)	38	36	37	56
T_m value (°C)	80	78	84	92
Contour length (μm)	44		39	
Mol. wt (daltons)				
calc. from contour length	$9·2 \times 10^7$	20×10^7	9×10^7	
calc. from analytical complexity	$5–6 \times 10^9$	5×10^9	2×10^9	
calc. from kinetic complexity	$1·8 \times 10^8$	$1·9 \times 10^8$	$0·9 \times 10^8$	$2·2 \times 10^7$
Number of repetitive copies/molecule	25–35	26	22	

Data adapted from Bastia *et al.*, 1971; Edelman *et al.*, 1967; Kolodner and Tewari 1972; Kung *et al.*, 1972a, b; Manning *et al.*, 1971; Rochaix, 1972; Stutz, 1970; Wells and Birnstiel, 1969.

which can differ markedly from one species to another (Kirk, 1971a). In spite of the striking similarities, it is too premature to assume that the cDNA is identical in all green plants. DNA–DNA hybridization studies have provided evidence for the similarity of cDNAs from different higher plants; however, hybridization has also been observed between the cDNA of broad bean and the DNA of *Anacystis nidulans* (Kung, 1973), which are quite different in GC content. Heterogeneity in DNA melting curves and renaturation suggests not only differences in the properties of various DNA species but also heterogeneity within one type of chloroplast (Bard and Gordon, 1969).

Comparison (Table II) of the mol. wts, as calculated from the contour length and the kinetic complexity (renaturation studies), with the data of the analytical complexity (gradient centrifugation) reveals large differences in cDNA preparations from the same organism. The non-coincidence of *E. gracilis* cDNA in the data from kinetic complexity and from contour length might be due to the unknown contribution of the base distribution to the renaturation kinetics, which can introduce an uncertainty factor of two. The mol. wts of the cDNAs from *Chlamydomonas* (Manning *et al.*, 1971) and pea (Kolodner and Tewari, 1972), as determined by the two methods, agree excellently and suggest that the real mol. wt of cDNA is around 9×10^7 daltons. Taking the data from kinetic measurements, we can conclude that the cDNA from both algae and higher plants is extensively reiterated. Every chloroplast "chromosome" contains DNA with 20–30 repetitive nucleotide sequences, which are copies of a genome responsible for the total information content. The informational capacity of each of the 20–30 copies would then correspond to that of a virus (bacteriophage T_4) rather than to that of a bacterium (Bastia *et al.*, 1971). The reiteration of the cDNA is also supported by the estimate of 20–30 cistrons for the rRNA in *E. gracilis* cDNA (Rawson and Stutz, 1969; Scott and Smillie, 1967; Tewari and Wildman, 1970). In *Chlamydomonas* 10% of the cDNA (Rochaix, 1972), and in lettuce chloroplasts 25% (Wells and Birnstiel, 1969) reassociate much faster than the mass of cDNA, suggesting the existence of repetitive units in the range of $1–3 \times 10^6$ daltons. Furthermore, chloroplasts occur in various degrees of polyteny or polyploidy (Granick and Gibor, 1967; Herrmann, 1970).

B. CDNA SYNTHESIS AND DNA POLYMERASE

Semiconservative DNA replication within *Chlamydomonas* chloroplasts has been demonstrated using $^{15}N/^{14}N$ transfer experiments (Chiang and Sueoka, 1967), although the possibility that the nucleus might control the DNA synthesis is not excluded. This objection is emphasized by observations with inhibitors of 80 S ribosome protein synthesis which block the DNA polymerase formation (Giles and Taylor, 1971; Richards *et al.*, 1971; Surzycki *et al.*, 1970). Nevertheless, labelled deoxynucleotides have been incorporated into well characterized DNA in a DNA-dependent enzymatic process (Scott *et al.*, 1968; Spencer and Whitfeld, 1969; Tewari and Wildman, 1967). The DNA polymerase activity could not be washed off from the chloroplasts using hypotonic solutions. The enzyme is assumed to exist tightly bound to membranes, an assumption confirmed by *in vitro* DNA synthesis in toluene-treated *Chlamydomonas* cells (Howell and Walker, 1972). Under these conditions the internal morphology of the cell is disrupted, leaving the outer membranes of the chloroplast intact. A soluble, readily leached chloroplast DNA polymerase has likewise been observed (Spencer and Whitfeld, 1969).

C. CODING CAPACITY OF CDNA

The total amount of cDNA per organelle does not reflect correctly the coding capacity since this depends on the non-repetitive base sequences only. From estimates of the mol. wts based on kinetic experiments, the information content of cDNA is contained in 20–30 genome copies of mol. wts between 0·9 and $1·9 \times 10^8$ daltons (Table II). This size is large enough to code for

150–300 protein species of mol. wts of 40 000 (Kirk, 1971b). The proportion of the cistrons for chloroplast rRNA and tRNA is less than 3% of the chromosome. On the other hand, only a limited number of the proteins localized in the organelle seem to be controlled or coded for by the cDNA (see Table IV). Even in an optimistic estimation, the genes for chloroplast proteins synthesized inside the plastid would account for not more than 10–20% of the genome. Thus the dilemma arises of how to uncover the functions of the other three-quarters of the cDNA genome. It is suggested that the cDNA genes are mainly concerned with synthesis and assembly of subcellular membranes, one of the main ways in which prokaryotic and eukaryotic cells differ in their organization. However, the interwoven control processes of intracellular co-operation in the plant cell would suggest the need for regulator genes situated in the cDNA. Genetic mapping experiments will certainly help us to make a step forward in this field (Sager and Ramanis, 1971).

A quantitative estimation of the cDNA coding capacity for the chloroplast ribosomal RNAs has been made. Newly synthesized rRNA was annealed with both cDNA and nDNA. The hybridization rates of 1–1·5% found with cDNA hybrids are much higher than those with nDNA (Ingle et al., 1970; Rawson and Stutz, 1969; Scott and Smillie, 1967; Tewari and Wildman, 1968, 1970) and indicated that 20–30 rRNA cistrons are present in E. gracilis cDNA (Kirk, 1971b). It was observed that a chloroplast-associated DNA with a buoyant density of 1·701 g/cm^3 predominantly hybridizes with chloroplast rRNA (Stutz and Vandrey, 1971). In a more recent paper (Rawson and Haselkorn, 1973), it has been suggested that the cistrons for the 23 + 16 S rRNAs in tandem could be attached to approximately 10% of that satellite cDNA. The authors found a DNA–rRNA hybridization of 1·9% which corresponds exactly with 1·9% that is occupied by the $1·65 \times 10^6$ daltons of the $1·1 + 0·56 \times 10^6$ rRNAS on a cDNA of 90×10^6 daltons, as calculated from the contour length and kinetic complexity (Manning and Richards, 1972; Manning et al., 1971). Thus, every cDNA genome contains one "polycistron" for the large plus small high-mol. wt chloroplast rRNA. The 5 S rRNA is not transcribed at the same cistron. Nothing is known about a possible gene amplification during chloroplast development; however, no gross amplifications or deletions of rRNA genes have been observed during the developmental phases of whole plants (Ingle and Sinclair, 1972).

III. ELEMENTS OF INFORMATION TRANSFORMATION

A. RNA POLYMERASE

The enzymology of the transcription of cDNA information into RNA has made appreciable progress and the results can be summarized as follows (see also section IIID). Chloroplast RNA polymerase together with the cDNA is firmly attached to chloroplast membranes (Bottomley et al., 1971a). The enzyme is sensitive to the rifamycins (Munsche and Wollgiehn, 1973; Schweiger, 1970; Surzycki, 1969) and thus resembles prokaryotic DNA-dependent RNA polymerase (Hartmann et al., 1967; Herzfeld and Zillig, 1971). Contrasting reports, i.e. those showing that rifamycin does not suppress RNA synthesis in chloroplasts (Bottomley, 1970; Polya and Jagendorf, 1971a; Sprey, 1972), obviously mean that the initiation region of the cDNA was inaccessible to the antibiotic during these particular experiments. The chloroplast enzyme is not different from the nuclear and soluble enzymes in the absolute dependency of its activity on divalent cations, in pH and temperature optima, K_m values, but it is different in the template-specificity of the enzyme subunits (Polya and Jagendorf, 1971b). The chloroplast RNA polymerase of maize seeds has a molecular weight of 50 000, is unstable if highly purified, and exhibits an unusually high temperature optimum of 48 °C (Bottomley et al., 1971b). The enzyme from tobacco chloroplasts was found to be more active than the respective nuclear enzyme (Tewari and Wildman, 1969).

B. POLYSOMES AND MESSENGER RNA

In all biological systems proteins are synthesized on polyribosomes. The occurrence of polysomes in isolated chloroplasts was first described by Clark (1964) using density gradient centrifugation. This observation was confirmed in many other laboratories (Avadhani and Buetow, 1972; Chen and Wildman, 1967, 1970; Chua et al., 1973; Filippovitch et al., 1973; Harris and Eisenstadt, 1971; Heizmann et al., 1972; Lyttleton, 1967; Oparin et al., 1972; Stutz and Noll; 1967; Williams and Novelli, 1968). A massive formation of chloroplast polysomes from their monomeric precursors is dependent on the illumination of etiolated cells (Pine and Klein, 1972). The light response seems to be more directly connected with polysome-RNA formation than with new rRNA, suggesting an induced synthesis of mRNA with subsequent aggregation of monomeric units to polysome structures. Isolated polysomes are able to incorporate labelled amino acids into acid-precipitable proteins; this endogenous incorporation can not be stimulated by the addition of synthetic polynucleotides.

The localization of polysomes within chloroplasts has also been examined by electron microscopy (Falk, 1969). In Chlamydomonas chloroplast polysomes are found attached to the non-stacked regions of thylakoid membranes as penta- or hexamers and they can be removed with high KCl and puromycin. Both ionic interactions and nascent peptide chains are therefore suggested to be involved in the ribosome–membrane attachment. Such membrane-bound polysomes may be responsible for the synthesis of thylakoid proteins. In pea chloroplasts polysomes are attached to the thylakoids of the grana regions only.

To our knowledge, no datum dealing with mRNA separation from chloroplasts has been reported. Because of the sensitivity of this fraction to nucleolytic attack and physical forces, it is difficult experimentally to determine the presence of mRNA in organelles. However, the occurrence of active polysomes is indicative that mRNA is also present. RNA with the genetic information for chloroplast proteins may be transcribed at cDNA or nDNA; translation of either mRNA, however, can theoretically take place on chloroplast or cytoplasmic ribosomes. According to the class-specificity of initiation factors in the prokaryotic and eukaryotic protein synthesizing systems, one might expect two sets of mRNA-binding factors. Further studies in this field are necessary in order to decide which of the theoretical possibilities of information translation predominates in chloroplast biogenesis.

C. RIBOSOMES AND RIBOSOMAL PROTEINS

We know a lot about the physical properties of bacterial ribosomes and something about how they function in protein synthesis. There is no doubt that chloroplast ribosomes possess the same features although the experimental data are far less detailed than for bacterial ribosomes. The results have been extensively reviewed (Kirk and Tillney-Bassett, 1967; Parthier, 1970; Smillie and Scott, 1969; Wittmann, 1970).

Together with the ribosomes of bacteria, blue-green algae and mitochondria, chloroplast ribonucleoprotein particles belong to the 70 S class of ribosomes (Table I). They differ from the 80 S cytoplasmic counterparts in the mol: wts and in content and composition of the RNA and protein constituents. Further differences exist in the amount of Mg^{2+} required to keep the two subunits attached together and in their sensitivity to inhibitors of protein synthesis. Chloroplast ribosomes contain three types of rRNA molecules which are present in a molar ratio of $1:1:1$. The large subunit (50 S) harbours a rRNA molecule of $1 \cdot 1 \times 10^6$ daltons mol. wt (23 S) and additionally $0 \cdot 04 \times 10^6$ rRNA (5 S); the small subunit (30 S) contains a rRNA of $0 \cdot 56 \times 10^6$ daltons mol. wt (16 S). In contrast, the rRNAs from cytoplasmic ribosomes of the same cells account for $1 \cdot 3 \times 10^6$ (25 S) and $0 \cdot 7 \times 10^6$ (18 S), show distinct differences in chromatographic and electrophoretic properties and also in base compositions. These types of data have been obtained in many laboratories (Arglebe and Hall, 1969; Boardman et al., 1966; Delihas et al., 1972; Hoober and Blobel, 1969; Rawson and Stutz, 1969; Rossi and Gualerzi, 1970; Ruppel, 1969; Scott et al., 1970).

The ribosomal proteins (ca. 20 species in the 30 S subunit and approx. 35 species in the 50 S subunit) have mol. wts around 20 000 daltons and behave quite differently during gel electrophoresis (Gualerzi and Cammarano, 1969; Jones et al., 1972; Lyttleton, 1968; Odintsova and Yurina, 1969) and show immunological specificity (Wittmann, 1970, 1972). These results also point to differences in the composition of ribosomal proteins of bacteria, blue-green algae and chloroplasts, indicating the dissimilarity of various 70 S ribosomes. The protein composition of chloroplast and cytoplasmic ribosomes of the same plant cells are expectedly different, but most striking is the dissimilarity of ribosomal proteins in the two subunits of the chloroplast ribosome (Gualerzi and Cammarano, 1969, 1970).

As mentioned above, rRNA is transcribed at cDNA regions. The proteins of chloroplast ribosomes, however, to a great extent seem to be synthesized on cytoplasmic ribosomes and to be coded for by nDNA (Bourque and Wildman, 1973; Honeycutt and Margulies, 1973; Mets and Bogorad, 1972; Schweiger *et al.*, 1972). Some of them may be synthesized on chloroplast ribosomes (Ellis and Hartley, 1971; Surzycki and Gillham, 1971). The mode of association between rRNA and ribosomal proteins to form a functional ribosome is still obscure. The highest proportion of the chloroplast ribosomes are localized in the stroma but membrane-bound particles have been observed (Chen and Wildman, 1970; Chua *et al.*, 1973).

D. RIBOSOMAL RNA

Electron microscopic autoradiography (Gibbs, 1967) and biochemical studies with isolated chloroplasts (Bottomley, 1970; Hartley and Ellis, 1973;

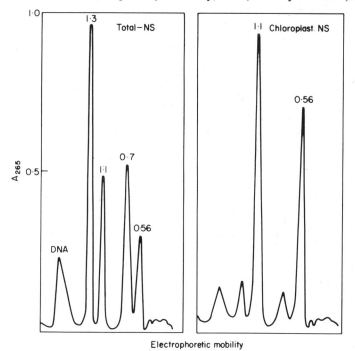

Fig. 2. Gel electrophoretic pattern of ribosomal RNAs from total homogenate of tobacco leaves (*left*) and from purified chloroplasts (*right*). The mol. wts are designated as million daltons (Munsche and Wollgiehn, 1973).

180   B. PARTHIER ET AL.

Spencer and Whitfield, 1967; Spencer *et al.*, 1970; Wollgiehn and Munsche, 1972) or enucleated *Acetabularia* (Schweiger, 1970) have confirmed that organelles have the ability to synthesize their own RNA. Chloroplast rRNA in particular has been extensively studied and this has been facilitated by the fact that more than 80% of the total chloroplast RNA consists of rRNA. Polyacrylamide gel electrophoresis will separate the $1{\cdot}1 \times 10^6$ and $0{\cdot}56 \times 10^6$ chloroplast rRNAs from the corresponding cytoplasmic rRNAs (Ingle *et al.*, 1970); Fig. 2). RNA thus prepared essentially contain 23 S and 16 S rRNAs (Fig. 2). The 23 S rRNA is fairly labile and can only be obtained in an intact form by using Mg^{2+}-containing buffers of high ionic strength at low temperatures. By contrast, the 16 S rRNA is very stable.

In vivo synthesis of chloroplast rRNA is stimulated by light (Gibbs, 1970; Ingle, 1968; Munns *et al.*, 1972; Zeldin and Schiff, 1967). The rate of synthesis is rapid and parallels chlorophyll formation in greening higher plants (Pollack and Davies, 1970; Smith *et al.*, 1970). During the greening process in *Euglena gracilis*, however, radioactivity in chloroplast rRNAs is not detected until 3 h after the exposure of dark-grown cells to light (Brown and Haselkorn, 1971; Heizmann, 1970). It is suggested that the synthesis of chloroplast rRNA is controlled by light in an indirect way.

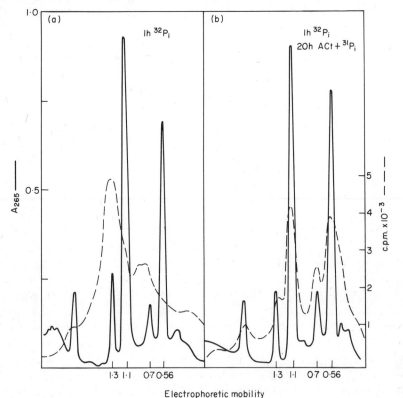

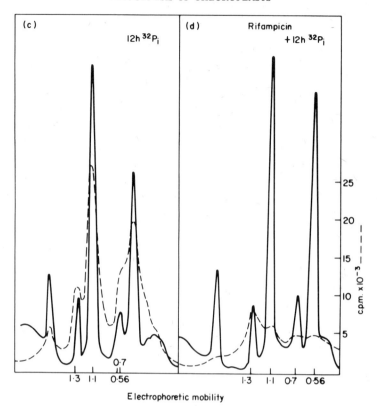

FIG. 3. Gel electrophoretic pattern of chloroplast rRNAs. (a) After 1 h [^{32}P]-pulse labelling of intact leaves; (b) 1 h [^{32}P]-pulse plus 20 h [^{31}P]-chase in the presence of 50 µg/ml actinomycin D; (c) 12 h [^{32}P] long-term incubation; (d) as (c) but preincubated for 10 h in the presence of 300 µg/ml rifampicin (Munsche and Wollgiehn, 1973).

The mode of rRNA synthesis in chloroplasts resembles that in prokaryotic organisms and differs from that of cytoplasmic rRNA in eukaryotes. Whereas in the latter case a high-mol. wt precursor molecule common to the two rRNAs is processed to the mature molecules in a complicated manner (Burdon, 1971; Grierson and Loening, 1972), in the prokaryotic blue-green algae two precursor molecules exist which have only a slightly higher mol. wt than the mature rRNAs (Szalay *et al.*, 1972). An identical situation is found in the synthesis of chloroplast rRNA (Hartley and Ellis, 1973; Munsche and Wollgiehn, 1973).

After ^{32}P pulse labelling to tobacco leaves the highest amount of radio-activity in the chloroplast RNA is found in the regions of 1.3×10^6 and 0.7×10^6 rRNAs, which are assumed to represent rRNA precursor molecules. Mature chloroplast rRNAs are scarcely labelled at all (Fig. 3a). After ^{32}P incorporation for 1 h in the absence of inhibitors, subsequent treatment with

actinomycin D for 20 h, or a chase period for 12 h, causes a shift of the radioactivity to the region of the mature molecules (Fig. 3b, c). Preincubation for
10 h of the leaves with 300 µg/ml rifampicin following 12 h ^{32}P incorporation
results in an inhibition of chloroplast rRNA synthesis (Fig. 3d), leaving the
synthesis of cytoplasmic rRNA unaffected (Munsche and Wollgiehn, 1973).
Participation of bacterial contamination in the synthesis of the 16 S and 23 S
rRNAs could be excluded (Wollgiehn *et al.*, 1974).

Synthesis of intact rRNA can be observed in isolated chloroplasts from
tobacco leaves, provided RNase activity is suppressed by the addition of
bentonite (Munsche and Wollgiehn, 1973). The label of [^{14}C]-ATP after gel
electrophoresis of the synthesized products is distributed over the regions of

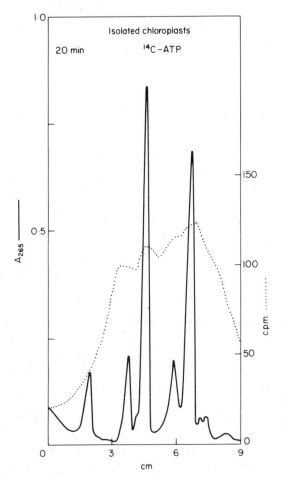

FIG. 4. Gel electrophoretic pattern of RNA prepared from tobacco chloroplasts after *in
vitro* [^{14}C]-ATP incubation for 20 min. After electrophoresis gels were sliced in 3 mm discs
and counted for radioactivity (Wollgiehn and Munsche, 1972).

precursor and mature molecules (Fig. 4). This *in vitro* rRNA synthesis is inhibited by addition of RNase, actinomycin D or rifampicin, but not by the presence of α-amanitine (Wollgiehn and Munsche, 1972).

The lability of the large chloroplast rRNA as mentioned above was studied in more detail in experiments using elevated temperatures. The results (Fig. 5) demonstrate that this lability apparently depends on the age of the molecules within the ribosome. Precursor and newly synthesized mature $1 \cdot 1 \times 10^6$ rRNA appeared stable (Ingle *et al.*, 1970), but the accumulated mature molecules after heating at 60° for 1 min are fragmented into smaller but reproducible pieces (major fractions 0·46, 0·40, 0·30 and 0·20 $\times 10^6$ mol. wts, as extrapolated from the electrophoretic mobilities of marker RNAs). The $0 \cdot 90 \times 10^6$ rRNA, which is a stable fragment after heating or Mg^{2+} removal from the $1 \cdot 1 \times 10^6$ rRNA of blue-green algae (Szalay *et al.*, 1973), is only transiently noted. In other experiments rooted leaf cuttings were fed with ^{32}P for 10 h and the radioactive phosphorus subsequently chased with ^{31}P. After 22 h the newly synthesized radioactive proportion of the $1 \cdot 1 \times 10^6$ rRNA is completely heat-stable while the unlabelled, i.e. older material, has decayed, as shown by the disappearance of absorption at 260 nm. The fragmentation of labelled rRNA can be observed, however, in preparations from cuttings labelled 5 or more days earlier (Fig. 5). From the kinetics of breakdown and from the use of RNase inhibitors such as proteinase K, it follows that the fragmentation is not caused by RNase during heating. This is supported by experiments on the breakdown of the rRNA with high-molar urea or dimethylsulphoxide in the cold.

We suggest that the specific cleavage of the $1 \cdot 1 \times 10^6$ rRNA is due to hidden breaks caused by specific endogenous RNases present during the ageing process in the chloroplast ribosomes. The breaks only become apparent after separation of the hydrogen bonds by heat or other treatment. The previously broken polynucleotide chains are held together by hydrogen bonds in the secondary structure of the RNA, and are released during the heat treatment.

E. TRANSFER RNA

The binding of an amino acid to its tRNA species by the cognate aminoacyl-tRNA synthetase is a highly selective reaction which secures the genetically correct insertion of the amino acid into the polypeptide chain synthesized at the ribosomes. As already shown in Table I, a class specificity exists between tRNAs and their cognate acylating enzymes that can also be observed for the chloroplast and cytoplasmic components.

The occurrence of isoaccepting tRNA species has been demonstrated in the green cells of higher plants and algae by means of the improved separation methods in benzoylated DEAE-cellulose (BD-cellulose) or by chromatography on a reversed phase system. The presence in *Phaseolus* chloroplasts of *N*-formyl-met tRNA was demonstrated (Burkard *et al.*, 1969); 2 tRNA met

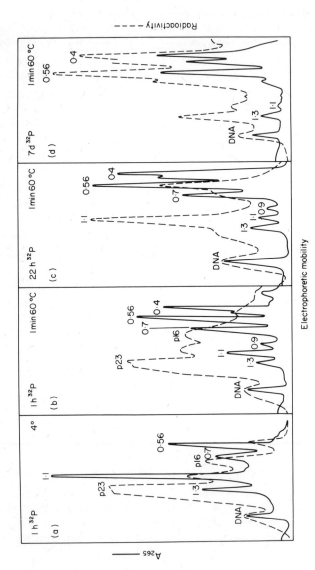

FIG. 5. Gel electrophoretic pattern of chloroplast rRNA labelled *in vivo* with ^{32}P. The isolated rRNA fractions were heat treated at 60 °C for 1 min, cooled to 0 °C and applied to electrophoresis. (a, b) 1 h pulse-labelling; (a) rRNA not heated; (b) rRNA heated at 60°; (c) 10 h ^{32}P labelling + 12 h chase; (d) 10 h ^{32}P labelling plus 7 days chase. The experiments (c) and (d) were performed with leaf cuttings, ^{32}P was administered via the root system.

species exist in cotton seedling chloroplasts, only one of them being formylated by endogenous or *E. coli* transformylase (Merrick and Dure, 1971). Five different forms of tRNA^{leu} and tRNA^{val} were found in bean plants (Burkard *et al.*, 1970, 1972) while there were seven of tRNA^{met} (Guillemaut *et al.*, 1973). In a comparison of the tRNAs of hypocotyls and green tissues or isolated organelles, two of each type were regarded as being attached to chloroplasts. Four to six tRNA^{leu} species were prepared from soyabean seedlings (Anderson and Cherry, 1969), bean leaves (Williams and Williams, 1970) and tobacco leaves (Guderian *et al.*, 1972); in each case, two were chloroplast-specific. A number of tRNA species are bound to chloroplasts of cotton cotyledons (Merrick and Dure, 1972); however, the increase in their amounts during seedling development is not dependent on the light-induced greening process.

Chloroplasts of green algae (*Euglena*) contain single tRNAs for phenylalanine, glutamic acid and isoleucine (Barnett *et al.*, 1969; Kislev *et al.*, 1972); the tRNA^{phe} preparation lacks the fluorescent base Y and thus resembles the tRNA^{phe} of *E. coli* or of *Neurospora* mitochondria but differs from the cytoplasmic tRNA^{phe} of *Euglena* (Fairfield and Barnett, 1971). Using BD-cellulose chromatography, we have observed at least two tRNA^{leu} peaks (III and IV in Fig. 6) in the total tRNA preparation of *E. gracilis* and regard them as being chloroplast-specific (Krauspe and Parthier, 1973;

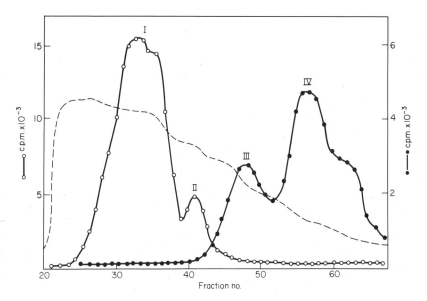

FIG. 6. BD-Cellulose chromatography elution pattern of total tRNA preparation from 3-day-old photoheterotroph grown *E. gracilis* and the distribution of the cytoplasmic tRNA^{leu} (○—○) and chloroplast tRNA^{leu} species (●—●) as determined of [^{14}C]-leucine charging with separated cytoplasmic and plastid leu-RS of *E. gracilis* (Krauspe and Parthier, 1973; Parthier and Krauspe, 1974).

Parthier and Krauspe, 1974). If we separate the tRNA preparations of u.v.-bleached plastid mutants of *E. gracilis* by BD-cellulose chromatography and check the fractions for their ability to accept [^{14}C]-leucine, peaks I and II are exclusively acylated by cytoplasmic leu-tRNA synthetase. The chloroplast enzyme (see section III F) cannot act on any of the fractions of this tRNA preparation suggesting a lack of the tRNA^{leu} species III and IV (Fig. 7). Thus, one is justified in using total tRNA prepared from intact plastid mutant cells instead of cytoplasmic tRNA chromatographically separated from organelle tRNA. Similarly, we attempted to substitute chloroplast tRNA by tRNA prepared from blue-green algae (Parthier and Krauspe, 1974). Figure 8 shows the separation of *Anacystis nidulans* tRNA into five tRNA^{leu} fractions which are all acylated by the leu-tRNA synthetases prepared from *A. nidulans* and *Euglena* chloroplasts. There is only one peak that accepts, to a very small extent, [^{14}C]-leucine in the assay with the cytoplasmic enzyme of *Euglena*.

The aminoacylation specificity of isoaccepting tRNA species in the chloroplasts and in the cytoplasm is absolute for tRNA^{leu} and also probably for others. Chloroplast tRNA^{leu} and that from *A. nidulans* can be exchanged, but further studies on the primary structures of the isoaccepting tRNAs are necessary to determine whether or not differences in the tertiary structures

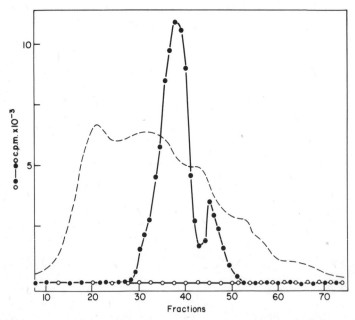

FIG. 7. BD-Cellulose chromatography elution pattern of total tRNA preparation from *E. gracilis* plastid mutant cells. [^{14}C]-leu acceptor activity as determined with separated cytoplasmic (●—●) and plastid leu-RS (○—○) of *E. gracilis* (Parthier and Krauspe, 1974).

exist which are responsible for the specificity in the tRNA-synthetase interaction.

Although the major tRNA^{phe} and tRNA^{Ile} species present in green *E. gracilis* cells could not be detected in dark-grown cells (Barnett *et al.*, 1969), no new isoaccepting tRNA species are induced during the light-dependent chloroplast biogenesis. Thus, we found an identical chromatographic elution pattern for tRNA^{leu} when prepared from dark-grown, mixotroph or auto-troph *E. gracilis*, the only difference being that the tRNA^{leu} peaks III and IV showed a much lower acceptor capacity in the tRNAs from dark-grown cells (Parthier and Krauspe, 1974). This observation is consistent with results obtained from etiolated and green tissues of higher plants (Burkard *et al.*, 1972; Merrick and Dure, 1972). Discrimination between the tRNAs in pro-plastids and in mature chloroplasts is, therefore, simply a quantitative matter.

Meagre evidence is available concerning the site of synthesis of chloroplast tRNA species. Tewari and Wildman (1970) annealed unfractionated [^{32}P]-tRNA with cDNA of tobacco and obtained hybridization rates between 0·4 and 0·7%. This value corresponds with 4·4 to 7·9 × 10^5 daltons of cDNA and was estimated to be sufficient to code for 20–30 tRNA species. Chloroplast tRNA^{leu} species of bean leaves hybridize with cDNA of the same plant with a rate of 0·026% while 0·010% annealed with nDNA. On the basis of these limited results, it is too premature to make any generalization. However,

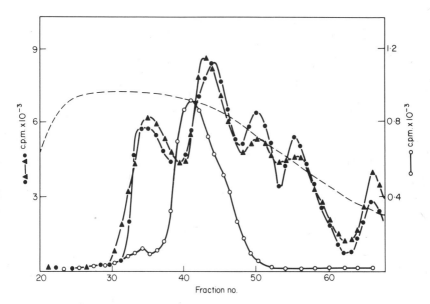

FIG. 8. BD-Cellulose chromatography elution pattern of total tRNA preparation of *A. nidulans* as charged with [^{14}C]-leucine by *A. nidulans* leu-RS (▲—▲), *E. gracilis* plastid leu-RS (●—●) or *E. gracilis* cytoplasmic leu-RS (○—○) (Parthier and Krauspe, 1974).

cDNA is most likely the transcription site for at least several chloroplast tRNA species.

F. AMINOACYL-tRNA SYNTHETASES

The presence of amino acid activating enzymes (aa–RS) has been demonstrated in chloroplasts of higher plants (Aliev and Filippovich, 1968; Burkard et al., 1970; Guderian et al., 1972; Kanabus and Cherry, 1971; Lanzani et al., 1969) and algae (Parthier and Krauspe, 1973; Parthier et al., 1972; Reger et al., 1970). Since these enzymes are easily dissolved in extraction buffers rendering it difficult to demonstrate their presence in isolated organelles, other ways of separation are of interest. Using hydroxyapatite chromatography, nine out of ten E. gracilis enzyme species we have studied could be separated into prokaryotic (plastid) and eukaryotic (cytoplasmic) enzymes (Krauspe and Parthier, 1973, 1974). The two leu-RS species in particular show marked differences of their properties.

Figure 9 demonstrates the separation of two major and one minor leu-RS activities from photoheterotroph-grown E. gracilis by chromatography on a hydroxyapatite column of the crude 100 000 × g supernatant. The enzyme activities of the fractions are determined by either ATP-[^{32}P]pyrophosphate exchange or [^{14}C]-leucine binding to the homologous tRNA. The minor peak eluted at fraction 11 is only occasionally observed and seems to be related to the stationary growth phase of the cultures. The two major peaks of activity,

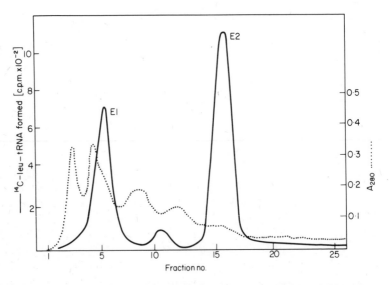

FIG. 9. Hydroxyapatite chromatography elution pattern of leu-RS activities of a crude enzyme extract prepared from 3 days old photoheterotroph grown E. gracilis. The activity of the fractions are determined by [^{14}C]-leucine binding to homologous tRNA (Krauspe and Parthier, 1974).

designated E1 and E2, are present in the cells grown under different conditions as in darkness or autotroph and also in the cells of the u.v.-bleached plastid mutant (Fig. 10). Exploiting the strict organelle specificity of the tRNA^{leu} species with their cognate synthetases, we can demonstrate that E1 exclusively acylates the tRNA^{leu} of *Euglena* chloroplasts or of *A. nidulans* cells. In contrast, E2 binds [^{14}C]-leucine to cytoplasmic tRNA^{leu} only. The attachment of the plastid leu-RS to the E1 activity is supported by the fact that this activity increases by a factor of ten or more during the greening process of dark-grown *Euglena*. However, a considerable E1 activity observed in the plastid mutant (Fig. 10c) indicates the involvement of mitochondrial leu-RS activity. Indeed, we were unable to separate the leu-RS after preparation from isolated chloroplasts and mitochondria by our methods (Krauspe and Parthier, 1974).

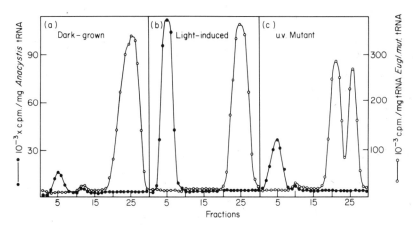

FIG. 10. Hydroxyapatite chromatography elution pattern of *E. gracilis* leu-RS activities as determined with total *A. nidulans* tRNA (●—●) or *E. gracilis* plastid mutant tRNA (○—○). (a) Cells grown in darkness for 3 days; (b) similar cells grown 3 days in the light at 1200 lux; (c) plastid mutant cells grown in the light for 3 days (Krauspe and Parthier, 1973).

The separation of chloroplast and cytoplasmic forms of other aa-RS species was partially successful. The enzymes activating lysine, serine, tyrosine and valine exhibited two activities and are as well separated as those for the leu-RS. Some of the aa-RS activities acting on *A. nidulans* tRNA were eluted in patterns that overlapped the activities of the corresponding cytoplasmic enzymes (e.g. gly-RS, Fig. 11). The thr-RS could never be separated into plastid- and cytoplasmic-specific forms (Fig. 11), suggesting that *E. gracilis* contains only one thr-RS which non-specifically acylates the tRNA of both prokaryotic and cytoplasmic origin. Another type of enzyme separation is represented by phe-RS (Fig. 11) and ile-RS. Here one form with non-specific acylating capacity is separated by hydroxyapatite chromatography

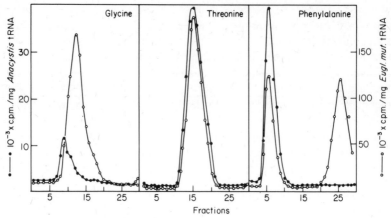

FIG. 11. Hydroxyapatite chromatography elution pattern of *E. gracilis* gly-RS, thr-RS and phe-RS as determined with total tRNA of *A. nidulans* (●—●) or tRNA of *E. gracilis* plastid mutants (○—○) (Krauspe and Parthier, 1973).

from a distinct cytoplasmic activity. The dual specificity of the enzyme eluted at low salt concentration results either from the identical elution pattern of two different forms (ile-RS) or from the non-specific acylation of one form, as shown for the phe-RS (Parthier and Krauspe, 1973).

TABLE III

Some properties of *Euglena gracilis* organelle and cytoplasmic leucyl-tRNA synthetases separated by chromatography on hydroxyapatite columns (Krauspe and Parthier, 1974)

Property	Organelle enzyme (E1)	Cytoplasmic enzyme (E2)
K_m for leucine	$2·5 \times 10^{-5}$ M	$2·5 \times 10^{-5}$ M
K_m for ATP	$8·0 \times 10^{-4}$ M	$8·0 \times 10^{-4}$ M
Optimal ATP	2·5 mM	2·5 mM
Optimal Mg^{2+}	7·5 mM	7·5 mM
Elution from hydroxyapatite column at	0·05 M PO_4^{3-}	0·18 M PO_4^{3-}
Molecular weight	105 000	110 000
Stimulation of ATP-PP-exchange reaction by tRNA	yes	no
Sensitivity to monovalent cations	weak	strong
Heat stability	high	low
Stabilization by ATP during heat treatment	no	yes
Decrease of heat stability by other purine nucleotides	no	yes

Besides having absolute recognition specificities to the cognate tRNAs, the prokaryotic (E1) and cytoplasmic (E2) leu-RS of *E. gracilis* differ in a number of properties (Krauspe and Parthier, 1973, 1974). These are summarized in Table III. While the catalytic parameters are identical, there are significant differences in physical properties. Such dissimilarities may reflect

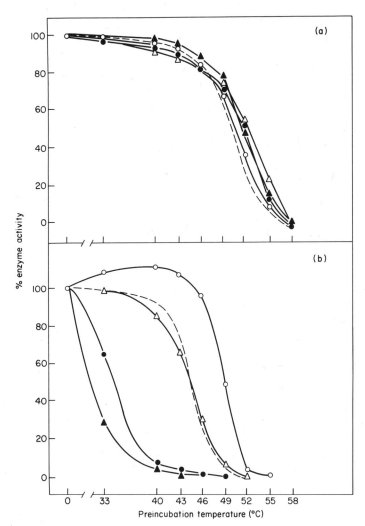

Fig. 12. Heat-inactivation during 15 min preincubation at various temperatures of the separated plastid (a) and cytoplasmic (b) leu-RS of *E. gracilis* and the effects of some nucleotides. Preincubation without addition (----); with 10 mM ATP (○—○); 10 mM UTP (△—△); 10 mM ITP (●—●); 10 mM NADP (▲—▲). After cooling to 0 °C, the [^{14}C]-leucine binding activity to *A. nidulans* tRNA (a) or *E. gracilis* plastid mutant tRNA (b) are assayed (Krauspe and Parthier, 1974).

structural variations in the polypeptide chains, suggesting that the two leu-RS forms are biosynthetically unrelated as well.

The different thermostabilities of E1 and E2 and the influence of certain purine nucleotides on this as a possible control mechanism on enzyme activity are of interest.

Heat inactivation curves of the two leu-RS of *E. gracilis* are obtained after preincubation of the enzymes between 33 and 58 °C for 15 min. The remaining activity is related to the reference activity of enzymes preincubated at 0°. After preincubation in Tris-HCl buffer, pH 7·5, containing 0·005 M $MgCl_2$, 0·01 M KCl, 0·002 M β-mercaptoethanol and 10% glycerol, 50% of the plastid enzyme activity is restored at 52 °C preincubation and 50% of the cytoplasmic enzyme activity at 44 °C preincubation (Fig. 12). Similar differences, indicating that the organelle enzymes are more heat-stable than the cytoplasmic counterparts, are likewise observed for lys-, ser-, val-RS, the cytoplasmic val-RS being the most heat-labile enzyme losing 50% of its activity after preincubation at 30 °C for 15 min (Krauspe and Parthier, 1974).

Preincubation of the cytoplasmic aa-RS in the presence of 5–10 mM ATP delays the inactivation process and stabilizes the enzyme (Fig. 12b). The plastid aa-RS are not affected (Fig. 12a). On the other hand, the presence of GTP, ITP, XTP, and other purine nucleotides during the preincubation period renders the cytoplasmic enzymes more heat-sensitive (Fig. 12b). Pyrimidine nucleotides are without influence. The effects of purine nucleotides again are not observed with plastid aa-RS. In these experiments, the activity in the control (preincubation at 0°) is measured in the presence of the added reagents.

ATP, if added simultaneously with equimolar amounts of ITP to the preincubation mixture with cytoplasmic leu-RS, abolishes the amplified heat-sensitivity caused by ITP (Fig. 13). Apparently ATP can displace ITP from the effector site of the enzyme. The same observation is made when ATP is combined with GTP or XTP. However, we could not observe a recovery of the enzyme activity after preincubation in the presence of NADP plus ATP (Fig. 13).

We have examined the effect of a number of other nucleotides and their derivatives on the heat-stability of the cytoplasmic leu-RS (Krauspe and Parthier, 1974). The following conclusions can be drawn: (i) thermostability is affected by purine nucleotides only, pyrimidines being ineffective; (ii) a triphosphate compound is a prerequisite for activity, di- or monophosphates, nucleosides and bases being ineffective; and (iii), ATP (and to a much less extent ADP) stabilizes molecular changes caused by heat, whereas other purine nucleotide triphosphates or the nicotinamide adenine dinucleotides increase heat-sensitivity. Preliminary results indicate that the cytoplasmic enzymes from tobacco leaves or rat liver behave the same way during heat treatment. The leu-RS from *Anacystis nidulans*, however, exhibited a very similar heat-inactivation curve to that of plastid leu-RS from *E. gracilis*.

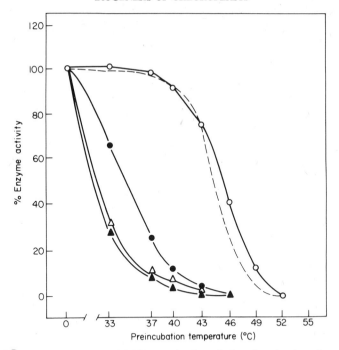

Fig. 13. Effect of ATP together with ITP or NADP on the heat-inactivation of cytoplasmic leu-RS of *E. gracilis*. Preincubation without addition (----); in the presence of 10 mM ITP (●—●); 10 mM ITP + 10 mM ATP (○—○); 10 mM NADP (▲—▲); 10 mM NADP + 10 mM ATP (△—△) (Krauspe and Parthier, 1974).

IV. Proteins—Basis of Chloroplast Function

A. CONTENT, DISTRIBUTION AND PROPERTIES

In cells of green leaves the content of chloroplast proteins may account for 50–65% of the total cell protein and thus represent a major food source for herbivores. Most of the enzyme proteins localized in the chloroplasts are soluble in aqueous solutions and are assumed to constitute the stroma proteins (Table IV). These enzymes are engaged in the fixation and reduction of CO_2, in starch formation and in the synthesis of fatty acids and lipids, amino acids and proteins, nucleic acids, porphyrins, carotenoids and quinones (Goodwin, 1971).

Approximately 50% of the total chloroplast proteins belong to the group of "structural proteins" which are distinguished by their insolubility in aqueous solutions. A number of them are constituents of the thylakoid membranes, together with phospho- and galactolipids and pigments. These proteins have been little characterized as yet; the activities of the photosystems (PS) I and II are found in this fraction, and possibly some other enzymes involved in the

TABLE IV

Sites of synthesis or genetic control of chloroplast proteins (P, plastid; N, nucleus; C, cytoplasm)

Enzyme or protein	Plant source	Site of Synthesis	Control	References
STROMA PROTEINS				
Ribulose 1,5-diphosphate carboxylase	Euglena	P		Schiff, 1970; Smillie et al., 1967, 1971.
Ribulose 1,5-diphosphate carboxylase	Chlamydomonas	P + C		Armstrong et al., 1971; Givan and Criddle, 1972; Margulies, 1971.
Ribulose, 1,5-diphosphate carboxylase (large + small subunit)	Nicotiana		P + N	Chan and Wildman, 1972; Kawashima and Wildman, 1972.
Ribulose 1,5-diphosphate carboxylase (large + small subunit)	Hordeum	P + C		Criddle et al., 1970.
Ribulose 1,5-diphosphate carboxylase	Phaseolus	P		Ireland and Bradbeer, 1971.
Ribulose 1,5-diphosphate carboxylase (large subunit)	Pisum	P		Blair and Ellis, 1972.
NADP-dept. triosephosphate DH	Euglena	P; C		Schiff, 1970; Smillie et al., 1967.
NADP-dept. triosephosphate DH	Phaseolus	C		Ireland and Bradbeer, 1971.
Fructose 1,6-diphosphate aldolase	Chlamydomonas		N	Surzycki et al., 1970.
Fructose 1,6-diphosphate aldolase	Pisum		N	Anderson and Levin, 1970.
Fructose 1,6-diphosphate aldolase	Phaseolus	C		Ireland and Bradbeer, 1971.
Ribulose 5-phosphate kinase	Chlamydomonas	C	N	Armstrong et al., 1971; Surzycki et al., 1970.
Ribulose 5-phosphate kinase	Phaseolus	P		} Ireland and Bradbeer, 1971.
Ribose 5-phosphate isomerase	Phaseolus	C		
Triose-phosphate isomerase	Chlamydomonas		N	Surzycki et al., 1970.
Transketolase	Phaseolus	C		Ireland and Bradbeer, 1971.
5'-aminolevulate dehydratase	Euglena	P		Smillie and Scott, 1969.
Nitrite reductase	Zea mays	P		Schrader et al., 1967.
Fatty acid synthetase	Euglena	P		Ernst-Fonberg and Bloch, 1971.
Adenosine 5'-triphosphatase	Phaseolus	P + C		Horak and Hill, 1972.
Alkaline DNase	Euglena	C	N	Egan and Carell, 1972.

Enzyme or protein	Plant source	Site of Synthesis	Site of Control	References
Malate dehydrogenase	*Acetabularia*		N	Schweiger *et al.*, 1972.
Lactate dehydrogenase	*Acetabularia*		N	
Ferredoxin	*Chlamydomonas*	C	N	Armstrong *et al.*, 1971; Surzycki *et al.*, 1970.
Ferredoxin-NADP reductase	*Euglena*	C + P		Smillie *et al.*, 1967.
Ferredoxin-NADP reductase	*Chlamydomonas*	C	N	Armstrong *et al.*, 1971; Surzycki *et al.*, 1970.
DNA polymerase	*Chlamydomonas*	C		Surzycki *et al.*, 1970.
DNA polymerase	*Euglena*	C		Richards *et al.*, 1971.
DNA polymerase	*Funaria*	C		Giles and Taylor, 1971.
RNA polymerase	*Chlamydomonas*	C		Surzycki *et al.*, 1970.
aa-tRNA synthetases (10 species)	*Euglena*	C		Parthier, 1974.
phe-tRNA synthetase	*Euglena*		N	Reger *et al.*, 1970.
ile-tRNA synthetase	*Euglena*		P	
RIBOSOMAL PROTEINS	*Chlamydomonas*	C (+P)		Honeycutt and Margulies, 1973; Margulies,1971; Surzycki *et al.*, 1970.
	Acetabularia		N	Schweiger *et al.*, 1972.
	Nicotiana		N (+P)	Bourque and Wildman, 1973.
THYLAKOID PROTEINS				
Total fraction	*Chlamydomonas*	C + P		Hoober and Blobel, 1969.
Total fraction	*Acetabularia*		N + P	Apel and Schweiger, 1972; Schweiger *et al.*, 1972.
L-protein (major band)	*Chlamydomonas*	C		Eytan and Ohad, 1970, 1972.
"Activating" proteins	*Chlamydomonas*	P	N	Jennings and Ohad, 1972.
Polypeptide "c" (major band)	*Chlamydomonas*	C		Hoober and Stegeman, 1973.
Photosystem I complex	*Antirrhinum*	P	P	Herrmann, 1971.
Photosystem I complex	*Vicia*	P		Machold and Aurich, 1972.
Photosystem II complex	*Antirrhinum*		P	Herrmann, 1972.
Photosystem II complex	*Vicia*	C		Machold and Aurich, 1972.
Photosystem II complex	*Nicotiana*		N	Kung *et al.*, 1972b.
Plastocyanin	*Chlamydomonas*		N	Surzycki *et al.*, 1970.
Cytochrome 553	*Euglena*	P		Schiff, 1970; Smillie *et al.*, 1971.
Cytochrome 553	*Chlamydomonas*	P	P	Armstrong *et al.*, 1971; Levine and Armstrong, 1972.
Cytochrome 563	*Chlamydomonas*	P	P	

photosynthetic processes may be attached to and associated with the structural protein fraction. This fraction can be solubilized by using detergents and gel electrophoresis will separate these membrane proteins into approximately 15 bands (Fig. 14). Certainly, some of these bands contain 2 or more individual proteins. Other separation methods will resolve such bands, e.g. that of the chlorophyll-containing complexes of PS I and PS II (Apel and Schweiger, 1972; Eytan and Ohad, 1970, 1972; Herrmann, 1971, 1972; Hoober and Stegeman, 1973; Jennings and Ohad, 1972; Lagoutte and Duranton, 1972; Machold and Aurich, 1972; Remy, 1973).

One soluble chloroplast protein, which has been extensively studied, is the fraction I protein, synonymous with ribulose-1,5-diphosphate carboxylase (RuDPC). As RuDP oxygenase activity is associated with this protein, the term fraction I protein should be maintained. This protein is implicated in the key reactions of photosynthesis and photorespiration and comprises up to 60% of the total soluble proteins in chloroplasts (Ellis, 1973; Kawashima and Wildman, 1970). It consists of a population of subunits: 8 large subunits of each $52–60 \times 10^3$ daltons mol. wt are combined with 8 to 10 subunits of mol. wts of $12–18 \times 10^3$ daltons (Rutner, 1970). Thus the complete enzyme, irrespective of the plant source, has a mol. wt between $5 \cdot 1$ and $6 \cdot 6 \times 10^5$ daltons and sediments with 16–20 S in a centrifugal field. The catalytic centre seems to be localized in the large subunit fraction as demonstrated by antisera inhibition, although isolated large subunits have not yet exhibited activity. Large and small subunits of RuDPC from the same plant species show different amino acid compositions and immunological properties. The large subunits of different plant species are closely related although there are great dissimilarities between different subunits.

B. PROTEIN SYNTHESIS AND THE ROLE OF INHIBITORS

In earlier *in vivo* studies using photosynthetically fixed $^{14}CO_2$ as a precursor for chloroplast proteins, it was shown that the radioactivity was rapidly incorporated into crude membrane fractions of thoroughly washed tobacco chloroplasts (Heber, 1962; Parthier, 1964). These findings were confirmed more recently and interpreted to suggest that thylakoid membrane-bound polysomes synthesize lamellar proteins (Chua *et al.*, 1973). Such experiments cannot be assessed quantitatively for two reasons. Firstly, chloroplasts isolated in aqueous media have lost an appreciable amount of stroma proteins which contaminate the cytoplasmic protein fractions. Secondly, protein synthesis in chloroplasts *in situ* cannot provide clear evidence whether the synthesized proteins have been made inside the organelles or whether the polypeptides have been formed on cytoplasmic ribosomes and subsequently been transported into the chloroplasts.

The second objection can be overcome by the determination of protein synthesis in isolated chloroplasts. This has been done and the results (reviewed in Kirk, 1970, 1971b; Parthier, 1970; Smillie and Scott, 1969) suggest that isolated chloroplasts can indeed incorporate labelled precursors into the chloroplast protein fraction. However, there is no conclusive evidence available as yet showing that isolated chloroplasts perform the translation of the genetic information of cDNA to a well-defined chloroplast protein. Experimental difficulties such as the leaching-out of soluble components necessary for polypeptide synthesis is one reason for this failure. Thus, the observation that isolated chloroplasts almost exclusively synthesize lamellar proteins (Goffeau, 1969; Hearing, 1973) might reflect the real situation found *in vivo* regarding the site of synthesis or the genetic control of the thylakoid proteins (Table IV). However, the *in vitro* results can likewise represent a residual synthetic activity due to a loss of necessary components that might be available, in a limited amount, for the synthesis of insoluble proteins at membrane-bound polysomes (Chen and Wildman, 1967, 1970; Chua *et al.*, 1973). Both the translation of viral mRNA on plastid ribosomes resulting in the formation of viral proteins (Sela and Kaesberg, 1969) and the function of chain initiation at chloroplast ribosomes (Schwartz *et al.*, 1967) favour the assumption of an active translation machinery in isolated plastids. These experiments, of course, do not *a priori* preclude the translation of mRNA transcribed at cDNA. The latter, however, is supported by studies on the synthesis of chloroplast proteins in enucleated *Acetabularia* (Craig and Gibor, 1970).

Finally, the use of specific inhibitors of protein synthesis on 70 S ribosomes, preferably chloramphenicol (CAP), provides evidence on the suppression of amino acid incorporation into polypeptides. Together with cycloheximide (CHI), a specific inhibitor of 80 S ribosomal protein synthesis, CAP is a valuable tool for the determination of the intracellular site of synthesis of chloroplast proteins in intact cells or tissues. Although a number of metabolic side-effects caused by these antibiotics have been reported, their use is still the method of choice for deciding whether or not a chloroplast protein is synthesized inside the organelle and, together with genetic studies on plastid mutants, for helping in understanding the process of chloroplast biogenesis.

C. SITES OF SYNTHESIS

Table IV lists data on the distribution inside the chloroplast of enzymes and structural proteins, together with the presumed sites of their synthesis. Treatment of cells or plants with the inhibitors of protein synthesis, CAP and CHI, and the use of mutants has given coincident results in most but not all cases. However, there may well be real differences between various plant species in these respects. The stroma enzymes are probably generally coded for by nDNA and synthesized on cytoplasmic ribosomes. Whether or not the exceptions listed in Table IV remain following further experiments is an open question. There is ample evidence that RuDPC consists of two different types of subunits. This enzyme is produced by the two ribosomal systems. The small subunits are synthesized and genetically controlled by the nucleo-cytoplasm, the large ones are coded for by the cDNA and synthesized on chloroplast ribosomes, since CAP is an inhibitor. The *Euglena* enzyme seems to be an exception; however, we have found its light-induced synthesis is inhibited by CHI (Table V) in relation to the concentration of the antibiotic. Similarly, most of the ribosomal proteins appear to be formed outside the chloroplast (Bourque and Wildman, 1973; Ellis and Hartley, 1971; Honeycutt and Margulies, 1973; Mets and Bogorad, 1972; Schweiger *et al.*, 1972; Surzycki and Gillham, 1971).

Several proteins of the thylakoid membrane complex of *Euglena gracilis* chloroplasts are either not synthesized or are only formed in traces if the cells have been grown in the presence of CAP or nalidixic acid. Those bands which are not present after separation by gel electrophoresis (Fig. 14) likewise show an inhibited incorporation of [^{14}C]-leucine in pulse experiments. The incorporation of label into other bands is inhibited by CHI (unpublished results). Some of the protein bands are affected by the two antibiotics suggesting the participation of the two sites of synthesis or, more likely, that at least two individual proteins can be aggregated in one band. According to Table IV, a predominant thylakoid protein in *Chlamydomonas* and the PS II complex are synthesized on cytoplasmic ribosomes (Eytan and Ohad, 1970, 1972; Hoober and Stegeman, 1973; Kung *et al.*, 1972b; Machold and Aurich, 1972); the PS I complex and "activating proteins" of the thylakoid membranes seem to be formed inside the chloroplasts. The synthesis of these "activating proteins" is inhibited by CAP but not by rifampicin. Thus, it has been suggested that the mRNA transcribed at nDNA is translated on chloroplast ribosomes (Jennings and Ohad, 1972). In conclusion, the results from a number of laboratories demonstrate that there is close co-operation between plastid and cytoplasm in the formation of thylakoid proteins. The assembly of the individual species to thylakoid membranes is a multistep process which is light-dependent and can be regulated by both chloroplast and cytoplasmic factors (Eytan and Ohad, 1970, 1972; Hoober *et al.*, 1969).

Of interest is the site of synthesis of the enzymes involved in the replication, transcription and translation of the genetic material of the chloroplast. Although it is too premature to make conclusive statements, the results at present favour a cytoplasmic origin of DNA polymerase and RNA polymerase.

We have performed experiments to determine the site of synthesis of chloroplast aa-RS in *Euglena*. The aa-RS activities of dark-grown cells increase considerably upon illumination (Krauspe and Parthier, 1973, 1974; Parthier *et al.*, 1972) and precede chlorophyll formation. This increase of activity can be suppressed using the inhibitors of protein synthesis, CAP and CHI. In accordance with the finding that the chloroplast phe-RS in *Euglena* is a constitutive enzyme (Reger *et al.*, 1970) and consequently controlled by the nucleus, we were able to show that the synthesis of 10 aa-RS species tested

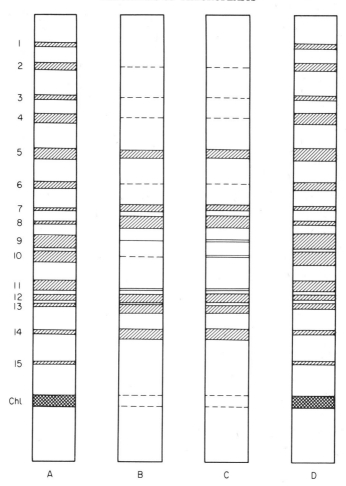

FIG. 14. Gel electrophoretic separation of the membrane protein fraction of *E. gracilis* chloroplasts. The cells are grown for 5 generations (3 days) under photoheterotrophic conditions with the following additions: (A) none; (B) 1·5 mg/ml CAP; (C) 40 µg/ml nalidixic acid; (D) 4 µg/ml CHI. Bands are stained with amido black; Chl, chlorophyll.

are effectively inhibited by CHI in appropriate concentrations. Figure 15 demonstrates that in a kinetic experiment the light-increased leu-RS synthesis is suppressed by 6 µg/ml CHI, adding the antibiotic at any time. It is concluded from this experiment that the light-dependent development of a protein synthesizing system in the plastid is not a prerequisite for leu-RS synthesis. A resumption of the enzyme synthesis is always observed. Using CAP alone chloroplast leu-RS synthesis is reduced to 60% of the activity of untreated cells, in contrast to nearly 100% inhibition of chlorophyll and RuDPC synthesis (Table V). Addition of CAP does not interfere with the

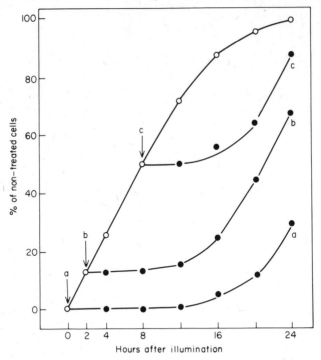

Fig. 15. Inhibition of plastid leu-RS synthesis by 6 μg/ml CHI added at various times (arrows) after illumination of dark-grown *E. gracilis*. CHI-treated (●—●), non-treated cells (○—○) (Parthier, 1973).

TABLE V

Effects of chloramphenicol (CAP) and cycloheximide (CHI) alone or in combination on the light-dependent increase of chlorophyll, RuDP carboxylase, plastid and cytoplasmic leucyl-tRNA synthetases of *Euglena*. The figures are given as % of non-treated controls (Parthier, 1973)

		Antibiotic concn/ml			
	2 mg CAP no CHI	No CAP		2 mg CAP	
		4 μg CHI	6 μg CHI	+4 μg CHI	+6 μg CHI
Chlorophyll	< 5	62	17	24	5
RuDP carboxylase	< 5	130	55	16	10
Plastid leu-RS	60	58	28	65	30
Cytoplasmic leu-RS	110	100	75	115	85

CHI-inhibition of plastid leu-RS synthesis. Thus, we suggest (Parthier, 1973) that the aa-RS are formed on cytoplasmic ribosomes; the inhibition by CAP, however, might be due to the suppressed synthesis of a derepressor, which is synthesized on 70 S ribosomes and regulates the transcription of mRNA for plastid aa-RS at nDNA.

V. Regulatory Aspects of Chloroplast Biogenesis

A. PHOTOREGULATION OF PLASTID TRANSFORMATION

Light plays a dual role in the life of chloroplasts. Firstly, it provides the electromagnetic energy for the conversion into chemical energy; and secondly, the development of a functioning chloroplast from its precursor organelle as judged by the greening process and the appearance of photosynthesis is strictly light-dependent in many plants. However, most algae, the cryptogams and some phanerogams are exceptional in being able to develop chloroplasts in darkness. This indicates that the light-dependency of the chlorophyll formation and lamellar protein synthesis is not a *conditio sine qua non*. This property might be due to repression mechanisms acquired during the past periods of evolution of the plant kingdom. Light, whether primarily or secondarily, might act as derepressor of chloroplast development. The formation of regulatory proteins as well as the massive synthesis of chloroplast proteins needs the continuous synthesis of RNA and proteins, as has been shown by inhibitor studies. Mutants have also been used to elucidate pathways which have to operate in the formation of a functional chloroplast from the proplastid, provided the specific genetic blocks can be characterized (Börner *et al.*, 1972; Goodenough and Levine, 1971; Sprey, 1972; Wettstein *et al.*, 1971).

In determining the photoregulation of chloroplast biogenesis, there are three problems to be solved: (i), what is the primary photoreceptor; (ii), which macromolecular syntheses are induced, increased or unaffected by light, and (iii), what is the sequential order of the light-stimulated processes?

Although the photochemical reduction of protochlorophyllide to chlorophyllide has an important effect on subsequent plastid development, protochlorophyllide for some reason cannot be regarded as the primary photoreceptor (Smillie and Scott, 1969). In angiosperms the photoregulation of chloroplast biogenesis is mediated by phytochrome, a system however which is absent from *Euglena*. In the case of the angiosperms, there is no conclusive evidence that phytochrome regulates chloroplast formation, but it could act on other cellular processes which promote the plastid transformation to chloroplasts.

In the light-induced sequence protochlorophyllide → chlorophyllide → membrane formation → RNA polymerase activation → rRNA synthesis → massive protein formation (Harel and Bogorad, 1973), the synthesis of new protochlorophyllide may be a limiting step. The onset of photosynthesis, as measured by oxygen production parallel with chlorophyll formation, shows a lag of several hours (Lagoutte and Duranton, 1972; Nadler *et al.*, 1972; Schiff, 1970; Schiff and Zeldin, 1968). An uncoupling of the two processes can be achieved by addition of δ-aminolevulinic acid which accelerates chlorophyll synthesis but not O_2 evolution, and by CAP which inhibits O_2 evolution and not chlorophyll synthesis (Nadler *et al.*, 1972). These results suggest that the formation of PS II enzymes is independent of chlorophyll and also that there is a continuous requirement of light for the synthesis of the

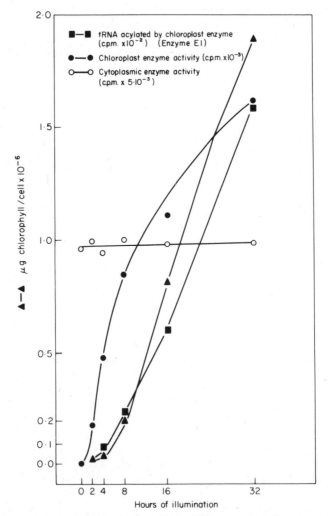

FIG. 16. Light-induced increase of chlorophyll content (▲—▲), plastid leu-RS activity (●—●), plastid tRNA^{leu} (■—■) during the greening process of dark-grown *E. gracilis*. Cytoplasmic leu-RS activity (○—○) (Parthier and Krauspe, 1974).

PS II complex. Which of the two photosystems is synthesized first is still a matter of controversy (Machold and Aurich, 1972; Nadler *et al.*, 1972; Remy, 1973). It is suggested that the protein components of the two photosystems may be already present in wheat etioplasts, which form "primary thylakoids" upon illumination, by the reorganization of preexisting proteins (Remy, 1973). The few structural proteins already present in etioplasts can become the main structural proteins in chloroplasts (Lagoutte and Duranton, 1972); other proteins may be readily labelled with light onset but do not increase during further illumination (Kaveh and Harel, 1973). These "early structural proteins" seem to be synthesized in the cytoplasm.

In *Euglena*, chlorophyll synthesis is probably controlled by a cytoplasmic repressor and by a derepressor synthesized on chloroplast ribosomes (Perl, 1972). The light-stimulated increase in the synthesis of certain chloroplast proteins preclude an activation of the genetic material in turn with RNA synthesis. After illumination of etiolated maize leaves a rapid labelling of chloroplast rRNA relative to mRNA (polysome) formation was observed (Harel and Bogorad, 1973). In *Euglena*, however, assembly of polysomes clearly precedes the rRNA synthesis (Brown and Haselkorn, 1971; Heizmann, 1970; Heizmann *et al.*, 1972). Light not only stimulates polysome formation in the chloroplasts but also the synthesis of cytoplasmic polysomes (Pine and Klein, 1972; Travis *et al.*, 1970).

We have studied the kinetics of light-stimulated synthesis of plastid tRNA^{leu} and the cognate leu-RS during the greening process in *Euglena*. As can be seen in Fig. 16, the accumulation of plastid leu-RS in terms of enzyme activity clearly precedes tRNA^{leu} and chlorophyll synthesis which synchronously increase with a lag phase of 6 h after illumination with 1200 lux. Since this period corresponds to the appearance of thylakoid membrane attachment forming chloroplast lamellae (Neumann and Parthier, 1973), we suggested (Parthier and Krauspe, 1974) that the assembly of chloroplast lamellae is a prerequisite for the synthesis or the function of RNA polymerase bound to membranes (Bottomley *et al.*, 1971a, b; Polya and Jagendorf, 1971; Spencer *et al.*, 1971; Tewari and Wildman, 1969). The aa-RS as stroma enzymes are synthesized on cytoplasmic ribosomes (Parthier, 1973) and are readily available for chloroplast protein synthesis. The mechanism of intracellular control of these processes are unknown but the extent and mode of photoregulation in chloroplast biogenesis may vary in different plant species.

B. INTERRELATIONS BETWEEN CHLOROPLAST AND CYTOPLASM

As was shown in Table IV and can be seen in Fig. 17, only a few proteins synthesized in the chloroplast, seem to be under the entire control of cDNA. The greater proportion of the chloroplast proteins or their non-functional polypeptide precursors, therefore, should be imported from the cytoplasm. The transport from the nucleus to chloroplasts of informational RNA for the

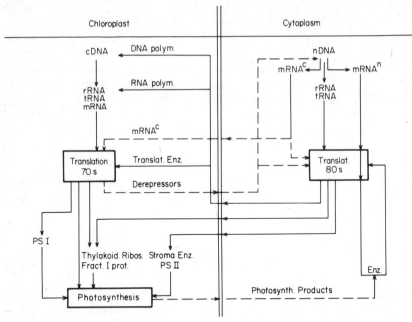

Fig. 17. Tentative scheme of possible interrelations between chloroplast and nucleo cytoplasm. PS, photosystem; cDNA, chloroplast DNA; nDNA, nuclear DNA; mRNA^c messenger RNA carrying genetic information for the synthesis of chloroplast proteins.

synthesis of chloroplast proteins (mRNA^c) and its subsequent translation on chloroplast ribosomes has been suggested for certain thylakoid proteins in *Chlamydomonas* (Jennings and Ohad, 1972). More experimental evidence is needed to confirm this possibility of chloroplast information translation. Data have been reported indicating that a precursor of chlorophyll can regulate the transcription of the mRNA for a thylakoid protein ("polypeptide c") that is synthesized on cytoplasmic ribosomes (Hoober and Stegeman, 1973). This regulation seems to be mediated by a protein synthesized on chloroplast ribosomes. A similar assumption has been made for the synthesis of chloroplast aa-RS (Parthier, 1973).

Synthesis of rRNA and possibly tRNA together with a very limited number of chloroplast proteins adjusts the synthetic capacity of chloroplasts to that of mitochondria. It indicates that most of the processes of chloroplast metabolism are directly or indirectly controlled by the cytoplasm, particularly during the early periods of chloroplast formation following illumination. Chloroplast functions intimately linked with the photosynthetic electron transport and the very processes of plastid inheritance are "autonomous" features of the chloroplasts. By means of various plastom mutants of barley, 86 nuclear genes have been identified that control chloroplast development (Wettstein *et al.*, 1971). A hypothesis suggesting that cDNA harbours regulator genes of chloroplast biogenesis while the structural genes are situated in nDNA is an interesting view (Kirk, 1971b) but experimental data are lacking as yet to prove this. The rapid increase of plastid aa-RS synthesis (Fig. 16) or of some other chloroplast proteins whose syntheses take place in the cytoplasm would presume the formation of components that derepress or activate transcription or translation processes involved in the cytoplasmic synthesis of chloroplast proteins (Fig. 17). A control of the activity of cytoplasmic enzymes by products of the photosynthesis is likewise suggestive.

Besides its limited reproductive autonomy, the chloroplast may be even more dependent on the surrounding cytoplasm for its nutrition (Givan and Leech, 1971). Very little is known about the capacity of the plastid to synthesize nucleic acid precursors or about which amino acids, except those directly derived from the photosynthetic carbon cycle, are synthesized in the plastids. As the requirement of chloroplasts for amino acids could be wholly or partially met from outside, the chloroplast–cytoplasm interrelationship has become a truly mutual symbiosis (Givan and Leech, 1971).

Regulatory correlations also exist between the organelles within a cell. After stress conditions or treatment with inhibitors etiolated *E. gracilis* cells (grown in the presence of CAP or nalidixic acid) upon illumination cannot transform proplastids to morphological and functionally mature chloroplasts (Neumann and Parthier, 1973). Since the growth rates of the cultures are unaffected by the antibiotics the energy for the metabolism must be produced in the mitochondria, which are increased either in size or in number. In contrast, growth in the presence of CHI results in the amplification of chloroplast structures and photosynthetic activity (Neumann and Parthier, 1973). This plasticity of the unicellular *E. gracilis* cells to break down and rebuild their organelles as a response to nutritional changes must be the result of a subtle regulation of gene expression at the molecular level.

REFERENCES

Aliev, K. A. and Filippovich, I. I. (1968). *Molec. Biol.* (Moscow) **2**, 364.
Anderson, M. B. and Cherry, J. H. (1969). *Proc. natn. Acad. Sci. U.S.A.* **62**, 202.
Anderson, L. E. and Levin, D. A. (1970). *Pl. Physiol., Lancaster* **46**, 819.
Apel, K. and Schweiger, H. G. (1972). *Eur. J. Biochem.* **25**, 229.
Arglebe, C. and Hall, T. C. (1969). *Pl. Cell Physiol.* **10**, 171.
Armstrong, J. J., Surzycki, S. J., Moll, B. and Levine, R. P. (1971). *Biochemistry* **10**, 692.
Avadhani, N. G. and Buetow, D. E. (1972). *Biochem. J.* **128**, 353.
Bard, S. A. and Gordon, M. P. (1969). *Pl. Physiol., Lancaster* **44**, 377.
Barnett, W. E., Pennington, C. J. and Fairfield, S. A. (1969). *Proc. natn. Acad. Sci. U.S.A.* **63**, 1261.
Bastia, D., Chiang, K., Swift, H. and Siersma, P. (1971). *Proc. natn. Acad. Sci. U.S.A.* **68**, 1157.
Blair, G. E. and Ellis, R. J. (1972). *Biochem. J.* **127**, 42P.

Boardman, N. K., Francki, R. I. B. and Wildman, S. G. (1966). *J. molec. Biol.* **17**, 470.

Boardman, N. K., Linnane, A. W. and Smillie, R. M. (eds), (1971). "Autonomy and Biogenesis of Mitochondria and Chloroplasts". North-Holland, Amsterdam.

Börner, T., Knoth, R., Herrmann, F. and Hagemann, R. (1972). *Theoret. appl. Genetics* **42**, 3.

Bottomley, W. (1970). *Pl. Physiol.*, *Lancaster* **45**, 608.

Bottomley, W., Spencer, D., Wheeler, A. M. and Whitfeld, P. R. (1971). *Archs Biochem. Biophys.* **143**, 269.

Bottomley, W., Smith, H. J. and Bogorad, L. (1971b). *Proc. natn. Acad. Sci. U.S.A.* **68**, 2412.

Bourque, D. P. and Wildman, S. G. (1973). *Biochem. biophys. Res. Comm.* **50**, 532.

Boulter, D., Ellis, R. J. and Yarwood, A. (1972). *Biol. Rev.* **47**, 113.

Brown, R. D. and Haselkorn, R. (1971). *Proc. natn. Acad. Sci. U.S.A.* **68**, 2536.

Burdon, R. H. (1971). *Progr. Nucleic Acid Res. molec. Biol.* **11**, 33.

Burkard, G., Eclancher, B. and Weil, J. H. (1969). *FEBS Letters* **4**, 285.

Burkard, G., Guillemaut, P. and Weil, J. H. (1970). *Biochim. biophys. Acta* **224**, 184.

Burkard, G., Vaultier, J. P. and Weil, J. H. (1972). *Phytochemistry* **11**, 1351.

Chan, P. and Wildman, S. G. (1972). *Biochim. biophys. Acta* **277**, 677.

Chen, J. L. and Wildman, S. G. (1967). *Science, N.Y.* **155**, 127.

Chen, J. L. and Wildman, S. G. (1970). *Biochim. biophys. Acta* **209**, 207.

Chiang, K. S. and Sueoka, N. (1967). *Proc. natn. Acad. Sci. U.S.A.* **57**, 1506.

Chua, N. H., Blobel, G., Siekevitz, P. and Palade, G. E. (1973). *Proc. natn. Acad. Sci. U.S.A.* **70**, 1554.

Clark, M. F. (1964). *Biochim. biophys. Acta* **91**, 671.

Craig, J. W. and Gibor, A. (1970). *Biochim. biophys. Acta* **217**, 488.

Criddle, R. S., Dau, B., Kleinkopf, G. E. and Huffaker, R. C. (1970). *Biochem. biophys. Res. Comm.* **41**, 621.

Delihas, N., Jupp, A. and Lyman, H. (1972). *Biochim. biophys. Acta* **262**, 344.

Edelman, M., Swinton, D., Schiff, J. A., Epstein, H. T. and Zeldin, B. (1967). *Bact. Rev.* **31**, 315.

Egan, J. M. and Carell, E. F. (1972). *Pl. Physiol.*, *Lancaster* **50**, 391.

Ellis, R. J. (1973). *Comment. Pl. Sci.* **4**, 29.

Ellis, R. J. and Hartley, M. R. (1971). *Nature New Biol.* **233**, 193.

Ernst-Fonberg, M. L. and Bloch, K. (1971). *Archs Biochem. Biophys.* **143**, 392.

Eytan, G. and Ohad, J. (1970). *J. biol. Chem.* **245**, 4297.

Eytan, G. and Ohad, J. (1972). *J. biol. Chem.* **247**, 122.

Fairfield, S. A. and Barnett, W. E. (1971). *Proc. natn. Acad. Sci. U.S.A.* **68**, 2972.

Falk, H. (1969). *J. Cell Biol.* **42**, 582.

Filippovich, I. I., Bezsmertnaya, I. N. and Oparin, A. I. (1973). *Expl Cell Res.* **79**, 159.

Gibbs, S. P. (1967). *Biochem. biophys. Res. Comm.* **28**, 653.

Gibbs, S. P. (1970). *J. Cell Biol.* **46**, 599.

Giles, K. L. and Taylor, A. O. (1971). *Pl. Cell Physiol.* **12**, 437.

Givan, A. L. and Criddle, R. S. (1972). *Archs Biochem. Biophys.* **149**, 153.

Givan, C. V. and Leech, R. M. (1971). *Biol. Rev.* **46**, 409.

Goffeau, A. (1969). *Biochim. biophys. Acta* **174**, 340.

Goodenough, U. W. and Levine, R. P. (1971). *J. Cell Biol.* **50**, 50.

Goodwin, T. W. (ed.) (1966). "Biochemistry of chloroplasts". Academic Press, London and New York.

Goodwin, T. W. (1971). *In* "Structure and Function of Chloroplasts" (M. Gibbs, ed.), p. 215. Springer, Berlin-Heidelberg-New York.

Granick, S. and Gibor, A. (1967). *Progr. Nucleic Acid Res. molec. Biol.* **6**, 143.

Green, B. R. and Burton, H. (1970). *Science, N.Y.* **168**, 981.

Grierson, D. and Loening, U. E. (1972). *Nature New Biol.* **235**, 80.

Gualerzi, C. and Cammarano, P. (1969). *Biochim. biophys. Acta* **190**, 170.

Gualerzi, C. and Cammarano, P. (1970). *Biochim. biophys. Acta* **199**, 203.

Guderian, R. H., Pulliam, R. L. and Gordon, M. P. (1972). *Biochim. biophys. Acta* **262**, 50.

Guillemaut, P., Burkard, G., Steinmetz, A. and Weil, J. H. (1973). *Plant Sci. Letters* **1**, 141.

Hagemann, R. (1967). *Biol. Rdsch.* **5**, 97.

Harel, E. and Bogorad, L. (1973). *Pl. Physiol., Lancaster* **51**, 10.

Harris, E. H. and Eisenstadt, J. M. (1971). *Biochim. biophys. Acta* **232**, 167.

Hartley, M. R. and Ellis, R. J. (1973). *Biochem. J.* **134**, 249.

Hartmann, G., Hornikel, K. O., Knüsel, F. and Nüesch, J. (1967). *Biochim. biophys. Acta* **145**, 843.

Hearing, K. J. (1973). *Phytochemistry* **12**, 277.

Heber, U. (1962). *Nature, Lond.* **195**, 91.

Heizmann, P. (1970). *Biochim. biophys. Acta* **224**, 144.

Heizmann, P., Trabuchet, G., Verdier, G., Freyssinet, G. and Nigon, V. (1972). *Biochim. biophys. Acta* **277**, 149.

Herrmann, F. (1971). *FEBS Letters* **19**, 267.

Herrmann, F. (1972). *Expl Cell Res.* **70**, 452.

Herrmann, R. G. (1970). *Planta* **90**, 80.

Herzfeld, F. and Zillig, W. (1971). *Eur. J. Biochem.* **24**, 242.

Honeycutt, R. C. and Margulies, M. M. (1973). *J. biol. Chem.* **248**, 6145.

Hoober, J. K. and Blobel, G. (1969). *J. molec. Biol.* **41**, 121.

Hoober, J. K., Siekevitz, P. and Palade, G. E. (1969). *J. biol. Chem.* **244**, 2621.

Hoober, J. K. and Stegeman, W. J. (1973). *J. Cell Biol.* **56**, 1.

Horak, A. and Hill, R. D. (1972). *Pl. Physiol., Lancaster* **49**, 365.

Howell, S. H. and Walker, L. L. (1972). *Proc. natn. Acad. Sci. U.S.A.* **69**, 490.

Ingle, J. (1968). *Pl. Physiol., Lancaster* **43**, 1850.

Ingle, J., Possingham, J. V., Wells, R., Leaver, C. J. and Loening, U. E. (1970). *Symp. Soc. exp. Biol.* **24**, 303.

Ingle, J. and Sinclair, J. (1972). *Nature, Lond.* **235**, 30.

Ireland, H. M. M. and Bradbeer, J. W. (1971). *Planta* **96**, 254.

Iwamura, T. (1966). *Prog. Nucleic Acid Res. molec. Biol.* **5**, 133.

Jennings, R. C. and Ohad, I. (1972). *Archs Biochim. Biophys.* **153**, 79.

Jones, B. L., Nagabushan, N., Gulyas, A. and Zalik, S. (1972). *FEBS Letters* **23**, 167.

Kanabus, J., and Cherry, J. H. (1971). *Proc. natn. Acad. Sci. U.S.A.* **68**, 873.

Kaveh, D. and Harel, E. (1973). *Pl. Physiol., Lancaster* **51**, 671.

Kawashima, N. and Wildman, S. G. (1970). *Rev. Pl. Physiol.* **21**, 325.

Kawashima, N. and Wildman, S. G. (1972). *Biochim. biophys. Acta* **262**, 42.

Kirk, J. T. O. (1970). *A. Rev. Pl. Physiol.* **21**, 11.

Kirk, J. T. O. (1971a). *In* Boardman *et al.* (1971), p. 267.

Kirk, J. T. O. (1971b). *A. Rev. Biochem.* **40**, 161.

Kirk, J. T. O. and Tillney-Bassett, R. A. E. (1967). "The Plastids". Freeman, San Francisco.

Kislev, N., Selsky, M. J., Norton, C. and Eisenstadt, J. M. (1972). *Biochim. biophys. Acta* **287**, 256.

Kolodner, R. and Tewari, K. K. (1972). *J. biol. Chem.* **247**, 6355.

208 B. PARTHIER ET AL.

Krauspe, R. and Parthier, B. (1974). *Biochem. Physiol. Pflanz.* **165**, 18.
Krauspe, R. and Parthier, B. (1973). *Biochem. Soc. Symp.* **38**, 111.
Kung, S. D. (1973). *FEBS Letters* **29**, 259.
Kung, S. D., Moscarello, M. A. and Williams, J. P. (1972a). *Pl. Physiol.* **49**, 331.
Kung, S. D., Thornber, J. P. and Wildman, S. G. (1972b). *FEBS Letters* **24**, 185.
Kung, S. D. and Williams, J. P. (1969). *Biochim. biophys. Acta* **195** 434.
Lagoutte, B. and Duranton, J. (1972). *FEBS Letters* **28**, 333.
Lanzani, G. A., Manzocchi, A., Galante, E. and Menegus, F. (1969).*Enzymologia* **37**, 97.
Levine, R. P. and Armstrong, J. (1972). *Pl. Physiol.*, *Lancaster* **49**, 661.
Lyttleton, J. W. (1967). *Biochim. biophys. Acta* **149**, 598.
Lyttleton, J. W. (1968). *Biochim. biophys. Acta* **154**, 145.
Mache, R. and Waygood, E. R. (1970). *Can. J. Bot.* **48**, 173.
Machold, O. and Aurich, O. (1972). *Biochim. biophys. Acta* **281**, 103.
Manning, J. E. and Richards, O. C. (1972). *Biochim. biophys. Acta* **259**, 285.
Manning, J. E., Wolstenholme, D. R., Ryan, R. S., Hunter, J. A. and Richards, O. C. (1971). *Proc. natn. Acad. Sci. U.S.A.* **68**, 1169.
Margulies, M. M. (1971). *Biochem. biophys. Res. Comm.* **44**, 539.
Menke, W. (1961). *Z. Naturf.* **16b**, 334.
Merrick, W. C. and Dure III, L. S. (1971). *Proc. natn. Acad. Sci. U.S.A.* **68**, 641.
Merrick, W. C. and Dure III, L. S. (1972). *J. biol. Chem.* **247**, 7988.
Mets, L. and Bogorad, L. (1972). *Proc. natn. Acad. Sci. U.S.A.* **69**, 3779.
Mühlethaler, K. (1971). *In* "Structure and Function of Chloroplasts" (M. Gibbs, ed.), p. 7. Springer, Berlin–Heidelberg–New York.
Munns, R., Scott, N. S. and Smillie, R. M. (1972). *Phytochemistry* **11**, 45.
Munsche, D. and Wollgiehn, R. (1973). *Biochim. biophys. Acta* **294**, 106.
Nadler, K. D., Herron, H. A. and Granick, S. (1972). *Pl. Physiol.*, *Lancaster* **49**, 388.
Nass, M. M. K. and Ben-Shaul, Y. (1972). *Biochim. biophys. Acta* **272**, 130.
Neumann, D. and Parthier, B. (1973). *Expl Cell Res.* **81**, 255.
Odintsova, M. S., Mikulska, E. and Turischewa, M. S. (1970). *Expl. Cell Res.* **61**, 423.
Odintsova, M. S. and Yurina, N. P. (1969). *J. molec. Biol.* **40**, 503.
Oparin, A. J., Filippovich, I. I. and Bezsmertnaya, I. N. (1972). *Fiziol. Rasten.* **19** 995.
Parthier, B. (1964). *Z. Naturf.* **19b**, 235.
Parthier, B. (1970). *Biol. Rundsch.* **8**, 289.
Parthier, B. (1973). *FEBS Letters* **38**, 70.
Parthier, B. and Krauspe, R. (1973). *Pl. Sci. Letters* **1**, 221.
Parthier, B. and Krauspe, R. (1974). *Biochem. Physiol. Pflanz.* **165**, 1.
Parthier, B., Krauspe, R. and Samtleben, S. (1972). *Biochim. biophys. Acta* **277**, 335.
Parthier, B. and Wollgiehn, R. (1966). *In* "Probl. Biol. Reduplikation" (P. Sitte, ed.), p. 244. Springer, Berlin–Heidelberg–New York.
Perl, M. (1972). *Biochem. J.* **130**, 813.
Pine, K. and Klein, O. (1972). *Dev. Biol.* **28**, 280.
Pollack, R. W. and Davies, P. J. (1970). *Phytochemistry* **9**, 471.
Polya, G. M. and Jagendorf, A. T. (1971a). *Archs Biochem. Biophys.* **146**, 635.
Polya, G. M. and Jagendorf, A. T. (1971b). *Archs Biochem. Biophys.* **146**, 649.
Rawson, J. R. Y. and Haselkorn, R. (1973). *J. molec. Biol.* **77**, 125.
Rawson, J. R. and Stutz, E. (1969). *Biochim. biophys. Acta* **190**, 358.
Reger, B. J., Fairfield, S. A., Epler, J. L. and Barnett, W. E. (1970). *Proc. natn. Acad. Sci. U.S.A.* **67**, 1207.

Remy, R. (1973). *FEBS Letters* **31**, 308.

Richards, O. C., Ryan, R. S. and Manning, J. E. (1971). *Biochim. biophys. Acta* **238**, 190.

Ris, H. and Plaut, W. (1962). *J. Cell Biol.* **13**, 383.

Rochaix, J. D. (1972). *Nature New Biol.* **238**, 76.

Rossi, L. and Gualerzi, C. (1970). *Life Sci.* **9**, 1401.

Ruppel, H. G. (1969). *Z. Naturf.* **24b** 1467.

Rutner, A. C. (1970). *Biochem. biophys. Res. Comm.* **39**, 923.

Sager, R. and Ramanis, Z. *In* Boardman *et al.* (1971), p. 250.

Schiff, J. A. (1970). *Symp. Soc. exp. Biol.* **24**, 277.

Schiff, J. A. and Zeldin, M. H. (1968). *J. Cell Physiol.* **72**, Suppl. **1**, 103.

Schnepf, E. and Brown, R. M. (1971). *In* "Origin and Continuity of Cell Organelles" (Reinert u. H. Ursprung, ed.). Springer, Heidelberg–Berlin–New York.

Schrader, L. E., Beevers, L. and Hageman, R. H. (1967). *Biochem. biophys. Res. Comm.* **26**, 14.

Schwartz, J. H., Meyer, R., Eisenstadt, J. E. and Brawerman, G. (1967). *J. molec. Biol.* **25**, 571.

Schweiger, H. G. (1970). *Sym. Soc. exp. Biol.* **24**, 327.

Schweiger, H. G., Apel, K. and Kloppstech, K. (1972). *Adv. Biosciences* **8**, 249.

Scott, N. S., Munns, R. and Smillie, R. M. (1970). *FEBS Letters* **10**, 149.

Scott, N. S., Shah, V. C. and Smillie, R. M. (1968). *J. Cell Biol.* **38**, 151.

Scott, N. S. and Smillie, R. M. (1967). *Biochem. biophys. Res. Comm.* **28**, 598.

Sela, J. and Kaesberg, P. (1969). *J. Virol.* **3**, 89.

Smillie, R. M., Bishop, D. G., Gibbons, G. C., Graham, D., Grieve, A. M., Raison, J. K. and Reger, B. J. *In* Boardman *et al.* (1971), p. 422.

Smillie, R. M., Graham, D., Dwyer, M. R., Grieve, A. and Tobin, N. F. (1967). *Biochem. biophys. Res. Comm.* **28**, 604.

Smillie, R. M. and Scott, N. S. (1969). *Progr. molec. subcell. Biol.* **1**, 136.

Smith, H., Stewart, G. R. and Berry, D. R. (1970). *Phytochemistry* **9**, 977.

Spencer, D. and Whitfeld, P. R. (1967). *Archs Biochem. Biophys.* **121**, 336.

Spencer, D. and Whitfeld, P. R. (1969). *Archs Biochem. Biophys.* **132**, 477.

Spencer, D., Whitfeld, P. R., Bottomley, W. and Wheeler, A. M. *In* Boardman *et al.* (1971), p. 372.

Sprey, B. (1972). *Z. Pflanzenphysiol.* **67**, 223.

Stutz, E. (1970). *FEBS Letters* **8**, 25.

Stutz, E. and Noll, H. (1967). *Proc. natn. Acad. Sci. U.S.A.* **57**, 774.

Stutz, E. and Vandrey, J. P. (1971). *FEBS Letters* **17**, 277.

Surzycki, S. J. (1969). *Proc. natn. Acad. Sci. U.S.A.* **63**, 1327.

Surzycki, S. J. and Gillham, N. W. (1971). *Proc. natn. Acad. Sci. U.S.A.* **68**, 1301

Surzycki, S. J., Goodenough, U. W., Levine, R. P. and Armstrong, J. J. (1970). *Symp. Soc. exp. Biol.* **24**, 13.

Szalay, A., Munsche, D., Wollgiehn, R. and Parthier, B. (1972). *Biochem. J.* **129**, 135.

Szalay, A., Munsche, D., Wollgiehn, R. and Parthier, B. (1973). *Biochem. physiol. Pflanz.* **164**, 1.

Tewari, K. K. (1971). *A. Rev. Pl. Physiol.* **22**, 141.

Tewari, K. K. and Wildman, S. G. (1967). *Proc. natn. Acad. Sci. U.S.A.* **58**, 689.

Tewari, K. K. and Wildman, S. G. (1968). *Proc. natn. Acad. Sci. U.S.A.* **59**, 569.

Tewari, K. K. and Wildman, S. G. (1969). *Biochim. biophys. Acta* **186**, 358.

Tewari, K. K. and Wildman, S. G. (1970). *Symp. Soc. exp. Biol.* **24**, 147.

Travis, R. L., Huffaker, R. C. and Key, J. L. (1970). *Pl. Physiol., Lancaster* **46**, 800.

Walles, B. (1971). *In* "Structure and Function of Chloroplasts" (M. Gibbs, ed.), p. 51. Berlin–Heidelberg–New York.

Wettstein, D. von, Henningsen, K. W., Boynton, J. E., Kannangara, G. C. and Nielsen, O. F. (1971). *In* "Autonomy and Biogenesis of Mitochondria and Chloroplasts" (N. K. Boardman *et al.*, eds), p. 205. North-Holland.

Wells, R. and Birnstiel, M. (1969). *Biochem. J.* **112**, 777.

Williams, G. R. and Novelli, G. D. (1968). *Biochim. biophys. Acta* **155**, 183.

Williams, G. R. and Williams, A. S. (1970). *Biochem. biophys. Res. Comm.* **39**, 858.

Wittmann, H. G. (1970). *Symp. Soc. gen. Microbiol.* **2**, 55.

Wittmann, H. G. (1972). *Biochem. J.* **129**, 30.

Wollgiehn, R. and Munsche, D. (1972). *Biochem. Physiol. Pflanz.* **163**, 137.

Wollgiehn, R., Munsche, D. and Mikulovich, T. P. (1974). *Biochem. Physiol. Pflanz.* **165**, 37.

Woodcock, C. L. F. and Bogorad, L. (1971). *In* "Structure and Function of Chloroplasts" (M. Gibbs, ed.), p. 89. Springer, Berlin–Heidelberg–New York.

Woodcock, C. L. F. and Fernández-Morán, H. (1968). *J. molec. Biol.* **31**, 627.

Zeldin, M. H. and Schiff, J. A. (1967). *Pl. Physiol., Lancaster* **42**, 922.

CHAPTER 8

Plant Proteins and Phenolics

C. F. VAN SUMERE, J. ALBRECHT, ANDRÉE DEDONDER, H. DE POOTER

AND IRMA PÉ

Laboratorium voor Plantenbiochemie, Rijksuniversiteit, Gent, Belgium

I. INTRODUCTION

Although free phenolics, or their oxidation products, and proteins have long been known to react with each other, the outcome of the many possible interactions is still far from being completely understood. In certain cases, knowledge of the chemistry and biochemistry involved is scanty and certainly not commensurate with the important roles played by these interactions. Such reactions are involved during the extraction of proteins from plants and, in the mechanism of action of phenolic disinfectants, the inhibition or activation of enzymes, the uncoupling of oxidative phosphorylation, the resistance of fruits to microbial attack, the production of humus, the formation of beer or wine hazes, the storing of other food products and the tanning of hides (Swain, 1965).

In addition to the reactions between the free, oxidized and (or) polymerized phenolics and proteins, certain phenolics such as ferulic acid may also be directly or indirectly bound to plant polypeptides and proteins by means of ester and amide linkages (Fausch et al., 1963; Neukom et al., 1967; Van Sumere et al., 1973).

Thus it seems that the reactions between phenolics and proteins are of great importance in many fields. The present review is devoted to the most recent work on the relation between phenols and plant proteins. Aminophenols and their interaction with proteins and enzymes have been deliberately omitted from the present discussion.

II. THE INTERACTION BETWEEN PHENOLICS AND PROTEINS

A. POSSIBLE REACTIONS BETWEEN PHENOLICS AND PROTEINS

The theoretical and practical importance of molecular interactions between proteins and phenolics has been increasingly recognized. Phenols combine with proteins reversibly by hydrogen bonding, and irreversibly by oxidation followed by covalent condensations (Loomis and Battaile, 1966). In the case of the tannins, weak ionic bonds between suitable charged anionic groups on the phenolics and cationic groups on the protein must also be considered (Swain, 1965). The hydrogen bond formed between phenols and N-substituted amides is relatively strong and the equilibrium in aqueous solution favours complex formation (Loomis, 1969). According to Loomis, the amount of phenolic material bound to protein by hydrogen bonding may be equal to more than 33% dry weight of the protein. This may be the reason for the inactivation and precipitation of enzymes during their isolation from plants. The presence of phenolic compounds often makes it impossible to isolate active enzymes by conventional techniques (Loomis and Battaile, 1966; Anderson, 1968; Neish, 1968; Loomis, 1969). Experiments on the tanning of modified collagen (Gustavson, 1956) and of synthetic polymers (Gustavson and Holm, 1952; Batzer, 1952; Batzer and Weissenberger, 1952; Gustavson, 1954, 1956) have further shown that peptide or amide linkages are very important in the formation of hydrogen bonded complexes between protein and tannin (see also Grassmann, 1937). Indeed, from the tanning experiments, it follows that only the —CO— group is required for the formation of complexes with vegetable tannins and that tannins form hydrogen bonds with the peptide linkages, most probably through the peptide oxygen, the tannins furnishing the hydrogen (Loomis and Battaile, 1966). This last conclusion is further supported by the more recent studies of Shuttleworth et al. (1968), according to whom tanning appears to be a solvent reversible, multipoint cross-linking mechanism in which hydrogen bonds provide again the dominant forces, and peptide links are the most powerful binding sites (see also Cannon, 1955; Van Buren and Robinson, 1969; Loomis, 1969). In order to be regarded as a tannin, a naturally occurring polyphenol must further possess a high enough molecular weight (between 500 and 3000) and a sufficiently large number of phenolic hydroxyl or other suitable groups (1–2 per 100 mol. wt) to enable it to form effective cross-links between proteins and (or) other macromolecules (Swain, 1965). According to Swain, phenolic compounds of lower molecular weight are too small in size to form effective cross-links, and although they may be absorbed onto proteins and other polymers, the stability constant of the complex is usually low. On the other hand, compounds of higher molecular weight may be ineffective as tannins because they are too large to penetrate between the collagen fibrils.

There are two types of tannin—condensed and hydrolysable—and, according to Gustavson (1956, 1963) and Loomis and Battaile (1966), they show very different pH responses in their reactions with polypeptides. Indeed, condensed tannins are bound almost independently of pH below 7–8, the binding decreasing rapidly above pH 8 because an ionized phenol cannot serve as proton donor for hydrogen bonding. Hydrolysable tannins, on the other hand, are bound very strongly at pH 3–4 but the binding decreases above pH 5. Again according to Loomis and Battaile (1966), the pH effects in the case of condensed tannins indicate that the binding involves un-ionized phenolic hydroxyl groups. However, in the case of hydrolysable tannins, the pH effects suggest that rather strong hydrogen bonds are formed by un-ionized carboxyl groups of the tannins and weaker hydrogen bonds by un-ionized phenolic hydroxyl groups. Hydrolysable tannins, as a result of possible internal hydrogen bonding, have a relatively low affinity for polyamide (see Endres and Hörmann, 1963; Egger, 1964). In a report by Van Buren and Robinson (1969) it was further shown that tannic acid (a mixture of phenolics) and protein interact to form soluble and insoluble complexes, the latter being favoured by a pH near the isoelectric point of the protein and an excess of tannic acid (see also Feeny, 1969). In addition, the formation of complexes may also be affected by the type of protein used (Hrazdina et al., 1969; Schneider and Hallier, 1970). According to Loomis and Battaile (1966), Batzer (1952), Batzer and Weissenberger (1952), Gustavson and Holm (1952) and Gustavson (1954), hydrated nylon also precipitates tannins. The same can be said of polyvinylpyrrolidone (PVP), since Gustavson (1954, 1963) established that both soluble and insoluble PVP form stable insoluble complexes with tannin material. Apparently, hide power, hydrated nylon and insoluble PVP all have nearly the same capacity for binding tannins.

The above reactions between tannins and proteins or polyamide are at least partially reversed by certain organic solvents (e.g. ketones, esters, dimethylformamide, N-methylpyrrolidone and dimethylsulfoxide) (Loomis, 1969), detergents (Goldstein and Swain, 1965), alkali (Pierpoint, 1970), tannase (Nishira, 1963) or, as shown by Lang (1956) for the tannin–albumin complex, by ethanol–water (l/l). In addition, caffeine (Mejbaum-Katzenellenbogen, 1959) and polyethylene glycol-4000 diluted with water (Toyama and Kamiyama, 1972) have been used to recover the protein from a tannin complex.

Although hydrogen bonding plays a very important role in associating phenolics to protein, there are also numerous indications that other and more stable bonds are rapidly formed (Gustavson, 1956; Loomis and Battaile, 1966; Pierpoint, 1970). Indeed, o-quinones, produced from o-dihydroxyphenols by reaction with phenolases, are compounds (Mason, 1955a, 1959; Mason and Peterson, 1965; Brown, 1967) which react rapidly and nonenzymatically with many components. According to Pierpoint (1970), they may polymerize, be reduced, or suffer nucleophilic attack by substances possessing amino, thiol and "activated methylene groups" (see also Hathway, 1958). Very frequently, therefore, they are liable to undergo a variety of reactions.

The reaction scheme shown in Fig. 1 is derived from Pierpoint's review (1970) and is a simple summary of the initial reactions that the first oxidation products of o-dihydroxyphenols, such as chlorogenic acid or flavanols, undergo. The scheme is based on reactions suggested for chlorogenoquinone. Their formulation (Pierpoint, 1966, 1969a, b) apparently relies heavily on analogies with reactions of simple quinones (Mason, 1955a). Furthermore, the reactions between the quinone of chlorogenic acid and amines, amino acids, thiols (see also Webb, 1966; Rich, 1969; Loomis, 1969) and compounds with reactive methylene groups (the reactions proceed much less readily with ε-amino groups of lysine: Mason, 1955b; Loomis and Battaile, 1966) are formulated as if the point of attack was the 6′-position. According to Pierpoint this seems probable and in the reaction with benzene-sulphinic acid, a reaction of no biological significance, almost certain. The monosubstituted products are aromatic and probably colourless; some are even relatively stable. Other primary products such as aminohydroquinones are further oxidized by an excess of o-quinone to coloured quinoids. With secondary amines, such as proline, the reaction stops. The quinone substituted with a primary amine, however, is further substituted and oxidized to give a disubstituted quinone. This type of reaction can continue until all the available positions on the chlorogenoquinone are occupied. Proteins possess several substituents (amine, α-amino, imino and thiol) which are potentially able to react with o-quinones. Crosslinks may thus be formed between different parts of the polypeptide chain(s) as reactive groups are attacked by the same quinone molecule (e.g. —NH—Q—NH—) or different ends of a group of polmerizing quinone molecules (—NH—Q—(Q)$_n$—Q—NH—). This type of reaction would again alter the biological and physical properties of the molecule.

According to Mason and Peterson (1955), o-quinones, which are more

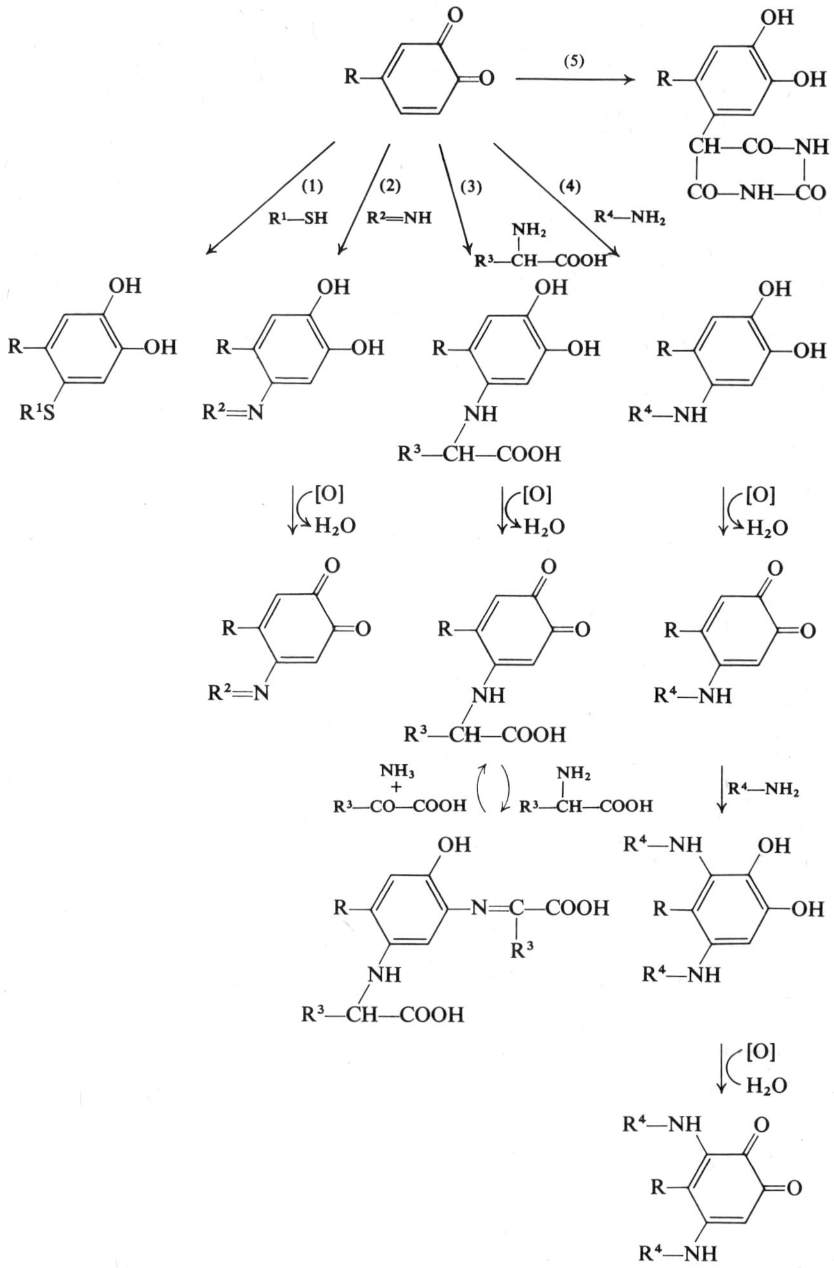

FIG. 1. Probable course of reactions of chlorogenoquinone with thiol compounds (1), amino acids with secondary amine groups (2), amino acids (3), amines (4), and the active methylene group of barbituric acid (5). (Reproduced with permission from Pierpoint (1970).)

reactive than p-quinones, may produce N- or S-catechol proteins which cannot further act as substrates for diphenolases. In the presence of excess enzyme-generated o-quinone, catecholic proteins may be oxidized to the corresponding quinoid proteins. Moreover, leucomelano-protein forms and persists, even in the presence of excess phenolase, when the molecular ratio of o-quinone to protein is low. When the ratio is high, coupled oxidations to melano-protein and to sclerotin take place readily. The enzymatic oxidation of phenols and the reaction of quinones with proteins have been reviewed by Mason (1955a, 1959), Gustavson (1956), Yasunobu (1959), Bouchilloux (1963) and Brunet (1967).

Thus a combination of hydrogen bonding and quinone reactions are involved in the products formed from protein and phenolics. Covalent bonds formed by the interaction of any quinone or semiquinone group that may be present in a tannin and any suitable reactive groups in the protein are important in the manufacture of leather (Swain, 1965). Indeed, since commercial tannin extracts are usually submitted to oxidative conditions during extraction and storage, they usually contain sufficient reactive groupings to form such bonds. According to Mason and Peterson (1955) the rate of reaction of enzyme-generated o-quinones with amino groups increases further within limits with increasing length of the peptide chain with which the amino group is directly associated. Furthermore, such complexes are usually harmful to the cell and under normal conditions, their formation is prevented. The oxidation and polymerization of polyphenols may be limited *in vivo* by the redox balance of the cell, by a combination of o-quinone reductase(s) and suitable reduced coenzymes, by correct compartmentation (Swain, 1965; Anderson, 1968), and by detoxification of polyphenols by conversion to more stable glycosides or esters. Phenolases may also exist as inactive proenzymes or in masked condition; they may be activated with $(NH_4)_2SO_4$, acid, alkali or various anionic wetting agents (Kenten, 1957). Under some circumstances o-quinones may play a beneficial role, e.g. in the defence mechanism of plants against microbial and fungal attack (hypersensitivity) and in reinforcing wall structures. They are almost certainly involved in the sclerotinization of exoskeletons and egg cases of arthropods and in the melanization of seed coats, spore walls and pigment cells of many other organisms (Brunet, 1967; Pierpoint, 1970). According to Mason and Peterson (1965), quinone–protein conjugates also pigment the teguments, feathers, hair and eyes of chordates.

B. PHENOL AS A SOLVENT FOR PROTEINS

The existence of a specific interaction between proteins and phenol(s) has been known since the discovery of phenol by Runge (1834a, b). Runge, one of the pioneers of chromatography (Van Sumere, 1969a, b), was also the first to notice the coagulating effect of phenol on aqueous solutions of proteins

(Pusztai, 1966a). However, phenol has also long been known to be a good and rather selective solvent for polypeptide material (for references see Pusztai, 1966a) and can therefore be used for extracting proteins from biological sources. Stanley *et al.* (1968) have further shown that phenol–acetic acid–water (1:1:1; w/v/v) is a very effective solvent for dissociating proteins, although it does not cleave peptide bonds. The use of phenol for the preparation of active proteins, however, has until recently been limited (Pusztai, 1973), since it was generally assumed that phenol irreversibly denatured proteins and abolished enzymatic activity (Grassmann and Deffner, 1953; Pusztai, 1966b). However, the demonstration of the reversible inhibition of ribonuclease by phenol (Huppert and Pelmont, 1962; Hsu and Wang, 1964) and the successful isolation of the active enzyme from various sources by extraction with the same organic compound have again increased interest in phenol as an extractant for protein (Synge, 1967). It should be particularly applicable to investigations of the nature and properties of water insoluble protein constituents of cell membranes. Although a variety of agents (e.g. aqueous solutions of urea or guanidinium salts, high concentrations of detergents such as sodium dodecyl sulphate and certain organic solvents such as pyridine and chloroethanol) are known to solubilize proteins, Pusztai (1973) claims that most of these solvents lead to serious difficulties during the more common protein fractionation methods. Because phenol is a powerful dissociating agent and decreases the molecular interaction between proteins and other materials (Craig, 1962; Bagdasarian *et al.*, 1964), Pusztai (1973) has used it during purification of a range of proteins.

The strong solvent action of phenol on protein is further indicated by the fact that it shows very little capacity to dissolve carbohydrates and nucleic acids (Gierer and Schramm, 1956; Schuster *et al.*, 1956; Kirby, 1956, 1957; Ralph and Bergquist, 1967; Synge, 1968). Indeed, phenol treatments are usually included in extraction schemes for nucleic acids. Nevertheless some interactions with these other macromolecules may occur in solvent mixtures containing phenol (Brattsten *et al.*, 1964, 1965). Pusztai (1973) has pointed out that, because of their absolute preference for phenol and because the extent of interaction between proteins and nucleic acids is determined mainly by pH, proteins can be selectively extracted from natural sources at pHs where this interaction is at a minimum. Indeed, at sufficiently high pHs, where the positive charges on the proteins have largely been eliminated and the nucleic acids and the protein carboxyl groups are highly charged, all interactions between proteins and nucleic acids are abolished, the proteins being recovered quantitatively from the phenol-rich phase and the nucleic acids from the aqueous phase. If the isolation of the nucleic acids has to be carried out at acidic pH values, the use of dissociating agents such as sodium bromide can help to keep any potential interaction between them and any protein present to a minimum. Some of the results obtained by Pusztai (1966a) are presented in Table I.

TABLE I

Partitioning of various polyelectrolytes between phenol and aqueous buffers at different pHs. The results are expressed as percentages (w/w) of the original material recovered from (a) the aqueous and (b) the phenol-rich phases calculated on the basis of N (Kjeldahl), neutral sugar, deoxypentose, P etc. values. Occasional low recoveries are due to the formation of small amounts of insoluble material under certain conditions.

Material subjected to standard extraction procedure	Recovery of materials (%, w/w)													
	At pH 1·5		At pH 2·5		At pH 4·0		At pH 6·0		At pH 8·0		At pH 9·2		At pH 10·0	
	(a)	(b)	(a)	(b)	(a)	(b)	(a)	(b)	(a)	(b)	(a)	(b)	(a)	(b)
Bovine serum albumin	0	100	0	100	0	100	0	100	0	100	0	100	0	100
Cytochrome c	0	100	0	100	0	100	0	100	0	100	0	100	0	100
RNA	95	7	98	2	100	0	100	0	100	0	100	0	100	0
DNA	74	21	89	13	99	1	100	0	100	0	100	0	100	0
Polygalacturonic acid	12	8	14	6	16	0	33	0	82	0	92	0	94	0
Chondroitin sulphate	69	22	69	7	71	2	74	0	80	0	84	0	90	0
Starch	94	2	98	0	99	0	99	0	99	0	102	0	103	0

Reproduced with permission from Pusztai (1966a).

The preferential solubility of protein in phenol seems to be poorly understood, although it has been attributed to the strong affinity of phenols for the amide linkage and resulting disruption of the intramolecular hydrogen bonds (Dawydoff, 1953; Mankash and Pakshver, 1953; Pakshver and Mankash, 1953, 1954). By contrast, Yang (1958) and Urnes and Doty (1961) found that several synthetic polypeptides dissolved in m-cresol gave viscosity and optical-rotatory-dispersion data consistent with their being in a complete helical form, thus excluding the possibility of any interaction between the peptide linkage and cresol. In addition, several proteins, including enzymes, have been recovered from phenol-containing solvents and shown to be practically identical with the native protein, on the basis of physical measurements, immunochemical behaviour and enzyme activity (Pusztai, 1973). More information on this point may be obtained from physico-chemical studies. Indeed Hansch et al. (1965), who investigated the role of substituents in the hydrophobic bonding of phenols by serum and mitochondrial proteins, came to the conclusion that binding depends on the lipophilic character of the substituent and that a linear free-energy relationship exists between the logarithm of the binding constant and the substituent constant π ($\pi = \log P_X/P_H$, where P_H is the partition coefficient of a parent compound between octanol and water and P_X is that of the derivative X).

Recently, the purification on a preparative scale of several proteins, among them the intracellular proteolytic enzyme from the seeds of kidney bean, has been shown to be possible with phenol as one of the solvents (Pusztai, 1973). In addition, a preparative high-voltage electrophoresis method for proteins in free-flowing buffer films (buffered phenol–ethanediol–water 3:2:3; w/v/v) has been developed (Pusztai and Watt, 1971).

C. THE BINDING OF PHENOLS TO PROTEIN

Proof that phenol and phenolic disinfectants may effectively bind to proteins comes from the work of Starr and Judis (1968), Weinbach and Garbus (1964, 1965, 1966a, b) (see also Jerchel and Oberheiden, 1955). A definite association between p-tert-amylphenol-^{14}C and dichlorophenol-^{14}C and serum proteins could be demonstrated by sucrose density gradient ultra-centrifugation and Sephadex gel filtration (Starr and Judis, 1968). Figures 2 and 3 show the results obtained in the case of p-tert-amylphenol-[14]C. p-tert-Amylphenol-^{14}C has a distinctly different rate of elution from the human serum proteins when subjected to gel filtration individually, but in a mixture a radioactive peak appeared at a position identical with that of the human serum albumin peak. The data obtained further suggest that phenol itself does not form as tight a complex with serum proteins as in the case of the chloro or amyl derivatives. An explanation for the lack of correspondence between the protein and phenol derivative curves would be that the phenol–protein complex has a different mobility from that of the protein only (Starr

220 C. F. VAN SUMERE ET AL.

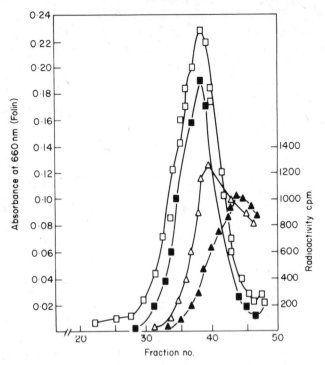

FIG. 2. Density gradient ultracentrifugation of human serum albumin, PTAP-C-14, and a mixture of serum albumin and PTAP-C-14. ■ albumin alone, □ albumin mixed with PTAP-C-14, ▲ PTAP-C-14 alone, △ PTAP-C-14 mixed with albumin. (PTAP-C-14 = *n-tert*-amylphenol-¹⁴C.) (Reproduced with permission from Starr and Judis (1968).)

and Judis, 1968). However, in the case of bacterial protein, sucrose density gradient ultracentrifugation shows that phenol-¹⁴C can also be associated with protein. These results, which could explain the interference by serum with the germicidal effects of phenolic disinfectants and enzyme inhibition and structural damage, may also account for their bactericidal action.

Protein–flavonoid interactions have also been described (Woof and Pierce, 1968) and differential spectroscopy in the u.v. and visible regions have been used to follow the reactions between flavonoids and proteins, peptides and amino acids. All the protein, peptide or amino acid–polyphenol pairs tested gave different spectra, except those involving catechin. The interaction was independent of pH between 3·0 and 7·5, but NaCl increased the spectral intensity. Dihydroquercetin combined further in a stepwise manner with α-amylase, forming definite complexes with flavonoid–protein molar ratios of 2:1, 6:1, 10:1. Similar complexes were also formed between rutin and gelatin. Further examples which prove the binding of phenolics to proteins may be found in the work of Moran and Walker (1968) and Nagashima *et al.*

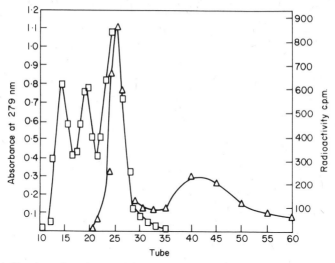

FIG. 3. Gel filtration of a mixture of human serum and PTAP-C-14. □ human serum, △ PTAP-C-14. (Reproduced with permission from Starr and Judis (1968).)

(1968) who have found that salicylates and dicoumarol bind to proteins in the blood.

From the point of view of the physiological activity of certain plant phenols, protein bonding plays an important part in the overall toxicity of certain of these compounds. The following examples are illustrative. The active substances produced by poison ivy (*Rhus toxicodendron radicans*) are a mixture of about 2% 3-*n*-pentadecylcatechol (1), 10% of the analogous Δ^8-mono-olefin, 64% of the $\Delta^{8,11}$-diolefin and 23% of the $\Delta^{8,11,14}$-triolefin (Mason and

3-*n*-pentadecylcatechol (1)

Lada, 1954; Singleton and Kratzer, 1969). These skin irritants or urushiols, which act like haptens, produce an allergenic response. According to Pierpoint (1970) the most likely mechanism for their conversion into antigens is that, once inside the animal, they are oxidized to *o*-quinones which react with skin and serum proteins. Such quinone–protein complexes have been made *in vitro* and their antigenicity has been demonstrated (Pierpoint, 1970). It seems thus that covalent bonding between the urushiols and protein may be involved in poison ivy dermatitis.

Gossypol, a yellow pigment, is an unusually toxic phenol. It causes loss of appetite and weight; pigs fed toxic levels may after a few weeks to a year die with severe anaemia (Kingsbury, 1964). In its occurrence, it seems to be confined to the genus *Gossypium* (Adams *et al.*, 1960). The aldehyde groups

gossypol (2)

of gossypol (2) are able to react with ammonia and the free amino group of lysine in protein (Cater and Lyman, 1969). At high levels of food protein, the phenolic proves to be less toxic and the lack of toxicity of gossypol to ruminants seems, according to Singleton and Kratzer (1969), to result from the indigestible binding of free gossypol to protein. The substance is thus not toxic if the binding to proteins occurs outside the body but the reverse is true inside the body (Harper, 1970; Smith and Clawson, 1970).

The last example with direct physiological importance comes from the field of the tannins. Indeed, tannic acid and other hydrolysable tannins may also at high levels produce various toxic effects (intestinal, liver and kidney damage and anaemia) and substances such as iron salts, calcium salts or high protein levels can prevent this toxicity (Singleton and Kratzer, 1969). The toxic effect of the tannins is much more pronounced when used intravenously (Spector, 1956) and may certainly be ascribed to the precipitation of blood proteins by hydrolysable tannins. It is further remarkable that condensed tannins are generally less toxic (Singleton and Kratzer, 1969).

D. COVALENTLY BOUND PHENOLICS IN PLANT PROTEINS

During the last decade, the isolation of proteins containing covalently bound phenolics such as ferulic acid has been described and the latter compounds must now be considered as being real, although minor constituents, of several plant proteins.

About 10 years ago, protein–chlorogenic acid complexes were isolated as soluble brown pigments from Burley tobacco by Wright (1963) and Wright *et al.* (1964). Amongst the dialysable pigments was a polymeric iron–protein–chlorogenic acid complex with a molecular weight of about 4000. Two minor reddish dialysable pigments were protein–chlorogenic acid polymers, whilst a soluble non-dialysable pigment proved to be an iron–protein–chlorogenic acid–rutin complex, with a molecular weight of 20 000–30 000. Unfortunately, the mode of linkage between the phenolic and protein moieties was not further investigated. Around the same time, Fausch *et al.* (1963) and

Neukom *et al.* (1967) reported ferulic acid as a component of a glycoprotein of wheat flour. The ferulic acid is most probably linked to one pentose residue in about 50, in the arabinoxylan moiety of the glycoprotein. Later, Geissmann and Neukom (1971) and Neukom (1972) suggested that such compounds might be responsible for the oxidative gelation of pentosans (see Fig. 4).

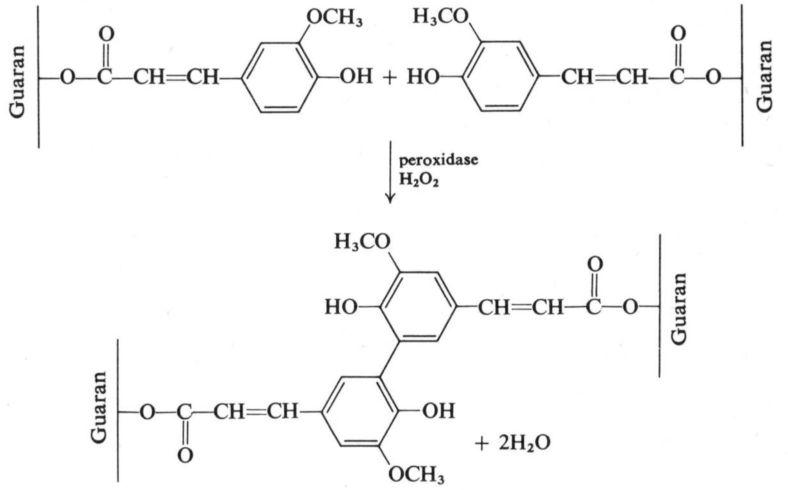

FIG. 4. Mechanism of the oxidative gelation of ferulic acid esters of guaran. Formation of diferulic acid by oxidative coupling. (Reproduced with permission from Neukom (1972).)

In 1964, El-Basyouni *et al.* suggested that an insoluble enzyme ester of a hydroxycinnamic acid could be an intermediate in lignin biosynthesis. According to these authors, cinnamic acid derivatives are not released from phenylalanine ammonia lyase (PAL) or tyrosine ammonia lyase (TAL) but undergo further reactions which eventually lead to the formation of lignin. New methods for the isolation and purification of certain of the ethanol-insoluble phenolic esters of *Mentha arvensis* have very recently been reported by Majak and Towers (1973). Insoluble conjugates of caffeic, ferulic and *p*-coumaric acids were purified and were shown to be electrophoretically and chromatographically homogeneous. A second pool of caffeic acid was associated with a high mol. wt fraction, the high absorbance of this fraction at 280 nm reflecting a protein-type association with caffeic acid. The rigorous method of isolation (i.e. chromatography in 7 M urea: 0·1% acetic acid) indicated the presence of a covalent bond between the caffeyl moiety and the protein. Further support for the existence of covalently linked phenolic acids in proteins may be found in the work of Alibert *et al.* (1968), who identified phenolic acids in a protein fraction from leaves of *Quercus pedunculata*, and that of Brieskorn and Mosandl (1970), who isolated a caffeic acid containing protein from fruits of Umbelliferae with the general formula (3). Finally, in a

Caffeoyl bound protein in umbellifer seed (3)

study of proteins allowed to react with phenolic compounds in the presence of
o-diphenol oxidase, Horigome and Kandatsu (1968) reported the binding of
caffeic and p-coumaric acids to casein.

Nevertheless, although phenolic acid containing proteins seem to be
widespread (Van Sumere, C. F., unpublished results) and ferulic acid and
p-hydroxybenzoic acid respectively have been found to be present in purified
Aspergillus niger cellulase and papain (Van Sumere, C. F. and De Brucker, J.,
unpublished results) there is little knowledge about the type of linkage by
which ferulic acid or any other phenolic acid may be bound to the amino acid
residues. In a recent paper, Van Sumere *et al.* (1973) found that N-ferulylglycyl-L-phenylalanine (4) can be obtained from barley globulins by partial
hydrolysis with 4 N HCl. The partial hydrolysis is difficult because the

(4)

N-ferulylglycyl-L-phenylalanine, as it is liberated from the protein, may be
further hydrolysed. Therefore, a procedure based on repeated short hydrolyses,
followed by ethyl acetate extractions or hydrolysis with HCl at room temperature for 4–10 h has proved most successful. The ferulic acid containing
peptide is subsequently obtained from the combined extracts by means of
preparative and "multiple elimination" TLC (METC) (Van Sumere, 1969a;
Van Sumere *et al.*, 1972a) and then identified by TLC comparison with a
synthetic sample (De Pooter *et al.*, 1973a, b). Additional proof for its identity

was obtained by u.v. fluorescence and i.r. spectroscopy and by the action of carboxypeptidase A.

Recently, both N-ferulylglycyl-L-phenylalanine and N-p-coumaryl-glycyl-L-phenylalanine have been found to be very useful substrates for the micro- and submicro determination of carboxypeptidase A (Van Sumere, C. F., De Pooter, H. and Pé, I., unpublished results). This compound (4) is the first N-acylamino acid derivative of the substituted benzoyl or cinnamyl type to have been found in plants, although small amounts of benzoylaspartic acid were formed when pea epicotyl sections were incubated in a solution of benzoic acid (Andreae and Good, 1957; Venis, 1972). In addition radio-activity from L-phenylalanine-carboxyl-^{14}C was incorporated specifically into the carboxyl group of cinnamamide by cultures of *Streptomyces verticillatus* (Bezanson *et al.*, 1970). Booth *et al.* (1957) and Armstrong *et al.* (1956) detected N-ferulylglycine and N-m-coumarylglycine as components of human urine. All these reports confirm the finding of Van Sumere *et al.* (1973).

It is further possible that N-ferulylglycine could play a role in protein biosynthesis in the seed or embryo of barley in the same way as N-formyl-methionine (HCO–Meth) (Marcker and Sanger, 1964; Adams and Capecchi, 1966; Webster *et al.*, 1966; Salas *et al.*, 1967) and unformylated methionyl-tRNA (for references see Marcus *et al.*, 1970) play this role respectively in *E. coli* and the cytoplasm of certain eukaryotes. Indeed, Boulter *et al.* (1972) indicate that (a) micro-organisms contain a transformylase and a formylatable initiating Meth-tRNA; (b) animals and yeast do not contain a transformylase, but the initiating Meth-tRNA is formylatable; (c) that in plants the initiating Meth-tRNA is not formylatable but the supernatant enzyme fraction appears to contain a transformylase which might, however, originate in the cell organelles. The initiation of protein synthesis may therefore be different in microbial, animal, yeast and plant systems. In the same context, Pearlman and Bloch (1963) and Krishna and Krishnaswamy (1966) have shown that N-acetylamino acids (e.g. MeCO–GlyOH) may be involved in protein synthesis in animal tissue and *E. coli*. Indeed Pearlman and Bloch (1963) have shown that certain N-acetylamino acids undergo the reactions which are generally believed to initiate protein synthesis from free amino acids. The results obtained demonstrate that the NH_2-group of the amino acid need not be free for carboxyl activation and transfer to tRNA. Furthermore, Verhoef *et al.* (1967) have shown that N-acetylphenylalanine can be incorporated into an N-terminal position of polypeptides synthesized under the direction of alfalfa mosaic virus (see also Lucas-Lenard and Lipmann, 1967). Moreover, after injection of radioactive methionine into the thorax of honey bees, Polz and Kreil (1970) were able to detect N-formyl- and N-acetyl-methionine in the proteins. However, the nature of the blocking group seems to be of some importance since N-carbobenzoxyamino acids and certain peptides are un-reactive in the amino acid activating system (Lipmann, 1958; Novelli, 1958).

The presence of N-acetylamino acids as N-terminal residues of a number of proteins or polypeptides is now recognized (Meister, 1965), e.g. N-acetylglycine has been identified in horse heart cytochrome c (Margoliash et al., 1961). In addition, in a number of proteins, free N-terminal amino groups cannot be detected, indicating perhaps that the amino terminal group is masked by substitution (Pearlman and Bloch, 1963). The utilization of N-acetylamino acids in early steps of protein synthesis raises the possibility that the acetylation reaction leading to acetylproteins occurs with the free amino acid and not at the tRNA stage or with the complete protein. It is theoretically possible that other N-acylamino acids such as N-ferulylglycine could be involved in protein biosynthesis. Investigations to prove or disprove this possibility are now in progress in our laboratory.

III. DIFFICULTIES CAUSED BY PHENOLIC COMPOUNDS DURING THE EXTRACTION OF PLANT PROTEINS, ENZYMES AND ORGANELLES

The presence of phenolic compounds frequently upsets the conventional techniques by which enzymes are isolated from plant material. In addition, problems due to reactions between plant proteins and phenolics are much more prevalent and much more complex than is generally recognized (Loomis and Battaile, 1966). The medium for extracting enzymes and subcellular organelles from plants thus requires careful formulation in order to obtain maximum activity in a cell-free extract (Anderson, 1968). In the early stages of enzyme extraction and purification, there are some problems peculiar to vascular plants, due to the presence of tannins, phenolic acids, esters, flavonoids and coumarins. When plant material is ground in a buffer the hydrolases and phenolases released can thus catalyse reactions with these phenolics and quinones, and brown pigments, formed from them by polymerization, may precipitate and (or) inhibit many enzymes and subcellular organelles (Hulme and Rhodes, 1971). Most successful work in the field of plant enzymes has been based on bland plant materials such as spinach or leguminous seedlings which are not rich in phenolic compounds (Neish, 1968).

More or less successful methods have been devised for overcoming the inhibition caused by phenolics and (or) their metabolites. First, the use of various polymers for preventing inactivation of enzymes and mitochondria during extraction by polyphenols and phenolase products has been suggested. Secondly, the employment of various reducing and copper chelating agents inhibiting browning and quinoid formation has been put forward by a number of investigators.

In connection with the use of various polymers, reference must first be made to lightly chromed hide powder. Indeed addition (about 10 mg/mg of tannin) of this material increased the phosphatase activity of suspensions of leaf powder of Kalanchoë (Ehrenberg, 1954) while addition of soluble proteins had no effect. The inhibition of β-glucosidase from bitter almonds by tannins could also be reversed by the same material (Friedrich, 1955). In later attempts to prevent loss of activity of different enzymes, PVP has been used. Goldstein and Swain (1965) have investigated the effect of a large number of simple molecules, detergents and polymers, including PVP, as reactivators of the β-glucosidase-tannic acid complex, in pH 6·0 phosphate buffer. Caffeine at 0·1 M gave only 50% activity, 0·1 M urea only 2%, whereas 0·1 M boric acid gave 53% recovery. Stronger solutions of urea (up to 5 M) gave appreciably higher recoveries but in view of the success obtained with detergents, reactivation with simple compounds was not investigated further. Six detergents were examined: three non-ionic, Ethylan (p-nonylphenyl ether of polyethyleneglycol), Lissapol N (p-iso-octylphenyl ether of polyethyleneglycol) and Tween 80 (dipolyethyleneglycol ether of sorbitan mono-oleate); two anionic, Tergitol 08 (sodium 7-ethyl-2-methyl-4-undecanol sulphate), Manoxal OT (bis(2-ethylhexyl)sodium sulphosuccinate); and one cationic, Cetavlon (cetyltri-methylammonium bromide). Only the non-ionic and cationic detergents proved to be effective, Tween 80 being the best, giving complete reactivation at a concentration of less than 1 mg/ml. According to Goldstein and Swain, the failure of the anionic detergents to reactivate the β-glucosidase–tannic acid complex is noteworthy, since these compounds are capable of altering the tertiary structure of proteins (Reithel, 1963).

From a number of high polymers then tested, the two samples of PVP (average mol. wt 11 000 and 25 000 respectively) and the highest molecular size polyethylene glycol (average mol. wt 20 000) were almost equally effective in restoring activity as was Tween 80 (see also Firenzuoli et al., 1969). The different methylcelluloses all gave about 80–85% reactivation at 0·5% (Goldstein and Swain, 1965). The effect of all these reactivation reagents strongly suggests that in the freshly prepared complex used in these experiments the protein and tannin are joined only by hydrogen bonds and coulombic forces. Furthermore, the reactivation reagents can all act by either altering the local charge on the tannin or the protein, or by competition for the tannin between the polymers used and the enzyme.

The results obtained by Goldstein and Swain (1965) with other enzymes, using both tannic acid and a condensed tannin, are shown in Table II; PVP was the most effective agent. The optimal conditions for binding plant phenols to insoluble PVP have been studied by Andersen and Sowers (1968) and the postulated bonding of plant phenol to PVP is presented in Fig. 5. A further important finding of Goldstein and Swain (1965) is that the condensed tannin fraction appears to give more stable complexes than does tannic acid since in only one case could more than half the activity be regained on reactivation. This indeed indicates that links other than hydrogen bonds are involved in the formation of these complexes. Further, the differences in the inhibition and reactivation of the various enzymes are presumably the result of differences in the fine structure of proteins. Loomis and Battaile (1966) have

TABLE II

Reactivation of enzyme–tannin complexes.

Enzyme[a]	Tannin[b]	Protein/tannin (mg/ml)	Activity of suspended precipitate	Activity in supernatant after treatment with						
				0·1 M Borate	0·1 M Caffeine	25% PEG 400	1% PVP	1% PVA	1% Me cellulose	25% Lissapol N
ADH	T	0·2/0·4	71	31	65	0[c]	94	54	0	—
	W	0·2/0·4	24	0	30	0[c]	35	24	0	24
LDH	T	0·5/1·0	0	0	2	3	65	6	66	—
	W	0·5/1·0	0	0	34	25	10	—	—	—
Peroxidase	T	0·4/1·0	59	14	27	29	23	35	42	31
	W	0·4/1·0	50	6	5	33	30	36	3	44
Catalase	T	0·1/1·0	58	0	24	19	79	26	0	—
	W	0·1/2·0	20	2	2	24	51	—	—	66
β-Glucosidase	T	1·0/2·0	52	40	50	45	74	71	22	—
	W	1·0/2·0	53	20	0	25	69	46	0	—

[a] ADH = alcohol dehydrogenase; LDH = lactate dehydrogenase.
[b] T = tannic acid; W = wattle tannin fraction.
[c] ADH is inhibited by PEG.

Anthocyanase did not form a precipitate with either tannin.
Reproduced with permission from Goldstein and Swain (1965).

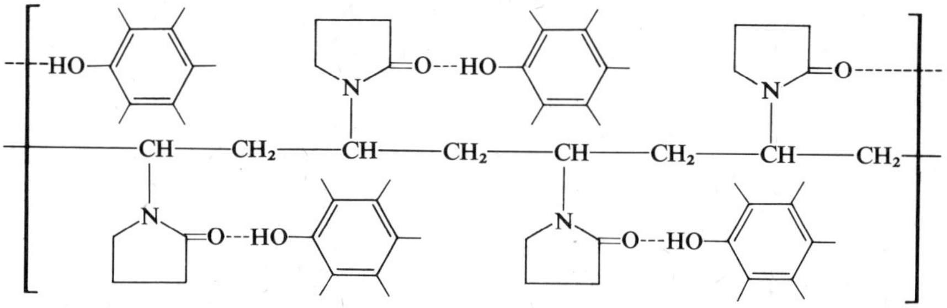

FIG. 5. Postulated hydrogen bonding of plant phenol to polyvinylpyrrolidone. (Reproduced with permission from Anderson (1968).)

similarly shown that phenols are effectively removed from hydrogen bonded complexes with protein by adding large amounts of polyvinylpyrrolidone; the authors were able to demonstrate mevalonic kinase, phosphomevalonic kinase, glutamyltransferase and alkaline phosphatase in peppermint leaves. Normal extracts prepared from such leaves by conventional techniques browned rapidly and no active enzyme, other than phenolase, could be found in them. More recently, Schneider and Hallier (1970) were able to reactivate the tannin-inhibited membrane bound photophosphorylation of the lamellar system from chloroplasts of leaves of Spinacia oleracea by means of PVP addition.

In certain cases, the extraction of cell organelles from plant tissues is also complicated by the production of quinones; the isolation of mitochondria of sweet potato tubers and chloroplasts from sugar cane leaves may be given here as examples (Pierpoint, 1970). While the evidence that phenolic compounds inhibit the mitochondrial system has often been indirect (Hulme and Jones, 1963), direct proof has also been advanced; for example, catechol and chlorogenic acid almost completely inhibit the major activities of sweet potato mitochondria (Lieberman and Biale, 1956). The use of water soluble polymers such as PVP reputed to cross link with tannins looked promising when incorporated into the extraction procedures. According to Hulme et al. (1964) using PVP, mitochondrial fractions from the peel of apples having 30 times the activity (higher rates of oxygen uptake) of the earlier preparations of Pearson and Robertson (1954) and comparable to those shown by mitochondria from leaves were thus obtained. These preparations showed the typical structure of plant mitochondria in the electron microscope. The method of preparation owed its success to the complete but gentle disintegration of the tissue cells in the presence of PVP which "protected" the mitochondria by preventing the oxidation of phenolic compounds and the formation of insoluble aggregates (Hulme and Rhodes, 1971). These oxidized phenolics (in young fruits there is a preponderance of simple leucoanthocyanins) not only directly inactivate the mitochondrial enzymes, but also cause considerable contamination of the mitochondrial preparations by co-sedimentation of non-mitochondrial protein–tannin complexes. Unfortunately, the use of PVP for extracting mitochondria for studies of oxidative phosphorylation seems less satisfactory, because, as pointed out by Rowan (1966), mitochondria prepared with PVP are uncoupled. Later work (Hulme et al., 1967) has further shown that mitochondria from apple prepared with PVP exhibit poor respiratory control. In addition, PVP inhibits certain enzymes such as purified preparations of o-diphenoloxidase from apple fruit (Harel et al., 1964; Walker and Hulme, 1965). Another more serious difficulty in connection with the use of PVP is the fact that the polymer has a pH optimum for the absorption of phenolics at 3·5.

Other polyamides such as polycaprolactam have been employed during the extraction of phenolase from tea leaves (Sanderson, 1964), while several proteins (e.g. collagen, gelatin, egg and bovine serum albumin (BSA)) have been used to prepare enzymes and mitochondria free from contamination by phenolics (Tager, 1958; Loomis and Battaile, 1966; Anderson, 1968; Pierpoint, 1970). In connection with BSA, there is no unequivocal evidence proving that it has a real effect on the efficiency of extraction of mitochondria, but Weinbach and Garbus (1966a, b) have shown that BSA powerfully binds phenolics and reverses their uncoupling effect. BSA is then also usually included in media for extracting plant mitochondria (Stokes et al., 1968).

Polymers thus bind rather powerfully tannins and polymerized oxidation products of phenolases. The most satisfactory agents seem to be soluble PVP for mitochondria (but poor respiratory controls may be expected) and insoluble PVP for enzymes (Loomis and Battaile, 1966). These polymers would thus appear to be very useful in tissues containing high amounts of high molecular weight phenolics and their derivatives, but since their ability to bind small phenols is rather limited, their use with tissues containing these compounds must be considered ineffective. The formation of oxidation products of o-diphenoloxidase activity can also be prevented or eliminated by other means such as the chelation of the copper from the active centre of the enzyme, the use of thiols, ascorbate and cyanide and the preparation of acetone powders. DIECA (diethyldithiocarbamate), a copper chelating reagent and powerful inhibitor of o-diphenolase, combines also with the quinoids formed. The importance of these reagents for the extraction of plant enzymes is shown in the work of Slack (1966), who was only able to prove the presence of sucrose synthetase in extracts of sugar cane after the addition of cysteine or DIECA to the extracting medium.

Thiols may also enhance the activity of enzymes and mitochondria from plant tissues. Most probably this is due to inhibition of o-diphenolase and the removal of quinone as it is formed. However, the requirement for thiols is far from universal (Palmer, 1967). The use of metabisulphite and dithionite, which prevent browning of freshly cut tissues, has also been recommended during the extraction of enzymes. Björk (1965) proved that dithionite prevented the formation of phenolic oxidation products during the preparation of nucleases from potato and Anderson and Rowan (1967) found metabisulphite more effective in shielding tobacco leaf peptidase from o-diphenolase products than thiols. Highly coupled mitochondria were also obtained from potato by addition of the same reductant to the extraction medium (Stokes et al., 1968).

In addition, successful isolations of enzymes may sometimes be expected by preparation of acetone powders. Indeed, cold acetone dissolves many of the phenols and separates them from the plant proteins before enzyme action can occur. Neish (1961) found the preparation of an acetone powder advantageous in the demonstration of L-tyrosine ammonia lyase in barley shoots and Sanderson and Roberts (1964) prepared peptidase from tea leaf, a tissue extremely rich in phenolics, by using the same solvent. However, Loomis and Battaile (1966) showed that acetone powders of peppermint leaves, which had been extracted with methylpyrrolidone, still browned rapidly when moistened with water.

Thus, it is impossible to recommend a single reagent for protecting enzymes or mitochondria from the effects of phenols. Guidelines for the isolation of enzymes and subcellular organelles have been published by Anderson (1968), Loomis (1969) and Pierpoint (1970). Table III indicates the methods that can be used for preventing o-quinones forming in plant extracts. For the isolation of tightly coupled mitochondria, Anderson (1968) suggests that perhaps a combination of the method of Stokes et al. (1968) with a rapid preparation of mitochondria as recommended by Palmer (1967) may be advantageous.

IV. Difficulties caused by Phenolic Compounds during the Extraction of Plant Viruses

Phenolic compounds may also interfere with the isolation of viruses from plants. Indeed, the precipitation of viruses by tannins is well known, but albumin, lead acetate, nicotine sulphate, quinine sulphate, caffeine, hide powder, soluble protein, dried egg white or skimmed milk powder (Cadman, 1960; Wessel-Riemens, 1965; Brakke, 1967) reverse the action of the phenolics. A rise in pH and the use of large amounts of buffer also decrease the precipitate formed (Brakke, 1967). Soluble proteins or hide powder are very frequently effective but, as indicated by Brunt and Kenten (1963), hide powder is preferred because of the ease of separating it from the extract. Although a phenolase–phenolic compound reaction appears to inactivate plant viruses in vitro and possibly in necrosing tissue (Kosuge, 1969), only a minority of known plant viruses seems to be inactivated by the o-quinones formed in leaf extracts. However, the fact that viruses are not inactivated does not necessarily mean that they do not react with o-quinones, because Pierpoint (1970) claims

TABLE III

Methods of preventing o-quinones forming in plant extracts

Principle	Technique	Example
Exclude oxygen	Disrupt tissue in N_2 atmosphere	Extraction of bulk leaf protein
Remove polyphenols	Extract tissue with solvents	Washing acetone powders of leaves
	Absorb phenols onto polymers:	
	polyvinylpyrrolidone (PVP)	Extraction of enzymes from apple fruit and
	insoluble PVP (polyclar AT)	leaves of peppermint (*Mentha piperita*)
	polyethylene glycol	
	albumin	
Inhibit polyphenoloxidase	Extract tissue in:	Extraction of amino acids from leaves
	trichloroacetic acid	
	sodium diethyldithiocarbamate (DIECA)	Extraction of viruses from leaves of tobacco
	potassium ethylxanthate	(*Nicotiana tabacum*) and enzymes from tubers
	thioglycollate	of potatoes (*Solanum tuberosum*)
	metabisulphite	
Reduce quinones	Extract tissue in ascorbate	Extraction of viruses from leaves of tobacco (*Nicotiana tabacum*) and enzymes from tubers of potatoes (*Solanum tuberosum*)
Trap quinones	Extract tissue in:	Extraction of enzymes, including polyphenol
	cysteine	oxidase, from acetone powders
	benzene sulphinic acid	

Reproduced with permission from Pierpoint (1970).

that southern bean mosaic virus retains its infectivity even after reaction with
o-quinones.

On the other hand, while cucumber mosaic virus (Harrison and Pierpoint,
1963; Pierpoint and Harrison, 1963) reacts with chlorogenoquinone (oxidized
chlorogenic acid), neither tobacco mosaic virus (TMV) nor its protein react.
This is in spite of the fact that each protein subunit of TMV contains one
thiol and two amino groups. Pierpoint (1970) suggests that, because of
differences in molecular structure, the amino and thiol groups of proteins
differ in accessibility to and affinity for *o*-quinones, as they do for other
reagents.

Other suggestions have also been made for these variations in reactivity.
On the basis of experiments with Tulare apple mosaic virus, Mink *et al.*
(1966) suggest that quinone inactivation of the virus occurs as a result of oxidation
of the viral nucleic acid. The fact that not all the viruses are inactivated by
quinones may be due to a difference in accessibility of the ribonucleic acid to
the oxidant. However, there is a contradictory report that infectious ribonucleic
acid from tobacco mosaic virus is not inactivated by exposure to oxidized
chlorogenic acid (Milo and Santilli, 1967). It seems therefore that the inactivation of plant viruses by the phenolase–phenolic reaction requires further
research.

Tannic acid and raspberry tannins probably act on most viruses by causing
the virus particles to form clumps (Cadman, 1960). These largely uninfective
complexes are a loose union between virus and tannin since they are also
readily dissociable by dilution or increase of pH. Dissociation may thus occur
when such complexes are injected into living cells.

In conclusion, there is no real evidence that a plant species which gives a
highly inhibitory sap, containing polyphenols, is in any way protected from
virus infection in nature (Matthews, 1960; Williams, 1963). However, unless
precautions are taken, the phenolics and their derivatives may complicate or
render the extraction of viruses from plant material very difficult.

V. The Inhibition and Activation of Enzymes by Phenolics or their Oxidation Products

Although phenolics and (or) their oxidation products may thus hamper the
isolation of plant proteins, enzymes and viruses, the inhibition or activation
of purified enzymes and enzyme systems by the above compounds has also
been reported. References are available in the reviews of Hoffmann-Ostenhof
(1963), Hulme and Jones (1963), Williams (1963), Mandels and Reese (1963,
1965), Loomis and Battaile (1966), Webb (1966), Goodman *et al.* (1967),
Wood (1967), Anderson (1968) and Kosuge (1969). However, no recent
survey is available; therefore, an attempt will be made here to discuss the

effects of simple phenolics on enzymes and the effect of tannins and quinones will also be referred to.

Although most work has been concerned with inhibition, it is equally likely that certain phenolics (especially before oxidation) will activate enzymes (Williams, 1963; Lyr, 1965; Wood, 1967). For example, with IAA-oxidase, kaempferol and kaempferol-3-(p-coumaryltriglucoside) activate the enzyme (Mumford et al., 1961; Furuya et al., 1962). In addition, Gortner and Kent (1958) and Gortner et al. (1958) have shown that o- and p-dihydric phenols inhibit pineapple IAA-oxidase, while monohydric phenols, particularly p-coumaric acid activate it. Other examples can be found in the papers of Rabin and Klein (1957), Sondheimer (1963), Varga and Köves (1962), Mumford et al. (1963), Engelsma and Meijer (1965a, b), Gaspar (1966) and Kosuge (1969).

The effect of several flavonoids on pea root IAA-oxidase has been investigated by Stenlid (1963, 1968) who found that all 4'-hydroxyflavonoids are cofactors for the oxidation of IAA, whereas 3',4'-dihydroxyflavonoids inhibit the destruction of the plant hormone. The aglycone naringenin (5) is a more active stimulator than is the glycoside naringin whereas the reverse is true for the pair phloretin (6)–phloridzin (7). Hydroxyl groups at the positions 2', 4, 4' in phloridzin and at 4', 5, 7 in naringenin give compounds which are very active cofactors (genistein, apigenin, and 2',4,4'-trihydroxychalcone (8) are all potent stimulators), whereas even small structural deviations may result in a considerable reduction of stimulatory properties. Indeed, sieboldin (9) a hydroxy analogue of phloridzin, is a strong inhibitor of IAA-oxidase (10^{-6} M gives about 50% inhibition).

Similar results have been obtained with phenolic compounds and their effect on the rate of oxidation of NADH by horseradish peroxidase and hydrogen peroxide (Gamborg et al., 1961; see also Akazawa and Conn, 1958). The relative effects of the aglycones phloretin and naringenin compared to their glycosides were determined with a purified preparation of this enzyme. In this case, phloridzin proved to be superior to phloretin in stimulating the oxidation of IAA, whereas naringenin was slightly more active than its glycoside. In addition, anthocyanins and anthocyanidins extracted from corn leaves have been reported to inhibit the oxidation of IAA by peroxidase (Voronkov, 1970).

According to Stenlid (1968) the relative efficiency of various flavonoids varies with the enzyme preparation used. Also the agreement between the effect of various flavonoids upon IAA-oxidase and upon growth is not very good. The possible implications of these findings for IAA-induced growth and host–parasite interactions are discussed by Zucker et al. (1967), Stenlid (1968) and Kosuge (1969).

At low concentrations (5×10^{-6} M) chlorogenic acid and caffeic acid almost completely inhibit the oxidative decarboxylation of amino acids by peroxidase (Mazelis, 1962), whereas at the same concentration phenol proved

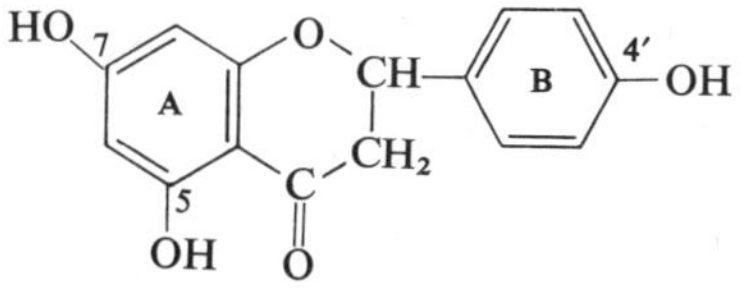

Naringenin (5)

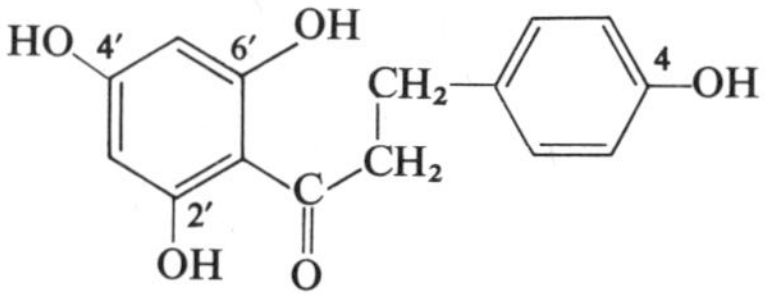

Phloretin (6)

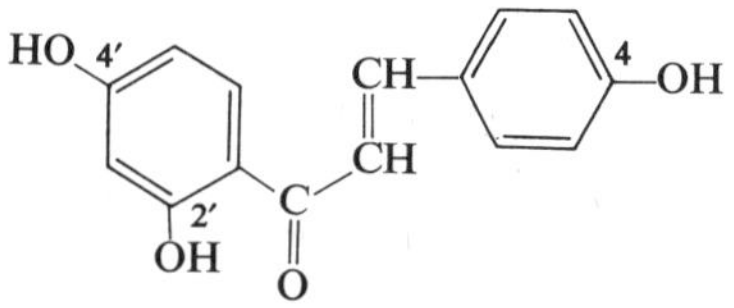

2′,4,4′-trihydroxychalcone (8)

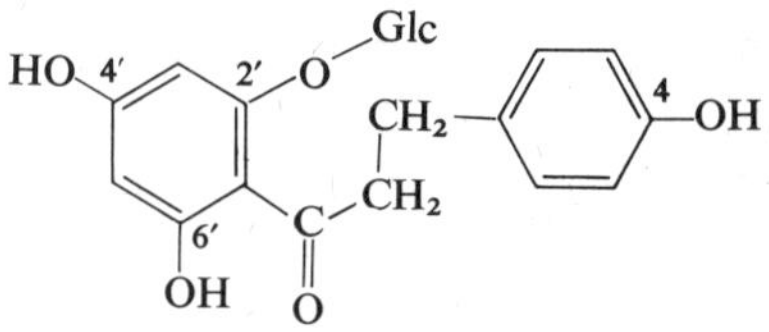

Phloridzin (7)

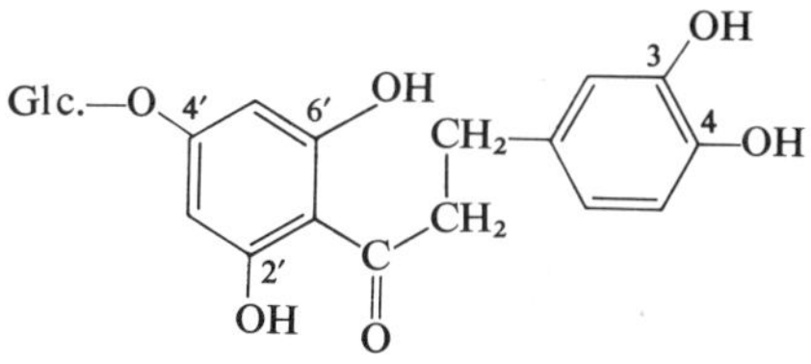

Sieboldin (9)

to be an activator. Quercetin was further shown by an isotopic method to be an effective inhibitor of non-specific and specific (induced) histidine decarboxylase, both *in vivo* and *in vitro* (Smyth *et al.*, 1964) and *trans*-cinnamic acid, *p*-coumaric acid and NADPH were found to be inhibitory to *o*-diphenoloxidase (Hyodo and Uritani, 1967). Oxidised polyphenols are supposed to inhibit glucose-6-phosphate dehydrogenase in cut-injured sweet potato tissue (Muto *et al.*, 1969).

The respiration of cells is also affected by phenolics and quinones (Lehninger and Schneider, 1958; Dedonder and Van Sumere, 1971a, b; Van Sumere *et al.*, 1972a). At 10^{-3} M most of the compounds tested by Van Sumere and coworkers proved to be stimulators of the oxygen consumption in *Saccharomyces cerevisiae* and *Chlorella vulgaris*, but quinones were inhibitory. However, at lower concentrations, quinones also stimulated yeast respiration. It seems that stimulation or inhibition is mainly determined by a suitable combination of both structure and concentration. In addition, pH is most probably connected with the dissociation of the compounds (Simon and Beevers, 1951). Moreover, the uptake of the substances, which is usually maximal at their pK value (Flaig, 1968), as well as a possible detoxification may have significant effects. It is also interesting that catechin, epicatechin and sodium epicatechin-2-sulphonate increase the respiratory activity of cytochrome oxidase in rat liver homogenate (Horn *et al.*, 1970).

Certain naturally occurring phenolics have further been reported to uncouple oxidative phosphorylation. Dinitrophenol (DNP), the classical uncoupler (Lehninger, 1964), at appropriate concentrations limits phosphorylation and either stimulates or does not affect respiration (Racker, 1965). Most of the naturally occurring simple phenolics, like DNP, are true uncouplers. Indeed, the results obtained in this laboratory with rat liver (Wolf, G. and Van Sumere, C. F., unpublished results) and yeast mitochondria (Van Sumere et al., 1972a) (see Tables IV and V) indicate that phenolics and related substances such as salicylaldehyde, β-resorcylaldehyde and naphthoquinone may uncouple oxidative phosphorylation and increase respiration. The same holds true for cinnamaldehyde and cinnamic acid (see also Tillberg, 1970; Stenlid and Saddik, 1962). However, coumarin and umbelliferone are ineffective in uncoupling oxidative phosphorylation in yeast mitochondria (see Table IV). This is a rather unexpected result, because several other authors (for references, see Van Sumere et al., 1972a) have shown the uncoupling

TABLE IV

Effect of phenolics and coumarins on oxidative phosphorylation in yeast mitochondria

Compound	Concentration (M)	% Inhibition
2-Methyl-1,4-naphthoquinone	10^{-4}	26·8
2,4-Dinitrophenol	10^{-4}	36·4
Salicylaldehyde	10^{-3}	20·7
β-Resorcylaldehyde	10^{-3}	24·8
Vanillin	10^{-3}	0
Cinnamaldehyde	10^{-3}	—
	10^{-4}	9·6
Cinnamic acid	10^{-3}	—
	10^{-4}	19·5
o-Coumaric acid	10^{-3}	—
	10^{-4}	13·8
p-Coumaric acid	10^{-3}	—
	10^{-4}	0
Caffeic acid	10^{-3}	—
	10^{-4}	19·5
Ferulic acid	10^{-3}	0
Coumarin	10^{-3}	0
Umbelliferone	10^{-3}	0
Esculetin	10^{-3}	—
	10^{-4}	0
Scopoletin	10^{-3}	30·2

Reproduced from Van Sumere (1972).

TABLE V

Effect of phenols, coumarins and related compounds on the oxidative
phosphorylation in rat liver mitochondria

Compound (5×10^{-4} M or as stated)		% Inhibition (uncoupling) after 30 min experiment
Phenol		19
Hydroquinone		12
p-Nitrophenol		89
2,4-Dinitrophenol		88
p-Benzoquinone		2
1,4-Naphthoquinone		74
Menadione		7
Benzaldehyde		0
Salicylaldehyde		31
β-Resorcylaldehyde	10^{-3} M[a]	91
	10^{-3} M[b]	54
	10^{-3} M[c]	43
	5×10^{-4} M	18
p-Nitrobenzaldehyde		0
Gallic acid	10^{-3} M	21
	5×10^{-4} M	0
Cinnamic aldehyde		14
p-Coumaric acid	10^{-3} M	32
	5×10^{-4} M	0
Caffeic acid		6
Hydrocaffeic acid	10^{-3} M	38
	5×10^{-4} M	14
Coumarin		14

Wolf, G. and Van Sumere, C. F. (unpublished results).

[a] Mitochondria 3 h ⎫
[b] Mitochondria 1·5 h ⎬ preincubated with the compound before start of the
[c] Mitochondria 0·5 h ⎭ experiment.

property of the compound and from Table V it follows that coumarin decreases the P/O ratio in rat liver mitochondria. It is possible that the speed with
which the compound is converted into o-coumaric acid (uncoupler of the
oxidative phosphorylation) and eventually into melilotic acid by the mitochondrial preparation, may be one of the reasons for this anomaly. The uncoupling effect of phloretin, 2′,4,4′-trihydroxychalcone and flavonoids was
studied by Stenlid (1963, 1968). Glycosides as a rule proved again to be less
active, although phloridzin influences mitochondrial activity and respiration
(Lehninger and Schneider, 1958; Lothspeich, 1961). In addition, certain
phenols such as 2,4-dinitrophenol and p-hydroxybenzaldehyde are uncouplers
of the cyclic photophosphorylation only (Dedonder and Van Sumere,
1971b), while others, e.g. o-hydroxybenzaldehyde, are active in the non-cyclic

but inactive in the cyclic energy production. β-Resorcylaldehyde and cinnamaldehyde, both powerful inhibitors of algal growth, have been shown to decrease both types of photophosphorylation. The same is true for phloridzin which also affects photosynthesis and the related energy production (for references, see Stenlid, 1968). The mechanism of action of DNP and other phenolics in oxidative (Wainio, 1970) and photosynthetic phosphorylation is not yet known. Binding between the phenolics and mitochondrial or chloroplast proteins may play an important part. Indeed, in addition to the classical hypotheses for explaining the effect of DNP on oxidative phosphorylation (Wainio, 1970), Hemker (1964) has suggested that nitrophenols may inhibit by a general destructive effect on mitochondrial structure. Hemker (1962) and Gladtke and Liss (1958) further concluded that the uncoupling activity of phenols is determined by the amount dissolved in mitochondrial lipid (this may also be true for other phenolics—Wolf, G. and Van Sumere, C. F., unpublished results) and Weinbach and Garbus (1965) presented results showing the extensive capacity of intact mitochondria to bind these reagents. In addition, the amounts found in rat liver mitochondria at uncoupling concentrations were greater than those calculated from empirically determined partition coefficients. Moreover, not only intact mitochondria but mitochondria depleted to various degrees of their lipid content bound pentachlorophenol, 2,4-dinitrophenol and other lipophilic uncoupling reagents. The extent of binding of the inhibitory reagents to this mitochondrial protein (bound pentachlorophenol and DNP migrated on Sephadex G25 with the protein moiety and not with the lipids) was similar to that with intact mitochondria as regards the effects of pH, concentration of the inhibitor or concentration of protein. The binding studies, in conjunction with additional experiments with intact mitochondria, suggest that the uncoupling phenols interact with the protein moiety of intact mitochondria and that the protein–phenol interaction may be an important factor in the uncoupling phenomenon. Such an interaction could induce configurational changes in the enzymes associated with oxidative phosphorylation or prevent the binding or rebinding of structural or functional factors essential for this bioenergetic process.

The concept of a general interaction of DNP with the mitochondrial protein (see also Hansch et al., 1965) gains further support from the observations of Grillo and Cafiero (1964) who found that the activity of hexokinase was 81% inhibited by $7 \cdot 5 \times 10^{-3}$ M DNP and that a number of plasma proteins, amongst which serum albumin was the most effective, counteracted the inhibition, presumably by binding DNP (Weinbach et al., 1963) through their amino groups. The binding to hexokinase was readily reversible, since dilution relieved the inhibition. In the case of the strong inhibition of oxidative phosphorylation by quinones such as the 1,4-naphthoquinones, Dedonder and Van Sumere (1971a) have suggested that inhibitory amounts of the quinones may well interfere with the electron flow in the respiratory chain by accepting electrons from reduced flavoproteins without passing them immediately to the cytochrome system. The quinones would then form semiquinones and, in this way, respiration would be hampered. In this connection, it is interesting that Michaëlis (see Fieser and Fieser, 1961) has shown that the reduction of quinone to hydroquinone, as well as the reverse reaction, proceeds via a semiquinone. The latter compound, possessing a lone electron and the character of a free radical, is relatively stable, due to resonance stabilization (mesomerism). Semiquinones are further known to be coloured compounds and during the growth and respiration experiments it has been found that the algal suspension, which contained quinones, showed a pronounced reddish colour (Dedonder and Van Sumere, 1971a). In addition, quinones may also react with proteins and enzymes in various different ways (see p. 214) and, as such, oxidative phosphorylation may also be inhibited through an inhibition of the catalytic activity.

Recently, the inhibition of pancreatic ribonuclease by simple phenolics and flavonoids has been reported (Mori and Noguchi, 1970; Van Sumere *et al.*, 1972b; Pé, I., Dedonder, A., Hrazdina, G. and Van Sumere, C. F., unpublished results). According to Mori and Noguchi, flavones and flavonols with three hydroxyl groups at positions 7, 3' and 4' inhibit the activity of the enzyme. The inhibitory action of the flavonoids is further partly attributed to the keto function at position 4 conjugated with the hydroxyl group at position 7 and partly to the keto group conjugated with the hydroxyl group at position 4'. A methoxyl group at positions 6 or 8 diminishes the inhibitory action of the 4-keto-7-hydroxyl system, while the hydroxyl groups at position 3' or 5' increase the action of the 4-keto-4'-hydroxyl system. Flavanones, flavanols and most of the simple phenols have no effect on the RNase activity.

The results obtained in our laboratory, shown in Table VI, support to a certain extent the findings of the Japanese workers, although the inhibition found by them is more pronounced. It is possible that the more extreme values recorded by Mori and Noguchi (1970) are at least partly due to the different technique employed for the determination of the RNase activity. However, the attribution by the same authors of the inhibitory effect of the 4-keto group cannot be correct, since we found that anthocyanidins, which lack this group, are also inhibitory. We agree that more highly oxidized flavonoids are most inhibitory, since reduction to flavanone, flavanonol or flavanol gave inactive compounds (see Table VI). In addition, the hydroxyl groups at the 5- and 7-positions as well as those at 3'- and 4'- are important; furthermore, the hydroxyl at position 3 should be free. In analogous experiments, pancreatic DNase activity was also found to be inhibited ($\pm 30\%$) by kaempferol (10^{-3} M) and quercetin (10^{-3} M) (Pé, I., Dedonder, A. and Van Sumere, C. F., unpublished results).

Phenylalanine ammonia lyase (PAL) is strongly inhibited by cinnamic acid (for references see Camm and Towers, 1973), but not by *trans*-cinnamamide (Emes and Vining, 1970). *p*-Coumaric acid and caffeic acid are also inhibitory

TABLE VI

Effect of flavonoids and simple phenols on RNase activity
(Pancreatic enzyme)

Compound (10^{-3} M unless otherwise stated)	% Inhibition
SIMPLE PHENOLS	
Catechol	17
Pyrogallol	7
Salicylic acid	16
β-Resorcylaldehyde	7
β-Resorcylic acid	34
Gallic acid	8
Syringic acid	5
Caffeic acid	35
Ferulic acid	7
Anthrarobin (3,4-diOH-9-anthrone)	35
4,4'-Bishydroxycoumarin	21
FLAVONES	
Chrysin (5,7-diOH)	15
Apigenin (5,7,4'-triOH)	26
Acacetin (5,7-diOH, 4'-OMe)	13
FLAVONOLS	
Galangin (3,5,7-triOH)	10
Kaempferol (3,5,7,4'-tetraOH)	84
5×10^{-4} M	63
10^{-4} M	0
Fisetin (3,7,3',4'-tetraOH)	55
Quercetin (3,5,7,3',4'-pentaOH)	100
10^{-4} M	19
10^{-5} M	11
Quercitrin (quercetin 3-rhamnoside)	23
FLAVANONE	
Hesperitin (5,7,3'-triOH,4'-OMe)	10
FLAVANONOL	
Dihydroquercetin (taxifolin) (3,5,7,3',4'-pentaOH)	21
ANTHOCYANINS and ANTHOCYANIDINS	
Pelargonidin (3,5,7,4'-tetraOH)	100
10^{-4} M	40
Peonidin (3,5,7,4'-tetraOH, 3'-OMe)	40
5×10^{-4} M	23
10^{-4} M	0
Peonidin 3,5-diglucoside	54
Malvidin (3,5,7,4'-tetraOH, 3,5'-diOMe)	22
Malvidin 3-glucoside	34

TABLE VI—*contd.*

Compound (10⁻³ M unless otherwise stated)	% Inhibition
Petunidin 3,5-diglucoside (7,3′,4′-triOH,5′-OMe,3,5-O-glc)	20
Cyanidin (3,5,7,3′,4′-pentaOH)	100
5×10^{-4} M	100
10^{-4} M	77
5×10^{-5} M	11
Cyanidin 3-O-rutinoside	22
DIHYDROCHALCONE	
Phloretin (4,2′,4′,6′-tetraOH)	12

Substances having no effect: phenol, hydroquinone, arbutin, *p*-benzoquinone, juglone, emodin, 1,8-dihydroxyanthraquinone, orcinol, 2,4-dinitrophenol, salicylaldehyde, protocatechuic acid, cinnamaldehyde, cinnamic acid, *o*-coumaric acid, *m*-coumaric acid, *p*-coumaric acid, sinapic acid, chlorogenic acid, coumarin, imperatorin; *flavone*: tectochrysin (5-OH; 7-OCH₃); *flavonol*: rutin (5,7,3′,4′-OH; 3-O-rutinoside); *flavanones*: flavanone (−), naringenin (5,7,4′-OH), naringin (5,4′-OH; 7-O-glc-rha), hesperidin (5,3′-OH; 4′-OCH₃; 7-O-rutinoside); *flavanonol*: dihydrofisetin = fustin (3,7,3′,4′-OH); *anthocyanin*: malvidin-3,5-diglucoside (7,4′-OH; 3′,5′-OCH₃; 3,5-O-glc); *dihydrochalcone*: phloridzin hydrate (4,4′,6′-OH; 2′-O-glc); *flavanols*: (+), (−) and (±)-catechin (2,3-*trans*-3,5,7,3′,4′-OH), (−)-epicatechin (2,3-*cis*-3,5,7,3′,4′-OH).

The RNase activity was determined by the increase in acid uranyl chloride soluble phosphates liberated from RNA by RNase action. Essentially a modification of "the acid-soluble phosphorus method" of Dubos and Thompson (1938), using a modified Fiske and Subbarow technique (1925) for the phosphorus determination, was employed for the analysis of soluble phosphate.

(Minamikawa and Uritani, 1965; Camm and Towers, 1973), but chlorogenic acid and isochlorogenic acid, the main phenolics of sweet potato root, showed no effect on the two PAL isozymes isolated from this plant tissue (Minamikawa and Uritani, 1965). In the case of PAL from *Pisum sativum* (Attridge *et al.*, 1971), quercetin and the *Pisum* flavonoids proved to be inhibitory.

The uptake of labelled substances by cells or tissues may also be affected by phenolics. The effect of phloridzin on the sugar absorption in renal and intestinal animal tissue is well known, because the compound has long been used for inducing artificial diabetes in test animals. Very low concentrations of phloridzin inhibit active sugar transport in intestinal tissues [10^{-6} M gives about 50% inhibition (Stenlid, 1968)], whereas phloretin proved much less efficient. However, in plant tissues phloridzin is a less potent inhibitor than phloretin; in addition, the required phloretin concentration is higher (10^{-4} M). Further information on the inhibition of active transport of sugars or glycine by phloridzin or phloretin in animals or plants may be found in the work of Lehninger and Schneider (1958), Cividanes *et al.* (1964) and Kotyk *et al.* (1965). Similarly, flavonoids inhibit the uptake of labelled sugars by excised wheat roots (Stenlid, 1968), although they do not appear to affect substantially the uptake of leucine by potato tuber tissues; the glycosides, as a rule, were less active than the corresponding aglycones (Parups, 1967). Several naturally occurring phenolics were further shown to decrease the uptake of phenylalanine-1-^{14}C by *Chlorella vulgaris* (Van Sumere and Dedonder, 1971), *Saccharomyces cerevisiae* cv. *ellipsoideus*, excised barley embryos and lettuce seeds (Van Sumere *et al.*, 1972a). The same authors have also studied the inhibitory effect of a series of simple phenolics and related compounds on the incorporation of phenylalanine-1-^{14}C into the proteins of yeast and *Chlorella*. In addition, they have reported on the inhibition of the incorporation of labelled phenylalanine caused by coumarin and ferulic acid in lettuce seeds and excised barley embryos. From the results obtained with yeast and *Chlorella* it may further be concluded that uncouplers of oxidative photophosphorylation such as β-resorcylaldehyde and *p*-hydroxybenzaldehyde act on the transport and incorporation system by depressing the energy formation. The germination and growth inhibitory action of these compounds may thus most probably, and at least in part, be ascribed to their inhibition of the energy production and protein biosynthesis. In the case of cinnamaldehyde (which, under aerobic conditions, is to a large extent converted to the corresponding acid), competition with the phenylalanine-1-^{14}C substrate and uncoupling of oxidative phosphorylation has been observed (Van Sumere and Dedonder, 1971; Van Sumere *et al.*, 1972a). It is further possible that ferulic acid, a derivative of cinnamic acid, may also directly produce an analogous inhibitory effect. Moreover, it cannot be excluded that exogenous ferulic acid may temporarily increase the cinnamic acid level in plants and seeds via feedback effects (Hanson *et al.*, 1967; Van Sumere *et al.*, 1972a).

Protein biosynthesis is also influenced by salicylate (Dawkins *et al.*, 1971) and by certain polyphenols (Parups, 1967) and the effect of a series of these compounds on the amino acid activating enzymes (amino acyl tRNA-synthetases) of barley embryos has recently been studied (Pé, I., Hrazdina, G. and Van Sumere, C. F., unpublished results) with the results shown in Table VII. Quinones and to a lesser extent certain flavonoids are powerful inhibitors of the *in vitro* amino acid activation, while catechins do not show any activity.

Parups (1967) has investigated the effect of some polyphenols on protein synthesis in potato tuber slices. From his results, it appears that the aglycones hesperitin, naringenin and phloretin are more effective inhibitors than their respective glycosides hesperidin, naringin and phloridzin. However, quercetin and quercitrin inhibited incorporation of leucine-^{14}C almost equally. Besides inhibiting the incorporation of the radioactive amino acid into the alcohol-insoluble fraction, hesperitin, naringenin and phloretin also decreased incorporation into the alcohol-soluble fraction. In addition, Parups (1967) has shown that certain of these compounds also inhibit synthesis in a cell-free extract from *E. coli* (see Table VIII). The same author suggests that differences

TABLE VII

Effect of phenols, cinnamic acids and flavonoids on the activation of amino acids

Compound* (10⁻³ M or as stated)		% Inhibition (% ATP ⇌ PP³²/mg prot. of the blank = 100%)
PHENOLS		
Salicylaldehyde		7
Hydroquinone	5×10^{-4} M	100
	10^{-4} M	50
	5×10^{-5} M	35
	10^{-5} M	10
p-Benzoquinone	5×10^{-4} M	100
	10^{-4} M	45
	5×10^{-5} M	16 (10^{-5} M:0)
Naphthoquinone	10^{-4} M	37
	5×10^{-5} M	16 (10^{-5} M:0)
Juglone		100
	5×10^{-4} M	86
	10^{-4} M	61
	5×10^{-5} M	41
	10^{-5} M	4
Anthrarobin		91
	5×10^{-4} M	61
Emodin		52
Caffeic acid at pH 8·3		23 (at pH 7·0 = 0)
Chlorogenic acid		6
trans-α-Phenylcinnamic acid		12
4,4′-Bishydroxycoumarin		31
FLAVONES		
Flavone		8
Apigenin		29
Acacetin		10
FLAVONOLS		
Galangin		36
Kaempferol		30
Fisetin		27
Quercetin		90
	5×10^{-4} M	33
	10^{-4} M	0
Quercitrin		13
FLAVANONES		
Flavanone		22
Naringenin		23
Hesperitin		32

* For substitution pattern of various compounds, see Table VI.

TABLE VII—*contd.*

Compound* (10^{-3} M or as stated)	% Inhibition (% ATP $\rightleftharpoons$ PP^{32}/mg prot. of the blank = 100%)
FLAVANONOLS	
Dihydrofisetin (fustin)	13
Dihydroquercetin (taxifolin)	—
ANTHOCYANINS and ANTHOCYANIDINS	
Pelargonidin	98
5×10^{-4} M	22
10^{-4} M	0
Peonidin	95
5×10^{-4} M	61
10^{-4} M	0
Peonidin 3,5-diglucoside	10
Malvidin 3,5-diglucoside	14
Cyanidin	65
Cyanidin 3-rutinoside	18
DIHYDROCHALCONE	
Phloretin	53

* For substitution pattern of various compounds, see Table VI.

Substances showing no effect: phenol, catechol, β-resorcylaldehyde, β-resorcyclic acid, arbutin, menadione, 1,8-dihydroxyanthraquinone, cinnamic acid, o- and p-coumaric acid, ferulic acid, coumarin, umbelliferone, esculetin, esculin, imperatorin; *flavones*: chrysin (5,7-OH), tectochrysin (5-OH; 7-OCH$_3$); *flavonol*: rutin (5,7,3',4'-OH; 3-rutinoside); *flavanones*: naringin (5,4'-OH; 7-O-glc-rha), hesperidin (5,3'-OH; 4'-OCH$_3$; 7-O-rutinoside); *flavanols*: (+) and (−)-catechin (2,3-*trans*; 3,5,7,3',4'-OH), (−)-epicatechin (2,3-*cis*; 3,5,7,3',4'-OH); *anthocyanins*: malvidin (3,5,7,4'-OH; 3',5'-OCH$_3$), petunidin-3,5-diglucoside (7,3',4'-OH; 5'-OCH$_3$; 3,5-O-glc).

TABLE VIII

Effect of various flavonoids (5×10^{-4} M) on the incorporation of leucine-1-^{14}C (100 000 cpm) into cell-free preparations of *Escherichia coli* (cpm/mg protein)[a]

Treatment	cpm	%	Treatment	cpm	%
Control	193	100	Phloridzin	71	37
Hesperitin	255	134	Phloretin	314	165
Hesperidin	346	181	Kaempferol	81	43
Naringenin	148	78	Coumarin	83	44
Naringin	94	50			

[a] Incubated for 30 min at 20°. Additives, except flavonoids, ribosome, and enzyme fractions, as outlined by Nirenberg.

Reproduced with permission from Parups (1967).

in protein biosynthesis as affected by the various flavonoids may be tissue specific and according to his view the plant growth regulating action of these compounds may also, at least in part, be mediated via their effect on protein formation.

Tannins and their oxidation products may also be powerful inhibitors of enzymes and mitochondria (Braunstein, 1947; Forsyth *et al.*, 1958; Byrde, 1963; Hulme *et al.*, 1963, 1964; Benoit, 1965; Loomis and Battaile, 1966; Wood, 1967; Anderson, 1968; Loomis, 1969; Schneider and Hallier, 1970). Inhibitors of cellulases have been reported from a large proportion of a variety of species (Bell *et al.*, 1962; Mandels and Reese, 1963, 1965). Thus Bell *et al.* investigated water soluble leaf extracts from 61 plant species in 32 families for their ability to inhibit fungal cellulase and pectinase. Leaf extracts from 29 species inhibited pectinase and extracts from 14 inhibited cellulase. In general, cellulase inhibition by the different plant species was less pronounced than that observed for pectinase. Only five species gave strong inhibition of cucumber flower cellulase, while muscadine grape and persimmon only slightly inhibited the commercial pectinase enzyme 19 AP. The water-soluble pectinase inhibitor in grape leaves is a tannin-like substance which is precipitated by caffeine, nicotine sulphate and gelatin; the other naturally occurring cellulase inhibitors are probably of the same nature. Moreover, cellulase activity is reduced by tannic acid, tannin and leucoanthocyanins (Mandels and Reese, 1963) and the same enzyme is further inhibited by DOPA-melanin in both a competitive and non-competitive way (Bull, 1970). The inhibition of pectolytic enzymes by naturally occurring compounds has been reviewed by Bell *et al.* (1958, 1960) and Wood (1967). In addition, the inhibition of pectic enzymes by tannins (Hathway and Seakins, 1958; Pollard *et al.*, 1958; Williams, 1963; Byrde, 1963; Grossmann, 1964; Bhatia *et al.*, 1972) and oxidation products of leucoanthocyanins and other phenolics has been reported (Cole, 1958; Pollard *et al.*, 1958).

An inactivator of polygalacturonase in grape leaves was described by Bell and Etchells (1958) and later studies revealed this inhibitor as a tannin-like substance (Bell *et al.*, 1962). However, Hobson (1964) did not find any inhibiting action of polyphenols, although the phenolic inactivation of pectin esterase has also been described (Hall, 1966). Coumarin can further inhibit the adaptive formation of polygalacturonase in germinating uredospores of *Puccinia graminis* cv. *tritici* (Van Sumere *et al.*, 1957b). Indeed, dormant spores which failed to germinate well, also failed to produce more than a trace of polygalacturonase activity. Addition of 30 μg/ml coumarin to the agar–pectin medium on which the spores were germinated was effective in stimulating germination (Van Sumere *et al.*, 1957a) but was unable to produce a corresponding increase in polygalacturonase production.

Several investigators have further shown that products formed by oxidation of *o*-dihydric phenols inactivate enzymes produced by plant pathogenic fungi (Byrde, 1963; Patil and Dimond, 1967; Kosuge, 1969). According to Patil and

Dimond (1967), inhibition of polygalacturonase by the oxidation products of dihydric phenols may be due to 1,4-addition of an amino or imino group in the enzyme to quinones. On the other hand, caffeic acid (250 µg/ml) increased cellulolytic and pectolytic activities in illuminated *Diplodia zeae* and ferulic acid (250 µg/ml) increased the pectolytic but decreased C_x activity (BeMiller *et al.*, 1969). Also *p*-coumaric acid caused a marked change in mycelial growth and in the amount of pectinase and cellulase secreted by the mycelia of *Fusarium roseum* cv. *graminearum* (Molot, 1970).

An inhibitory effect of oak leaf tannins on trypsin was shown by Feeny (1969). Indeed, when casein is complexed with oak leaf tannin the protein is almost completely protected from hydrolysis by the enzyme at pH 7·6. The pH of the mid-gut of larvae of the winter moth, *Operophtera brumata*, is 9·2 and at this pH, enzymic hydrolysis of complexes between casein and oak leaf tannins seem to be increasingly inhibited as the proportion of tannin in the initial complex is raised. With condensed oak leaf tannin, the inhibition is more pronounced than with the hydrolysable type. These observations may, according to Feeny, help to explain selection in some oak-feeding moth larvae for a high gut pH and for advance of the feeding period to avoid a high tannin content in the host plant. Indeed, in view of the substantial increase in digestibility of protein–tannin complexes at pH 9·2 in comparison with their digestibility at lower pHs, it is quite possible that one of the functions of the high gut pH in the larvae of many phytophagous insects is to increase the amount of nitrogen available from leaf protein–tannin complexes.

By analogy with the above results, Basaraba and Starkey (1966) found that the microbial hydrolysis of gelatin and gliadin was more thoroughly inhibited by tannins at pH 4·0 than pH 7·0. An explanation for this pH dependency may be found in the fact that the binding between proteins and tannins decreases as the pH value passes 7–8. Inhibition of β-amylase at pH 5 by tannins from the leaves of *Quercus petraea* Lieb. has been reported by Boudet and Gadal (1965a, b). However, above pH 6·7, inhibition decreased to zero, suggesting again that stable amylase–tannin complexes are most probably not formed under alkaline conditions. According to Feeny (1969) the tannins from *Q. robur* also almost completely inhibit the digestion of starch by α-amylase at pH 6·0. One final example of the effect of phenols on enzymes: it has been demonstrated by Baldry *et al.* (1970) that phenolics diminish photosynthesis in spinach chloroplasts, while chlorogenic acid and caffeic acid proved to be inhibitors for the reactions associated with CO_2-fixation by the photosynthetic carbon reduction cycle.

Finally, reference must also be made to the effect of quinones on enzyme activity, a subject reviewed by Hoffmann-Ostenhof (1963) and Webb (1966). The activity of quinones against diverse types of enzymes suggests that they are capable of disturbing cellular metabolism in many ways, the reactivity of the compounds usually being attributed to reaction with amino acids or proteins, alteration of the cellular redox potential, interference with co-factor and enzyme synthesis, and inhibition of specific enzyme systems (Webb, 1966; Rich, 1969; Kosuge, 1969; Levin, 1971). Quinones may inhibit enzymes by complexing with metal ions which participate in catalysis, by reaction with sulphydryl groups by 1,4-addition, by oxidation of the same groups, by reaction with the substrate or a co-factor, by production of H_2O_2 during oxidation of polyphenols, or by non-specific binding to enzymes through the aromatic ring and competition with the substrate.

According to Sissler and Cox (1960), quinones are effective inhibitors of pancreatic amylase, malt maltase, carboxylase and catalase. Certain quinones may also inhibit enzymes such as urease, pancreatic lipase, proteinase and phosphatase. Blank and Sondheimer (1969) (see also Schwimmer, 1958) have further proved that o-quinones inhibit potato phosphorylase. In addition Pé, I., Dedonder, A., Hrazdina, G. and Van Sumere, C. F. (unpublished results) have found that quinones may be powerful inhibitors of pancreatic RNase and amino acid activation (see Tables VI and VII). Inactivation of phenolase by reaction products formed from o-diphenols (chlorogenic acid and catechol) have also been described (Ingraham, 1955; Fling et al., 1963). The rapidity of inactivation frequently disturbs the exact measurement of the phenolase activity, the most accurate method being based on polarographic measurements (Mayer et al., 1966). In addition, Grossmann (1964) has proved that quinones also inhibit pectic enzymes and Pšenáková et al. (1972) have shown that benzoquinone decreases IAA-oxidase activity.

From the above discussion, it seems clear that phenolic inhibitors of enzymes are widespread in plants. These compounds can directly react and form complexes with the enzymes via hydrogen bonds following one of the different routes outlined in section II (p. 212). Tannins may further more or less reversibly precipitate enzymes, the frequently non-specific nature of the inhibition being suggested by the fact that the activity of certain of these polyphenol preparations is reduced or eliminated by gelatin. Presumably this protein combines with the inhibitors which therefore become unavailable to the enzymes when they are added later (Wood, 1967).

VI. Phenolics and Proteins in the Resistance of Plants against Phytopathogens

The possible importance of phenolics in disease resistance in plants has been thoroughly reviewed by Farkas and Király (1962), Rubin and Artsikhov-skaya (1963), Cruickshank and Perrin (1964), Wood (1967), Goodman et al. (1967), Kosuge (1969), Levin (1971) and Uritani (1971). Only one aspect of the subject will be considered here: the role that phenolics and (or) their oxidation products in combination with proteins may play in the defence mechanism of plants. Okasha et al. (1968), Levin (1971) and Hart and Hillis (1972) distinguish between constitutive and induced resistance, the former depending more upon the quality than upon the quantity of the phenolics, while the latter is mainly produced by an accumulation of various aromatic compounds in the surroundings of the lesion after infection (Farkas and Király, 1962).

The defence mechanisms may involve simple phenolics such as chlorogenic and caffeic acids, as well as more complex components such as the tannins. Usually the "active" phenols are stored in the plants or fruits as glycosides but, on infection, hydrolysis of these compounds by β-glycosidases may give rise to free phenolics. The oxidation of the liberated phenols is then accomplished by phenolases and peroxidases, both enzymes being widespread in nature and tending to accumulate in infected tissues (Farkas and Király, 1962; Seevers and Daly, 1970; Chant and Bates, 1970).

The inhibitory activity associated with the oxidation of phenolic compounds is frequently attributed to the reactivity of the quinones formed. These compounds can indeed, as indicated earlier, react in many ways with proteins, enzymes or intracellular amino acids. They can also interfere with energy production and enzyme synthesis. In addition, the cellular redox potential can be changed. Quinones may thus immediately be toxic to the pathogen or may reduce the nutritive capacity of the host. It is also evident that the plant itself may suffer from the quinones produced. Indeed, oxidized polyphenols may condense to polyquinoid structures which may, as indicated above, also inhibit plant enzymes or react with amino acids and proteins to form melanins. As a result dead tissue will be formed and the infected area will be sealed off due to the polymerizing and tanning action of the phenolase. The net effect of all these reactions may thus ultimately constitute part of the defence mechanism of the host by forming a physical barrier, which is stronger in resistant varieties.

VII. THE ROLE OF PHENOLICS AND PROTEINS IN THE FORMATION OF HUMUS

In soil humus, two types of polymer, humic acid and polysaccharides, are the major constituents since they may account for more than 90% of the total organic matter (Martin and Haider, 1971). "Humic acid" is a highly unspecific term which has been defined as that portion of "humus" which is soluble in aqueous sodium hydroxide, precipitated by acid (pH 2·0) and is brown to black in colour (Flaig, 1966; Haworth, 1971). Molecular weights from 1000–30 000 are reported. The material which does not precipitate is usually referred to as "fulvic acid". The humic acid and the analogous fulvic acid molecules appear to be very complex polymers of phenolics, amino acids, peptides and other organic constituents.

The complex aromatic core of humic acid is obtained by subsequent boiling of humic acid with water (20 h) and 6 N hydrogen chloride (20 h). Although it may contain o-benzoquinone polymers, its structure is still largely unknown

(Haworth, 1971). The polymers could further arise via enzymatic or autoxidative reactions to form highly active radicals or hydroxybenzoquinones which link with other phenolic units, peptides and amino acids to form the large humic acid molecules. Figure 6, derived from the work of Martin and Haider, gives a schematic representation of humic acid synthesis in soil.

It is quite possible that in humic acid some of the protein is hydrogen-bonded and that the remainder may be bound by the union of the free amino groups of the protein to a quinone nucleus. In the light of the findings of Van Sumere *et al.* (1973), it is also quite possible that some phenolic acid amide linkages may be present in the protein of humic acid. Soil polysaccharides (10–30% of the humus) from plant, animal and microbial origin, in combination with metal ions or clays may further form the resistant complexes that

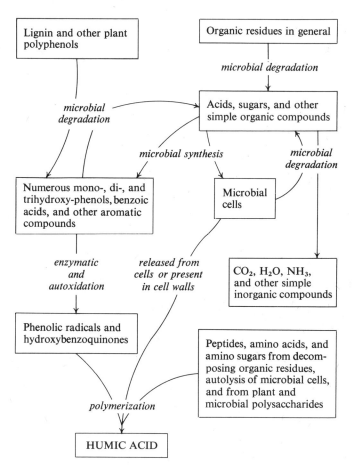

FIG. 6. Reproduced with permission from Martin and Haider (1971).

constitute the polysaccharide fraction of soil humus. How these poly-saccharides and phenolic acids are attached to the core is at this moment still difficult to indicate; glycosidic and ester linkages are most probably involved.

The relation of polyphenols to decomposition of organic materials in the soil has also recently been studied (Basaraba and Starkey, 1966; Davies, 1971). There seems to be little proof that lignin and mineral soil colloids are inhibitors of the microbial degradation of proteins (for references see Basaraba and Starkey, 1966), but there is some evidence that polypeptide material is preserved by complex formation with vegetable tannins (Davies et al., 1964a; Synge, 1972). Indeed, as suggested by Handley (1954), tanning of leaf proteins at senescence results in precipitates that are highly resistant to decomposition by soil organisms. Moreover the precipitates coat other naturally occurring compounds such as cellulose, protecting these energy-rich substrates from microbial and fungal attack. The resistance to decay may even be such that in base-deficient soils (mor site) the leaves, and especially their protein complexes, remain for many years largely undecomposed on the surface of mineral soil (Brown et al., 1966). It follows from the studies of Basaraba and Starkey (1966) that the inhibitive effect of the tannins on decomposition of the proteins increases with an increase in the ratio of tannin to protein from 1:4 to 4:1, the inhibitory effect being considerably greater at pH 4·0 than at neutrality. However the resistance of the tannin–protein complexes varied with the tannins and proteins (Lewis and Starkey, 1968). In addition, the effect of tannins would be greater in acid forest soils than in almost neutral grassland and cultivated soils, because forest vegetation contains greater amounts of tannin, and tannin–protein complexes are more resistant to decomposition at strongly acidic conditions. Coulson et al. (1960) have reported that the polyphenols of fresh, senescent and recently fallen leaves of plants grown on nutrient deficient (mor humus) sites are more diverse and greater in quantity as compared with the same species grown on nutrient rich (mull humus) sites. They observed a greater chance of tanning of the leaf proteins in fresh oak and beech green leaves from mor sites than those from mull sites. Later the same authors showed that the synthesis of leucoanthocyanins and tannins in plants changed from mull to mor sites and an increase of leucoanthocyanin and tannin was associated with deficiency of soil nitrogen and phosphorus (Davies et al., 1964b). However, the dead leaves of certain species such as willowherb (Chamaenerion angustifolium) containing a hydrolysable gallo-tannin do not give rise to organic layers of base poor mineral soils (Brown et al., 1966). Other observations made by Davies (1971) are consistent with a greater stability of litter at mor sites and with a less tanned litter at mull sites. However, it is not that mor humus is rich in tanned protein but rather that the initial tanning is an essential preliminary to the mode of degradation into mor. One must conclude that phenolics, proteins and suitable conditions for tanning of leaf proteins, may be very important in the formation of humus in the soil.

VIII. The Role of Phenolics and Proteins in the Formation of
Hazes in Beverages

Filtered beer, which should be clear, may be considered as a dilute alcoholic solution containing sugars, polypeptides, proteins, simple and complex
phenolics, bitter resins (De Keukeleire and Verzele, 1973), and other trace
materials. On storage at ambient temperature, beer will in due course become
hazy. This is due to haze induced by metals, protein–polyphenol (tannin)
hazes, polysaccharide sediments, oxalate hazes and sediments (Curtis, 1971).
In this review only protein–polyphenol turbidities will be discussed; these are
by far the most commonly encountered types of non-biological hazes.

At the stage of bottling beer, complex equilibria exist between the polypeptide material and the polyphenols (Chapon et al., 1961; Gramshaw, 1969;
Claesson and Sandegren, 1970). Subsequently, the beer polyphenols may
undergo a slow but continuous oxidative and (or) acid catalysed polymerization; such reactions increase their affinity for polypeptides, a process which
may be visually recognized by the formation of a reversible chill haze* or a
permanent haze (produced by further polymerization). Also, beer in contact
with air slowly deteriorates with the development of off-flavour, more intense
colour and the formation of hazes (Dadić et al., 1970; De Clerck, 1970).
Some of the more important review papers on this topic are those of Clark
(1960, 1963), Chapon (1965, 1971), Harris (1965), Pasfield (1968), Steiner
(1968) and Vancraenenbroeck (1968). According to most authors, 40–80% of
non-biological hazes are proteins or protein derivatives. In addition bitter
resins, carbohydrates (glucose and smaller amounts of arabinose and xylose),
simple phenols (vanillic, syringic, ferulic, sinapic, protocatechuic, gallic
acids), polyphenols and metals (iron, copper, tin, nickel, etc.) are also present.
Normally 2–4% carbohydrate, about 30–40% polyphenol and 1–3% ash are
found.

When the overall amino acid composition of barley is compared with
that of beer haze there appears to be a rather close correspondence. It has
therefore been concluded that no specific barley protein is responsible for
haze, all the proteins playing a part (Curtis, 1971). On the other hand,
Meredith and Tkachuk (1964) suggest that a water soluble protein–tannin
complex characteristic of chill haze is present in barley and persists relatively
unchanged through malting and brewing to beer. However, this protein is not
homogeneous and its heterogeneity varies during processing as has been
shown by the amino acid analysis for malt proteins. Haze proteins have been
very recently studied by isoelectric focusing on thin layers of polyacrylamide
gel (Savage and Thompson, 1972) and this has shown that hazes, whilst
containing most of the proteins found in beer, do differ appreciably from beer
in having a substantially higher proportion of acidic components.

* The European barley convention haze group has arbitrarily defined "chill haze" as
being the haze which forms in beer when it is chilled at 0° but redissolves at 20 °C.

The phenolic fraction of beer haze must be related to the phenolics of beer and thus of hops and barley. In beer the key anthocyanogens of hop and malt have been found (Harris, 1965), together with a large series of other phenolic compounds (Wye and McFarlane, 1957; Silbereisen and Kraffczyk, 1967; Gramshaw, 1967a, 1968; Steiner and Stoker, 1968; Vancraenenbroeck, 1968; Dadić et al., 1970).

Hop contains phenolic acids and derivatives, flavonols (kaempferol, quercetin and their glycosides), biflavans (procyanidin B), triflavans, catechins ((+)-catechin, (−)-epicatechin and gallocatechin) and the more complex products formed from them (condensed tannins) (Gramshaw, 1967b, 1969; Weinges et al., 1969a; Vancraenenbroeck et al., 1970).

With the exception of the flavonols, the same phenolic compounds have been identified in barley (Van Sumere et al., 1958; Harris and Ricketts, 1958; Gramshaw, 1967a, b, 1968, 1969; Vancraenenbroeck, 1968; Gorissen et al., 1970; Dadić et al., 1971). The main barley biflavans, isolated by Debeer et al. (1966), are "leucopelargonidin-(+)-catechin", "leucocyanidin-(+)-catechin" and "leucodelphinidin-(+)-catechin" (Vancraenenbroeck, 1968). However, the absence of leucocyanidins on the one hand, and the presence of catechin on the other in fruits containing dimeric procyanidins would, according to Weinges et al. (1969a) and Thompson et al. (1972), suggest that the latter substances were not formed by acidic condensation of a leucocyanidin with a catechin. Furthermore, the fact that the dimeric procyanidins* (10) always occur with catechins having the same configuration, seems to indicate that the former compounds are formed by enzymic dehydrogenation (or acid-catalysed polymerization) of the catechins. These arguments no doubt apply to the substances isolated by Debeer et al. (1966).

A method for the determination of the total polyphenol content of beer to cover all those polyphenols which may produce the red anthocyanidin colour in hot dilute acid has recently been proposed by Jerumanis (1973).

Although the first unequivocal proof of the presence of polyphenols in beer hazes was produced by Hartong (1937, 1949) and Sandegren (1947), the question as to whether the simple or the polymerized polyphenols of beer are the most important in haze formation has not yet been completely settled, but it seems that the latter are of greater importance (Curtis, 1971). Indeed, if isolated from beer they will, as shown by Vancraenenbroeck (1968), form haze instantly after their addition to beer. Gramshaw (1969) has further found

---

* Recently Weinges et al. (1969b) have proposed that flavan-3,4-diols and flavan-3-ol dimers and higher oligomers should be named respectively "leucoanthocyanidins" and "proanthocyanidins", the procyanidins being a part of the group of the latter compounds. This nomenclature has been used by Thompson et al. (1972), who have determined the structure of some plant proanthocyanidins in their free phenolic form (see also Weinges et al., 1969a). Biflavans and higher oligomers, which yield anthocyanidin pigments on acidic hydrolysis, have also been named "anthocyanogens" (Harris and Ricketts, 1958; see also Creasy and Swain, 1965; Geissman and Dittmar, 1965; Weinges and Freudenberg, 1965; Gramshaw, 1968; Vancraenenbroeck, 1968).

Procyanidin (10)

that the addition of synthetic 5,7,3',4'-tetrahydroxyflavan-3,4-diol at a rate of 12·5 ppm caused a rapid development of permanent haze under conditions of low air content, whilst use of 25 ppm resulted in a very rapid acceleration of haze formation. The question how beer hazes arise and how their formation may be prevented can, for the time being, only be partly answered. Beer phenolics undergo extensive oxidative and, possibly, although less likely, acid catalysed polymerization (Weinges et al., 1969a; Thompson et al., 1972). The flavans are less stable than the flavonols and oxidize and polymerize easily under the influence of acid, oxygen or light. Oxidative and acid-catalysed processes which can occur have also been summarized by Gramshaw (1969).

The final problem to be settled remained the mode of formation of the polyphenol–protein complexes. The reactions resulting in a combination between polymerized phenolics and proteins may certainly involve the classical tanning reactions discussed earlier. Indeed Roberts (1959) has shown that cysteine and glutathione react with the quinone derived from catechin. An increase in size of the existing polypeptide complexes by intermolecular hydrogen linkages, S–S-bridges and complexes with cations may also play a role (Vancraenenbroeck, 1968). However, oxygen reacts rapidly with beer (Curtis and Clark, 1959) and in an investigation using labelled oxygen (^{18}O) it was found (Owades and Jakovac, 1966) that about 65% reacted with phenolic materials and about 30% reacted with volatile carbonyls. The aqueous fraction showed no increase in ^{18}O, indicating that there had been no abstraction of hydrogen from organic compounds and this throws doubt on the role played by the oxidation of protein sulphydryl groups in haze formation. Nevertheless oxygen seems to increase haze formation, particularly chill haze production in agitated beer (Curtis, 1971), while treatment with metabisulphite has an opposite effect, indicating that haze is certainly not formed by hydrogen linkages only but that oxidative reactions are at least partly involved.

The formation of beer hazes may be eliminated by a variety of techniques. In his review, Curtis (1971) discusses the choice of barley, the effect of malting, mashing, boiling, cooling, fermentation, lagering, chilling and filtration on the stability of beer. Although it seems possible to take some steps at almost every stage of the brewing process to enhance stability, most brewers who wish to extend shelf life make use of stabilizing agents (Curtis, 1971). Stabilizing treatments include employment of reducing agents, the modification or partial removal of polypeptide material, the modification and partial removal of polyphenols. In this respect proteolytic enzymes (mainly papain) and protein adsorbents such as bentonite and activated silica gels (Stabifix and Stabiquick) have been used. Protein precipitants such as tannic acid (Vancraenenbroeck et al., 1972) have also been recommended. The removal of certain poly-phenols and (or) their oxidation products has also been described. Harris and Ricketts (1959) used adsorption onto nylon 66 and the same material was used industrially by Curtis and Clark (1960). However, nylon reduces foam stability; therefore polyvinylpyrrolidone (Polyclar AT) has been widely used in industry (McFarlane et al., 1955, 1961, 1964) especially since Polyclar AT may be incorporated into filter sheets (Keller, 1963). Vancraenenbroeck and Lontie (1964a, b) and Vancraenenbroeck et al. (1965) have also employed insoluble proteins like keratin and casein for the removal of beer polyphenols and the use of casein has been especially recommended (Vancraenenbroeck, 1968). According to Curtis (1971), combinations of stabilizing agents were more effective than single treatments.

The problem of clouding or haze formation is not strictly confined to beer but exists also both in wine (Kielhöfer, 1951; Wucherpfennig and Franke, 1963) and juices (Johnson et al., 1968). In both cases the importance of poly-meric phenolic–protein complexes has been recognized. Analogous techniques to those employed in the beer industry have been applied for the removal of wine hazes. Phenolics also play an important part in the browning of food products (Segal et al., 1970), and when non-enzymic the process is enhanced by heating, an increased pH (the reaction taking place at pH > 6) and amino acids such as lysine.

IX. The Interactions of Polyphenols with Proteins and Taste and Flavour in Foods

In addition to the ability of polyphenols to tan protein and produce unwanted precipitates or hazes in wines and beers, they also contribute to taste and flavour in foods (Synge, 1972). The impact of a simple phenolic such as chlorogenic acid on the taste of coffee and ciders has long since been recognized (J. Inst. Brew., 1958) and the astringency of tannins, i.e. the ability to cause a dry puckery sensation in the mouth, is well known (Joslyn and Goldstein, 1965). This sensation (Swain, 1965) is almost certainly due to the cross-linking action of phenolics of an appropriate size (tannins) with the proteins of the mouth, resulting in a reduction of the lubricant action of the glycoproteins in the saliva (Bate-Smith, 1954).

Astringency is very important in fruits and beverages. Indeed, many unripe fruits are unacceptable because of their astringency and varieties have generally been selected for low leucoanthocyanidin content (Harborne, 1967). The property of astringency is due to the presence of oligomers containing 4–8 flavan units and the loss of astringency on ripening in fruits seems to be directly attributable to the disappearance of the oligomers, the polymers being laid down in the cell wall or not contributing to the taste because of their insolubility.

Some astringency is necessary for avoiding insipidness and suitable blends of leucoanthocyanidin and sugar seem to be essential components of the palatability of wine, cider, beer, cocoa and tea. Recently, astringency (i.e. the efficiency of polyphenols as precipitants of proteins) has been determined by reacting polyphenols with the proteins of haemolysed blood and colorimetric determination of residual haemoglobin (Bate-Smith, 1973). Bate-Smith has further defined "relative astringency" as the ratio of the concentration of tannic acid to that of the tannin which effects the same degree of precipitation.

The astringent taste of black tea infusion has also been attributed to certain phenolics, namely the theaflavins (Bradfield, 1946) and the addition of milk to an infusion results in a reduction of the astringent taste (Brown and Wright, 1963), this reduction being due to a reaction between the polyphenols and certain milk proteins. Indeed, from the electrophoretic studies by Brown and Wright (1963), it is clear that when a tea infusion is mixed with milk the coloured tea polyphenols interact mainly with the α-casein complex and the β-casein of the milk to form soluble polyphenol complexes. β-Lactoglobulin and α-lactalbumin, the main whey proteins, appear to be unaffected by the polyphenols in the presence of casein at the concentration used in the investigation. It was further shown by membrane electrophoresis in phosphate buffer (pH 6·7) that 7 M urea inhibits the interactions between milk proteins and the coloured tea polyphenols, indicating that the interactions are due, at least initially, to the formation of hydrogen bonds.

X. CONCLUSION

It is evident that the interactions of phenolics and proteins are of great theoretical and practical importance, not only in physiology, biochemistry, phytopathology, pedology but also in the food sciences. More information on the oxidation and polymerization of phenolics and the subsequent reactions with proteins and enzymes is therefore required. It is also essential to investigate further how phenolics may be covalently bound to proteins and what function such compounds have in protein and enzyme chemistry. In addition, the possible role of N-ferulylglycine and related compounds in protein biosynthesis in plants should be investigated. Finally, food technology will benefit enormously from a better knowledge of the control of phenolic oxidation, polymerization and interaction with proteins.

256   C. F. VAN SUMERE ET AL.

ACKNOWLEDGEMENTS

The authors thank Professor R. L. M. Synge and Dr Pierpoint for valuable discussions. Financial support by the Belgian "Fonds voor Kollektief Fundamenteel Onderzoek" is gratefully acknowledged.

REFERENCES

Adams, J. M. and Capecchi, M. R. (1966). *Proc. natn. Acad. Sci. U.S.A.* **55**, 147.
Adams, R., Geissman, T. A. and Edwards, J. D. (1960). *Chem. Rev.* **60**, 555.
Akazawa, T. and Conn, E. E. (1958). *J. biol. Chem.* **232**, 403.
Alibert, G., Marigo, G. and Boudet, A. (1968). *C.r. hebd. Séanc. Acad. Sci., Paris* **267D**, 2144.
Andersen, R. A. and Sowers, J. A. (1968). *Phytochemistry* **7**, 293.
Anderson, J. W. and Rowan, K. S. (1967). *Phytochemistry* **6**, 1047.
Anderson, J. W. (1968). *Phytochemistry* **7**, 1973.
Andreae, W. A. and Good, N. E. (1957). *Pl. Physiol.* **32**, 566.
Armstrong, M. D., Shaw, K. N. F. and Wall, P. E. (1956). *J. biol. Chem.* **218**, 293.
Attridge, T. H., Stewart, G. R. and Smith, H. (1971). *FEBS Letters* **17**, 84.
Bagdasarian, M., Matheson, N. A., Synge, R. L. M. and Youngson, M. A. (1964). *Biochem. J.* **91**, 91.
Baldry, C. W., Bucke, C., Coombs, J. and Gross, D. (1970). *Planta* **94**, 107.
Basaraba, J. and Starkey, R. L. (1966). *Soil Sci.* **101**, 17.
Bate-Smith, E. C. (1954). *Food* **23**, 124.
Bate-Smith, E. C. (1973). *Phytochemistry* **12**, 907.
Batzer, H. and Weissenberger, G. (1952). *Makromol. Chem.* **7**, 320.
Batzer, H. (1952). *Makromol. Chem.* **8**, 183.
Bell, T. A. and Etchells, J. L. (1958). *Bot. Gaz.* **119**, 192.
Bell, T. A., Aurand, L. W. and Etchells, J. L. (1960). *Bot. Gaz.* **122**, 143.
Bell, T. A., Etchells, J. L., Williams, C. F. and Porter, W. L. (1962). *Bot. Gaz.* **123**, 220.
BeMiller, J. N., Tegtmeier, D. O. and Pappelis, A. J. (1969). *Phytopathology* **59**, 674.
Benoit, R. (1965). *In* "Effects of Tannin and Enzyme Activity", Ph.D. thesis, Rutgers University.
Bezanson, G. S., Desaty, D., Emes, A. V. and Vining, L. C. (1970). *Can. J. Microbiol.* **16**, 147.
Bhatia, I. S., Sharma, H. K. and Bafaf, K. L. (1972). *Zbl. Baker. (W. Germany)* **126**, 437.
Björk, W. (1965). *Biochim. biophys. Acta* **95**, 652.
Blank, G. E. and Sondheimer, E. (1969). *Phytochemistry* **8**, 823.
Booth, A. N., Emerson, O. H., Jones, F. T. and DeEds, F. (1957). *J. biol. Chem.* **229**, 51.
Bouchilloux, S. (1963). *In* "Plant Phenolics and their Industrial Significance", Proc. Symp. Plant Phenolics Group North America (Corvallis, Oregon 1962) (V. C. Runeckles, ed.), p. 1. Imperial Tobacco Company Canada, Montreal.
Boudet, A. and Gadal, P. (1965a). *C.r. hebd. Séanc. Acad. Sci., Paris* **260**, 4057.
Boudet, A. and Gadal, P. (1965b). *C.r. hebd. Séanc. Acad. Sci., Paris* **260**, 4252.
Boulter, D., Ellis, R. J. and Yarwood, A. (1972). *Biol. Rev.* **47**, 113.
Bradfield, A. E. (1946). *Chemy Ind.* 242.

Brakke, M. K. (1967). *In* "Methods in Virology" (K. Maramorosch and H. Koprowski, eds), Vol. II, p. 93. Academic Press, New York and London.
Brattsten, I., Synge, R. L. M. and Watt, W. B. (1964). *Biochem. J.* **92**, 1P.
Brattsten, I., Synge, R. L. M. and Watt, W. B. (1965). *Biochem. J.* **97**, 678.
Braunstein, A. E. (1947). *Adv. Protein Chem.* **3**, 1.
Brieskorn, C. H. and Mosandl, A. (1970). *Tetrahedron Letters* **1**, 109.
Brown, B. R., Brown, P. E. and Pike, W. T. (1966). *Biochem. J.* **100**, 733.
Brown, B. R. (1967). *In* "Oxidative Coupling of Phenols" (W. I. Taylor and A. R. Battersby, eds), p. 167. Marcel Dekker, New York.
Brown, P. J. and Wright, W. B. (1963). *J. Chromatogr.* **11**, 504.
Brunet, P. C. J. (1967). *Endeavour* **98**, 68.
Brunt, A. A. and Kenten, R. H. (1963). *Virology* **19**, 388.
Bull, A. T. (1970). *Enzymologia* **39**, 333.
Byrde, R. J. W. (1963). *In* "Perspectives of Biochemical Plant Pathology" (S. Rich, ed.), Conn. Agr. Expt. Sta. Bull. 663, New-Haven, Conn., p. 193.
Cadman, C. H. (1960). *In* "Phenolics in Plants in Health and Disease", Proc. Plant Phenolics Group Symp. (Bristol 1959) (J. B. Pridham, ed.), p. 101. Pergamon Press, Oxford.
Camm, E. L. and Towers, G. H. N. (1973). *Phytochemistry* **12**, 961.
Cannon, C. G. (1955). *Mikrochim. Acta* 555.
Cater, C. M. and Lyman, C. M. (1969). *J. Am. Oil Chem. Soc.* **46**, 649.
Chant, S. R. and Bates, D. C. (1970). *Phytochemistry* **9**, 2323.
Chapon, L., Chollot, B. and Urion, E. (1961). *Bull. Soc. Chim. Biol.* **43**, 429.
Chapon, L. (1965). *J. Inst. Brew.* **71**, 299.
Chapon, L. (1971). *Bios* 3.
Cividanes, I., Fernandez-Otero, P. and Larralde, J. (1964). *Rev. Espan. Fisiol.* **20**, 63.
Claesson, S. and Sandegren, E. (1970). "Proc. Eur. Brew. Conv. (Interlaken, 1969)", p. 339. Elsevier Publishing Company, Amsterdam.
Clark, A. G. (1960). *J. inst. Brew.* **66**, 318.
Clark, A. G. (1963). *Brewers Dig.* **38**, 47.
Cole, M. (1958). *Nature, Lond.* **181**, 1596.
Coulson, C. B., Davies, R. I. and Lewis, D. A. (1960). *J. Soil. Sci.* **11**, 20.
Craig, L. C. (1962). *Archs Biochim. Biophys.* Suppl. I, 112.
Creasy, L. L. and Swain, T. (1965). *Nature, Lond.* **208**, 151.
Cruickshank, I. A. M. and Perrin, D. R. (1964). *In* "Biochemistry of Phenolic Compounds" (J. B. Harborne, ed.), p. 511. Academic Press, London and New York.
Curtis, N. S. and Clark, A. G. (1959). *Brewers' Guild J.* 186.
Curtis, N. S. and Clark, A. G. (1960). *J. Inst. Brew.* **66**, 226.
Curtis, N. S. (1971). *In* "Modern Brewing Technology" (W. P. K. Findlay, ed.), p. 254. Macmillan, London.
Dadić, M., Van Gheluwe, J. E. A. and Valyi, Z. (1970). *J. Inst. Brew.* **76**, 267.
Dadić, M., Van Gheluwe, J. E. A. and Valyi, Z. (1971). *J. Inst. Brew.* **77**, 48.
Davies, R. I., Coulson, C. B. and Lewis, D. A. (1964a). *J. Soil Sci.* **15**, 299.
Davies, R. I., Coulson, C. B. and Lewis, D. A. (1964b). *J. Soil Sci.* **15**, 310.
Davies, R. I. (1971). *Soil Sci.* **111**, 80.
Dawkins, P. D., McArthur, J. N. and Smith, M. J. H. (1971). *Biochem. Pharmacol.* **20**, 1303.
Dawydoff, W. (1953). *Faserforsch. Text. Tech.* **4**, 412.
Debeer, L., Vancraenenbroeck, R. and Lontie, R. (1966). *Arch. Intern. Physiol. Biochim.* **74**, 312.

De Clerck, J. (1970). *Brauwelt* **110**, 685.
Dedonder, A. and Van Sumere, C. F. (1971a). *Z. Pflanzenphysiol.* **65**, 70.
Dedonder, A. and Van Sumere, C. F. (1971b). *Z. Pflanzenphysiol.* **65**, 176.
De Keukeleire, D. and Verzele, M. (1973). *In* "Proceedings of the FEBS Special
    Meeting of Industrial Aspects of Biochemistry, Dublin" (in press).
De Pooter, H., Haider Ali and Van Sumere, C. F. (1973a). *Bull. Soc. Chim. Belges*
    82, 243.
De Pooter, H., Haider Ali and Van Sumere, C. F. (1973b). *Bull. Soc. Chim. Belges*
    82, 259.
Dubos, R. J. and Thompson, R. H. S. (1938). *J. biol. Chem.* **124**, 501.
Egger, K. (1964). *Planta Med.* **12**, 265.
Ehrenberg, M. (1954). *Biochem. Z.* **325**, 102.
El-Basyouni, S. Z., Neish, A. C. and Towers, G. H. N. (1964). *Phytochemistry* **3**,
    627.
Emes, A. V. and Vining, L. C. (1970). *Can. J. Biochem.* **48**, 613.
Endres, H. and Hörmann, H. (1963). *Angew. Chem.* **75**, 288.
Engelsma, G. and Meijer, G. (1965a). *Acta bot. neerl.* **14**, 54.
Engelsma, G. and Meijer, G. (1965b). *Acta bot. neerl.* **14**, 73.
Farkas, G. L. and Király, Z. (1962). *Phytopath. Z.* **44**, 105.
Fausch, H., Kündig, W. and Neukom, H. (1963). *Nature, Lond.* **199**, 287.
Feeny, P. P. (1969). *Phytochemistry* **8**, 2119.
Fieser, L. F. and Fieser, M. (1961). *In* "Advanced Organic Chemistry", p. 851.
    Reinhold, New York.
Firenzuoli, A. M., Vanni, P. and Mastronuzzi, E. (1969). *Phytochemistry* **8**, 61.
Fiske, C. H. and Subbarow, Y. (1925). *J. biol. Chem.* **66**, 375.
Flaig, W. (1966). *In* "Handbuch der Pflanzenernährung und Düngung II: Boden und
    Düngemittel". (H. Linser, ed.). Springer-Verlag, Wien.
Flaig, W. (1968). *Pont. Acad. Scient. Scripta varia, Rome* **20**, 1.
Fling, M., Horowitz, N. H. and Heinemann, S. F. (1963). *J. biol. Chem.* **238**, 2045.
Forsyth, W. G. C., Quesnel, V. C. and Roberts, J. B. (1958). *J. Sci. Fd Agric.* **9**, 181.
Friedrich, H. (1955). *Arch. Pharm.* **288**, 583.
Furuya, M., Galston, A. W. and Stowe, B. B. (1962). *Nature, Lond.* **193**, 456.
Gamborg, O. L., Wetter, L. R. and Neish, A. C. (1961). *Can. J. Biochem. Physiol.*
    39, 1113.
Gaspar, T. (1966). *Congr. Coll. Univ. Liège* **38**, 41.
Geissman, T. A. and Dittmar, H. F. K. (1965). *Phytochemistry* **4**, 359.
Geissmann, T. and Neukom, H. (1971). *Helv. chim. Acta* **54**, 1108.
Gierer, A. and Schramm, G. (1956). *Nature, Lond.* **177**, 702.
Gladtke, E. and Liss, E. (1958). *Biochem. Z.* **331**, 65.
Goldstein, J. L. and Swain, T. (1965). *Phytochemistry* **4**, 185.
Goodman, R. N., Király, Z. and Zaitlin, M. (1967). "The Biochemistry and
    Physiology of Infectious Plant Disease", p. 205. D. Van Nostrand, Princeton.
Gorissen, H., Debeer, L., Vancraenenbroeck, R. and Lontie, R. (1970). *Bull. Ass.*
    Anc. Etud. Brass. Univ. Louvain **66**, 3.
Gortner, W. A. and Kent, M. J. (1958). *J. biol. Chem.* **233**, 731.
Gortner, W. A., Kent, M. J. and Sutherland, G. K. (1958). *Nature, Lond.* **181**, 630.
Gramshaw, J. W. (1967a). *J. Inst. Brew.* **73**, 258.
Gramshaw, J. W. (1967b). *J. Inst. Brew.* **73**, 455.
Gramshaw, J. W. (1968). *J. Inst. Brew.* **74**, 20.
Gramshaw, J. W. (1969). *J. Inst. Brew.* **75**, 61.
Grassmann, W. (1937). *Collegium* **809**, 530.
Grassmann, W. and Deffner, G. (1953). *Z. physiol. Chem.* **293**, 89.

Grillo, M. A. and Cafiero, M. (1964). *Biochim. biophys. Acta* **82**, 92.
Grossmann, F. (1964). *Z. Pflkrankh.* **71**, 148.
Gustavson, K. H. and Holm, B. (1952). *J. Am. Leather Chemists' Assoc.* **47**, 700.
Gustavson, K. H. (1954). *J. Polymer Sci.* **12**, 317.
Gustavson, K. H. (1956). "The Chemistry of Tanning Processes". Academic Press, New York and London.
Gustavson, K. H. (1963). *Leder* **14**, 27.
Hall, C. B. (1966). *Nature, Lond.* **212**, 717.
Handley, W. R. C. (1954). *Forestry Comm. Bull.* **23**. Imperial Forestry Institute, Oxford.
Hansch, C., Kiehs, K. and Lawrence, G. L. (1965). *J. Am. chem. Soc.* **87**, 5770.
Hanson, K. R., Zucker, M. and Sondheimer, E. (1967). *In* "Phenolic Compounds and Metabolic Regulation" (B. J. Finkle and V. C. Runeckles, eds), p. 69. Appleton-Century-Crofts, New York.
Harborne, J. B. (1967). *In* "Comparative Biochemistry of the Flavonoids", p. 300. Academic Press, London and New York.
Harel, E., Mayer, A. M. and Shain, Y. (1964). *Physiologia Pl.* **17**, 921.
Harper, G. A. (1970). *J. Am. Oil Chem. Soc.* **47**, 438.
Harris, G. and Ricketts, R. W. (1958). *Chemy Ind.* 686.
Harris, G. and Ricketts, R. W. (1959). *J. Inst. Brew.* **65**, 256.
Harris, G. (1965). *J. Inst. Brew.* **71**, 292.
Harrison, B. D. and Pierpoint, W. S. (1963). *J. gen. Microbiol.* **32**, 417.
Hart, J. H. and Hillis, W. E. (1972). *Phytopathology* **62**, 620.
Hartong, B. D. (1937). *Woch. Brau.* **54**, 321.
Hartong, B. D. (1949). "Proc. Eur. Brew. Conv. (Lucerne 1949)", pp. 56, 250. Elsevier, New York.
Hathway, D. E. (1958). *J. Soc. Leather Trades' Chemists* **42**, 108.
Hathway, D. E. and Seakins, J. W. T. (1958). *Biochem. J.* **70**, 158.
Haworth, R. D. (1971). *Soil Sci.* **111**, 71.
Hemker, H. C. (1962). *Biochim. biophys. Acta* **63**, 46.
Hemker, H. C. (1964). *Biochim. biophys. Acta* **81**, 9.
Hobson, G. E. (1964). *Biochem. J.* **92**, 324.
Hoffmann-Ostenhof, O. (1963). *In* "Metabolic Inhibitors" (R. M. Hochster and J. H. Quastel, eds), Vol. II, p. 145. Academic Press, New York.
Horigome, T. and Kandatsu, M. (1968). *Agric. biol. Chem.* **32**, 1093.
Horn, R., Vonder Mühll, M., Comte, M. and Grandroques, C. (1970). *Experientia* **26**, 1081.
Hrazdina, G., Van Buren, J. P. and Robinson, W. B. (1969). *Ann. J. Emol. Viticuii.* **20**, 66.
Hsu, Y. C. and Wang, T. P. (1964). *Acta biochim. Sin.* **5**, 413.
Hulme, A. C. and Jones, J. D. (1963). *In* "Enzyme Chemistry of Phenolic Compounds" (J. B. Pridham, ed.), p. 97. Pergamon Press, Oxford.
Hulme, A. C., Jones, J. D. and Wooltorton, L. S. C. (1963). *Proc. R. Soc. B*, **158**, 514.
Hulme, A. C., Jones, J. D. and Wooltorton, L. S. C. (1964). *Phytochemistry* **3**, 173.
Hulme, A. C., Rhodes, M. J. C. and Wooltorton, L. S. C. (1967). *Phytochemistry* **6**, 1343.
Hulme, A. C. and Rhodes, M. J. C. (1971). *In* "The Biochemistry of Fruits and their Products" (A. C. Hulme, ed.), Vol. II, p. 346. Academic Press, London and New York.
Huppert, J. and Pelmont, J. (1962). *Archs Biochem. Biophys.* **98**, 214.
Hyodo, H. and Uritani, I. (1967). *Archs Biochem. Biophys.* **122**, 299.

Ingraham, L. L. (1955). *J. Am. chem. Soc.* **77**, 2875.

Jerchel, D. and Oberheiden, H. (1955). *Angew. Chem.* **67**, 145.

Jerumanis, J. (1973). *Bull. Assoc. Anc. Etud. Brass. Univ. Louvain* **69**, 1.

Johnson, G., Donnelly, B. J. and Johnson, D. K. (1968). *J. Fd. Sci.* **33**, 254.

Joslyn, M. A. and Goldstein, J. L. (1965). *Wallerstein Lab. Commun.* **28**, 143.

Keller, K. (1963). *Brauwissenschaft* **16**, 252.

Kenten, R. H. (1957). *Biochem. J.* **67**, 300.

Kielhöfer, E. (1951). *Z. Lebensm. Untersuch. Forsch.* **92**, 1.

Kingsbury, J. M. (1964). *In* "Poisonous Plants of the United States and Canada". Prentice-Hall, Englewood Cliffs, New Jersey.

Kirby, K. S. (1956). *Biochem. J.* **64**, 405.

Kirby, K. S. (1957). *Biochem. J.* **66**, 495.

Kosuge, T. (1969). *A. Rev. Phytopath.* **7**, 195.

Kotyk, A., Kolínská, J., Vereš, K. and Szammer, J. (1965). *Biochem. Z.* **342**, 129.

Krishna, R. V. and Krishnaswamy, P. R. (1966). *Life Sci.* **5**, 2053.

Lang, W. (1956). *Planta Med.* **4**, 33.

Lehninger, A. L. and Schneider, M. (1958). *Z. physiol. Chem.* **313**, 138.

Lehninger, A. L. (1964). *In* "The Mitochondrion", p. 9. W. A. Benjamin, New York.

Levin, D. A. (1971). *Am. Nat.* **105**, 157.

Lewis, J. A. and Starkey, R. L. (1968). *Soil Sci.* **106**, 241.

Lieberman, M. and Biale, J. B. (1956). *Pl. Physiol.* **31**, 420.

Lipmann, F. (1958). *Proc. natn. Acad. Sci. U.S.A.* **44**, 67.

Loomis, W. D. (1969). *In* "Methods in Enzymology" (J. M. Lowenstein, ed.), Vol. 13, p. 555. Academic Press, New York and London.

Loomis, W. D. and Battaile, J. (1966). *Phytochemistry* **5**, 423.

Lothspeich, W. D. (1961). *Harvey Lecture Ser.* **56**, 63.

Lucas-Lenard, J. and Lipmann, F. (1967). *Proc. natn. Acad. Sci. U.S.A.* **57**, 387.

Lyr, H. (1965). *Phytopath. Z.* **52**, 229.

Majak, W. and Towers, G. H. N. (1973). *Phytochemistry* **12**, 1141.

Mandels, M. and Reese, E. T. (1963). *In* "Advances in Enzymic Hydrolysis of Celluloses and Related Materials" (E. T. Reese, ed.). Pergamon Press, New York.

Mandels, M. and Reese, E. T. (1965). *A. Rev. Phytopath.* **3**, 85.

Mankash, E. K. and Pakshver, A. B. (1953). *Zh. prik. Khim., Leningr.* **26**, 830.

Marcker, K. and Sanger, F. (1964). *J. molec. Biol.* **8**, 835.

Marcus, A., Weeks, D. P., Leis, J. P. and Keller, E. B. (1970). *Proc. natn. Acad. Sci. U.S.A.* **67**, 1681.

Margoliash, E., Smith, E. L., Kreil, G. and Tuppy, H. (1961). *Nature, Lond.* **192**, 1125.

Martin, J. P. and Haider, K. (1971). *Soil Sci.* **111**, 54.

Mason, H. S. (1955a). *Adv. Enzymol.* **16**, 105.

Mason, H. S. (1955b). *Nature, Lond.* **175**, 771.

Mason, H. S. and Lada, A. (1954). *J. Invest. Dermatol.* **22**, 457.

Mason, H. S. and Peterson, E. (1955). "Congr. Int. Biochim. Résum. Comm. (3rd Congr. Brussels)", p. 107.

Mason, H. S. (1959). *In* "Pigment Cell Biology" (M. Gordon, ed.), p. 563. Academic Press, New York and London.

Mason, H. S. and Peterson, E. W. (1965). *Biochim. biophys. Acta* **111**, 134.

Matthews, R. E. F. (1960). *Diss. Abstr.* **20**, 461.

Mayer, A. M., Harel, E. and Ben-Shaul, R. (1966). *Phytochemistry* **5**, 783.

Mazelis, M. (1962). *J. biol. Chem.* **237**, 104.

McFarlane, W. D., Wye, E. and Grant, H. L. (1955). "Eur. Brew. Conv. (Baden-Baden 1955)", p. 298. Elsevier Publishing Company, Amsterdam.
McFarlane, W. D. and Bayne, P. D. (1961). "Eur. Brew. Conv. (Vienna 1961)", p. 278. Elsevier Publishing Company, Amsterdam.
McFarlane, W. D., Sword, P. F. and Blinoff, G. (1964). "Eur. Brew. Conv. (Brussels 1963)", p. 174. Elsevier Publishing Company, Amsterdam.
Meister, A. (1965). In "Biochemistry of the Amino Acids", Vol. I, p. 558. Academic Press, New York.
Mejbaum-Katzenellenbogen, W. (1959). Acta biochim. polon. 6, 385.
Meredith, W. O. S. and Tkachuk, R. (1964). J. Inst. Brew. 70, 410.
Milo, G. E., Jr. and Santilli, V. (1967). Virology 31, 197.
Minamikawa, T. and Uritani, I. (1965). J. Biochem., Tokyo 58, 53.
Mink, G. I., Huisman, O. and Saksena, K. N. (1966). Virology 29, 437.
Molot, P. M. (1970). C.r. hebd. Séanc. Acad. Sci, Paris Ser. D, 270, 2097.
Moran, C. J. and Walker, W. H. C. (1968). Biochem. Pharm. 17, 153.
Mori, S. and Noguchi, I. (1970). Archs Biochem. Biophys. 139, 444.
Mumford, F. E., Smith, D. H. and Castle, J. E. (1961). Pl. Physiol. 36, 752.
Mumford, F. E., Stark, H. M. and Smith, D. H. (1963). Phytochemistry 2, 215.
Muto, S., Asahi, T. and Uritani, I. (1969). Agric. biol. Chem. 33, 176.
Nagashima, R., Levy, G. and Nelson, E. (1968). J. pharm. Sci. 57, 58.
Neish, A. C. (1961). Phytochemistry 1, 1.
Neish, A. C. (1968). In "Constitution and Biosynthesis of Lignin" (K. Freudenberg and A. C. Neish, eds), p. 14. Springer-Verlag, Berlin.
Neukom, H., Providoli, L., Gremli, H. and Hui, P. A. (1967). Cereal Chem. 44, 238.
Neukom, H. (1972). Getreide, Mehl Brot 26, 299.
Nishira, H. (1963). Mem. Hyogo Univ. Agr. 16, 33 pp.
Novelli, G. D. (1958). Proc. natn. Acad. Sci. U.S.A. 44, 86.
Okasha, K. A., Ryugo, K., Wilhelm, S. and Bringhurst, R. S. (1968). Phytopathology 58, 1114.
Owades, J. L. and Jakovac, J. (1966). Proc. Am. Brew. Chem. 180.
Pakshver, A. B. and Mankash, E. K. (1953). Zh. prikl. Khim., Leningr. 26, 835.
Pakshver, A. B. and Mankash, E. K. (1954). Zh. prikl. Khim., Leningr. 27, 182.
Palmer, J. M. (1967). Nature, Lond. 216, 1208.
Parups, E. V. (1967). Can. J. Biochem. 45, 427.
Pasfield, J. (1968). Process Biochem. 3, 49.
Patil, S. S. and Dimond, A. E. (1967). Phytopathology 57, 492.
Pearlman, R. and Bloch, K. (1963). Proc. natn. Acad. Sci. U.S.A. 50, 533.
Pearson, J. A. and Robertson, R. N. (1954). Aust. J. biol. Sci. 7, 1.
Pierpoint, W. S. and Harrison, B. D. (1963). J. gen. Microbiol. 32, 429.
Pierpoint, W. S. (1966). Biochem. J. 98, 567.
Pierpoint, W. S. (1969a). Biochem. J. 112, 609.
Pierpoint, W. S. (1969b). Biochem. J. 112, 619.
Pierpoint, W. S. (1970). "Rep. Rothamsted exp. Stn. 1970". Part 2, p. 199.
Pollard, A., Kieser, M. E. and Sissons, D. J. (1958). Chemy Ind. 952.
Polz, G. and Kreil, G. (1970). Biochem. biophys. Res. Commun. 39, 516.
Pšenáková, T., Kolek, J. and Pšenák, M. (1972). Experientia 28, 1424.
Pusztai, A. (1966a). Biochem. J. 99, 93.
Pusztai, A. (1966b). Biochem. J. 101, 265.
Pusztai, A. and Watt, W. B. (1971). Biochim. biophys. Acta 251, 158.
Pusztai, A. (1973). In "Methodological Developments in Biochemistry—Vol. II, Preparative Techniques" (E. Reid, ed.), p. 145. Longmans, London.

Rabin, R. S. and Klein, R. M. (1957). *Archs Biochem. Biophys.* **70**, 11.

Racker, E. (1965). *In* "Mechanisms in Bioenergetics", p. 145. Academic Press, New York and London.

Ralph, R. K. and Bergquist, P. L. (1967). *In* "Methods in Virology" (K. Maramorosch and H. Koprowski, eds), Vol. II, p. 463. Academic Press, New York and London.

Reithel, F. J. (1963). *Adv. Protein Chem.* **18**, 123.

Rich, S. (1969). *In* "Fungicides, an advanced treatise" (D. C. Torgeson, ed.), Vol. II, p. 447. Academic Press, New York and London.

Roberts, E. A. H. (1959). *Chemy Ind.* 995.

Rowan, K. S. (1966). *Int. Rev. Cytol.* **19**, 301.

Rubin, B. A. and Artsikhovskaya, Ye. V. (1963). "Biochemistry and Physiology of Plant Immunity". Pergamon Press, Oxford.

Runge, F. F. (1834a). *Ann. Physik.* **31**, 65.

Runge, F. F. (1834b). *Ann. Physik.* **32**, 324.

Salas, M., Hille, M. B., Last, J. A., Wahba, A. J. and Ochoa, S. (1967). *Proc. natn. Acad. Sci. U.S.A.* **57**, 387.

Sandegren, E. (1947). "Proc. Eur. Brew. Conv. (Scheveningen 1947)", p. 28.

Sanderson, G. W. (1964). *Biochim. biophys. Acta* **92**, 622.

Sanderson, G. W. and Roberts, G. R. (1964). *Biochem. J.* **93**, 419.

Savage, D. J. and Thompson, C. C. (1972). *J. Inst. Brew.* **78**, 472.

Schneider, V. and Hallier, U. W. (1970). *Planta* **94**, 134.

Schuster, H., Schramm, G. and Zillig, W. (1956). *Z. Naturf.* **11B**, 339.

Schwimmer, S. (1958). *J. biol. Chem.* **232**, 715.

Seevers, P. M. and Daly, J. M. (1970). *Phytopathology* **60**, 1322.

Segal, B., Sahlean, V. and Hopulele, L. (1970). *Ind. Aliment., Bucharest* **22**, 326.

Shuttleworth, S. G., Russell, A. E. and Williams-Wynn, D. A. (1968). *J. Soc. Leather Trades' Chemists* **52**, 486.

Silbereisen, K. and Kraffczyk, F. (1967). *Monatschr. Brau.* **20**, 217.

Simon, E. W. and Beevers, H. (1951). *New Phytologist* **2**, 163.

Singleton, V. L. and Kratzer, F. H. (1969). *J. agric. Fd Chem.* **17**, 497.

Sissler, H. D. and Cox, C. E. (1960). *In* "Plant Pathology" (J. G. Horsfall and A. E. Dimond, eds), Vol. II, p. 507. Academic Press, New York and London.

Slack, C. R. (1966). *Phytochemistry* **5**, 397.

Smith, F. H. and Clawson, A. J. (1970). *J. Am. Oil Chem. Soc.* **47**, 443.

Smyth, R. D., Lambert, R. and Martin, G. J. (1964). *Proc. Soc. exp. Biol. Med.* **116**, 593.

Sondheimer, E. (1963). *In* "Plant Phenolics and their Industrial Significance", Proc. Symp. Plant Phenolics Group North America (Corvallis, Oregon 1962) (V. C. Runeckles, ed.), p. 15. Imperial Tobacco Company Canada, Montreal.

Spector, W. S. (1956). *In* "Handbook of Toxicology", Vol. I. Saunders, Philadelphia.

Stanley, P. E., Jennings, A. C. and Nicholas, D. J. D. (1968). *Phytochemistry* **7**, 1109.

Starr, J. E. and Judis, J. (1968). *J. pharm. Sci.* **57**, 768.

Steiner, K. (1968). *Schweiz. Brau.-Rundsch.* **79**, 219.

Steiner, K. and Stoker, H. R. (1968). "Eur. Brew. Conv. (Madrid 1967)", p. 407. Elsevier, Amsterdam.

Stenlid, G. and Saddik, K. (1962). *Physiologia Pl.* **15**, 369.

Stenlid, G. (1963). *Physiologia Pl.* **16**, 110.

Stenlid, G. (1968). *Physiologia Pl.* **21**, 882.

Stokes, D. M., Anderson, J. W. and Rowan, K. S. (1968). *Phytochemistry* **7**, 1509.

Swain, T. (1965). *In* "Plant Biochemistry" (J. Bonner and J. E. Varner, eds), p. 552. Academic Press, New York and London.

Synge, R. L. M. (1967). *A. Rep. Rowett Res. Inst.* **23**, 75.
Synge, R. L. M. (1968). *A. Rev. Pl. Physiol.* **19**, 113.
Synge, R. L. M. (1972). *Z. Phys. Chem.* **353**, 128.
Tager, J. M. (1958). *S. Afr. J. Sci.* **54**, 324.
Thompson, R. S., Jacques, D. and Haslam, E. (1972). *J. chem. Soc. Perkin Transact.* I, 1387.
Tillberg, J. E. (1970). *Physiologia Pl.* **23**, 647.
Toyama, N. and Kamiyama, S. (1972). Japan. 7216,793 (Cl.C. 07g), 17 May 1972, 3 pp. (Chem. Abstr. 1972, 60039 u).
Uritani, I. (1971). *A. Rev. Phytopath.* **9**, 211.
Urnes, P. and Doty, P. (1961). *Adv. Protein Chem.* **16**, 401.
Van Buren, J. P. and Robinson, W. B. (1969). *J. agric. Fd Chem.* **17**, 772.
Vancraenenbroeck, R. (1968). *Bull. Assoc. Anc. Etud. Brass. Univ. Louvain* **64**, 195.
Vancraenenbroeck, R., Callewaert, W., Gorissen, H. and Lontie, R. (1970). "Eur. Brew. Conv. (Interlaken 1969)", p. 29. Elsevier, Amsterdam.
Vancraenenbroeck, R. and Lontie, R. (1964a). "Eur. Brew. Conv. (Brussels 1963)", p. 513. Elsevier, Amsterdam.
Vancraenenbroeck, R. and Lontie, R. (1964b). *Bull. Assoc. Anc. Etud. Brass. Univ. Louvain* **60**, 153.
Vancraenenbroeck, R., Lontie, R. and Eyben, D. (1965). *Bull. Assoc. Anc. Etud. Brass. Univ. Louvain* **61**, 113.
Vancraenenbroeck, R., Neo-Badin, M. and De Clerck, J. (1972). *Bull. Assoc. Anc. Etud. Brass. Univ. Louvain* **68**, 1.
Van Sumere, C. F., Van Sumere-De Preter, C. and Ledingham, G. A. (1957a). *Can. J. Microbiol.* **3**, 761.
Van Sumere, C. F., Van Sumere-De Preter, C., Vining, L. C. and Ledingham, G. A. (1957b). *Can J. Microbiol.* **3**, 847.
Van Sumere, C. F., Hilderson, H. and Massart, L. (1958). *Naturwiss.* **45**, 292.
Van Sumere, C. F. (1969a). *Rev. Ferm. Ind. Alim.* **24**, 91.
Van Sumere, C. F. (1969b). *Rev. Ferm. Ind. Alim.* **24**, 131.
Van Sumere, C. F. and Dedonder, A. (1971). *Z. Pflanzenphysiol.* **65**, 159.
Van Sumere, C. F., Cottenie, J., De Greef, J. and Kint, J. (1972a) *In* "Recent Advances in Phytochemistry" (V. C. Runeckles and J. E. Watkin, eds), Vol. IV, p. 165. Appleton-Century-Crofts, New York.
Van Sumere, C. F., Dedonder, A. and Pé, I. (1972b). "Symp. Phytochem. Soc. North America (Syracuse, New York 1972)", Abstr., p. 17.
Van Sumere, C. F., De Pooter, H., Haider Ali and Degrauw-Van Bussel, M. (1973). *Phytochemistry* **12**, 407.
Varga, M. and Köves, E. (1962). *Naturwissenschaften* **49**, 355.
Venis, M. A. (1972). *Pl. Physiol.* **49**, 24.
Verhoef, N. J., Reinecke, C. J. and Bosch, L. (1967). *Biochim. biophys. Acta* **149**, 305.
Voronkov, L. A. (1970). *Sel'skokhoz. Biol.* **5**, 58 (Chem. Abstr. (1970) 73, 42033a).
Wainio, W. W. (1970). *In* "The Mammalian Mitochondrial Respiratory Chain", p. 342. Academic Press, New York and London.
Walker, J. R. L. and Hulme, A. C. (1965). *Phytochemistry* **4**, 677.
Webb, J. L. (1966). "Enzyme and Metabolic Inhibitors", Vol. III, p. 435. Academic Press, New York and London.
Webster, R. E., Engelhardt, D. L. and Zinder, N. D. (1966). *Proc. natn. Acad. Sci. U.S.A.* **55**, 155.
Weinbach, E. C., Sheffield, H. and Garbus, J. (1963). *Proc. natn. Acad. Sci. U.S.A.* **50**, 561.
Weinbach, E. C. and Garbus, J. (1964). *Science, N.Y.* **145**, 824.
Weinbach, E. C. and Garbus, J. (1965). *J. biol. Chem.* **240**, 1811.

Weinbach, E. C. and Garbus, J. (1966a). *J. biol. Chem.* **241**, 169.

Weinbach, E. C. and Garbus, J. (1966b). *J. biol. Chem.* **241**, 3708.

Weinges, K. and Freudenberg, K. (1965). *Chem. Commun.* 220.

Weinges, K., Gorissen, H. and Lontie, R. (1969a). *Ann. Physiol. vég.* **11**, 67.

Weinges, K., Bahr, W., Ebert, W., Goritz, K. and Marx, H. D. (1969b). *Fortschr. Chem. org. Naturstoffe* **27**, 158.

Wessel-Riemens, P. C. (1965). *Virology* **27**, 566.

Williams, A. H. (1963). *In* "Enzyme Chemistry of Phenolic Compounds" (J. B. Pridham, ed.), p. 87. Pergamon Press, Oxford.

Wood, R. K. S. (1967). "Physiological Plant Pathology", p. 457. Blackwell Scientific Publications, Oxford.

Woof, J. B. and Pierce, J. S. (1968). *J. Inst. Brew.* **74**, 544.

Wright, H. E., Jr. (1963). *In* "Plant Phenolics and their Industrial Significance", Proc. Symp. Plant Phenolics Group North America (Corvallis, Oregon 1962) (V. C. Runeckles, ed.), p. 39. Imperial Tobacco Company Canada, Montreal.

Wright, H. E., Jr., Burton, W. W. and Berry, R. C., Jr. (1964). *Phytochemistry* **3**, 525.

Wucherpfennig, K. and Franke, I. (1963). *Z. Lebensm. Untersuch. Forsch.* **124**, 22.

Wye, E. and McFarlane, W. D. (1957). "Proc. Eur. Brew. Conv. (Copenhagen 1957)", p. 299. Elsevier, Amsterdam.

Yang, J. T. (1958). *J. Am. Chem. Soc.* **80**, 1783.

Yasunobu, K. T. (1959). *In* "Pigment Cell Biology" (M. Gordon, ed.), p. 583. Academic Press, New York and London.

Zucker, M., Hanson, K. R. and Sondheimer, E. (1967). *In* "Phenolic Compounds Metabolic Regulation" (B. J. Finkle and V. C. Runeckles, eds), p. 68. Appleton-Century-Crofts, New York.

CHAPTER 9

Protein Sweeteners

G. E. INGLETT

Northern Regional Research Laboratory, Agricultural Research Service, U.S. Department of Agriculture, Peoria, Illinois, U.S.A.

I. Introduction

Applications of synthetic sweeteners in dietetic foods have increased rapidly during the past two decades. Wider uses could be anticipated if improved taste and wholesomeness could be found in a new sweetener. In the United States, combinations of saccharin and cyclamates were helpful in creating numerous dietetic foods between 1950 and 1969. The largest market was found in low-calorie soft drinks, such as the diet colas. This blend was also used in canned foods, gum, candies, and table products which were intended primarily for sweetening coffee and tea. The era of cyclamate–saccharin blends in the U.S. was abruptly terminated when the Food and Drug Administration banned cyclamates in 1969. Saccharin remains as the only nonnutritive sweetener allowed in the U.S. food supply. It has a bitter aftertaste for many people, and its healthfulness has been questioned sporadically since its commercial introduction in 1900.

The need for a safe, nonnutritive sweetener for the diabetic and the diet-conscious has spurred research in many diverse areas, such as the physiological basis of sweetness and the wide variety of chemicals that invoke a sweet taste. Sweet taste is a gustatory response invoked by substances on sweet taste buds which transmit a message to the brain indicating sweetness. The chemical nature of various substances that excite a sweet taste has been extensively studied with no universally accepted explanation. More basic information is needed concerning the nature of sweetness. Although much has been said and written about sweetness, basic experimental data are still needed.

II. A Macromolecular Taste Modifier

An important approach to sweet taste perception is the study of the strange properties of the miracle fruit. Although miracle fruit (*Synsepalum dulcificum*, Sapindaceae) has been known in the literature since 1852 (Daniell, 1852) to cause sour foods to taste sweet, scientific investigations of the fruit were not made until Inglett and his associates (1964, 1965a) found some experimental evidence that the active principle was macromolecular and that it might be a glycoprotein. Subsequently, Brouwer *et al.* (1968) and Kurihara and Beidler (1968) confirmed that the active principle was a glycoprotein with a molecular weight of around 42 000. This taste-modifying substance revealed a new concept in taste perception of sweet taste. Until this time, only small molecules were considered sweet-invoking substances, so this was the first time that macromolecules were considered capable of participating in either taste perception or modification of taste (Inglett *et al.*, 1964, 1965a).

Preliminary studies on miracle fruit by Inglett *et al.* (1964, 1965a) were intended to isolate, characterize, and synthesize the taste-modifying substance. The fruit was processed to give a stable concentrate. Since the active principle appeared polymeric, the material was incompatible with the sponsoring organization's mission, so research was discontinued.

With the increasing pressure for an excellent nonnutritive sweetener in recent years, a new corporation, Mirlin, has launched a more than $5 million venture on dietetic foods using miracle fruit concentrate as its source of latent sweetness. The miracle fruit concentrate developed by Mirlin Corporation has the flexibility of being used in either of two ways: taken separately before eating unsweetened foods or added directly to foods. Foods are designed to be used with the miracle fruit concentrate to produce a sweetness and flavor comparable to sugar-containing products and with from 6 to 27% of the usual calories of the same foods formulated with sugar. The horticultural technology has also been developed to grow large numbers of trees for the production of sufficient quantities of the miracle fruit concentrate (Dastoli and Harvey, 1973).

III. Micro- and Macromolecular Sweeteners

Although there are a large number of substances that exhibit some resemblance to sucrose-type sweetness (Inglett, 1971), only a few can be considered suitable for human consumption. Almost all building blocks of foodstuffs—carbohydrates, proteins, or lipids—are among the substances generally recognized as safe. Sweeteners derived from these materials may be less hazardous in foods than totally foreign organic structures.

Nature appeared a logical place to find an intense sweetener that may be considered safe for food uses, particularly if the natural materials were consumed as food by some people. The naturally occurring sweeteners, stevioside (1) and glycyrrhizin (2), are candidates and are consumed as components of

(1) Stevioside

plant materials by a limited number of people. Stevioside is the intensely sweet $(1' \rightarrow 2)$-linked disaccharide-containing substance (1) found in the leaves of a small shrub, *Stevia rebaudiana* Bertoni (Compositae), that grows wild in Paraguay. Primarily the Indians use the leaves to sweeten their tea and other foods. The pure compound can be obtained in 6% yield from the dried leaves and is 300 times sweeter than sucrose.

The other edible intense sweetener of natural origin is glycyrrhizin (2) found in the licorice root (*Glycyrrhiza glabra*, L., Leguminosae). Licorice extracts have been used for flavoring candy, tobacco, and pharmaceuticals for many years. The sweetener is present in the root as the calcium and potassium salt of glycyrrhizic acid and isolated commercially as the ammonium salt. Glycyrrhizin is a saponin of glycyrrhetinic acid with an attached

(2) Glycyrrhizic acid (Glycyrrhizin)

$(1' \rightarrow 2)$-linked disaccharide. Another class of sweeteners, the dihydrochalcones (e.g. 3 and 4) were discovered by Horowitz and Gentili (1963) who hydrogenated the bitter flavonoids, naringin and neohesperidin. These substances also contain the $(1' \rightarrow 2)$-linked disaccharide as a portion of their molecules.

(3) Naringin dihydrochalcone (R = OH, R' = H)
(4) Neohesperidin dihydrochalcone (R = OMe, R' = OH)

During the course of studying various sweetener candidates in 1965, I discovered the intense sweetness of some pink berries from West Africa. The fruit was called serendipity berry, and its botanical name, *Dioscoreophyllum cumminsii* Diels (Menispermaceae), was established many months later.

IV. THE SERENDIPITY BERRY SWEETENER

A. THE PLANT SOURCE

Serendipity berries (Fig. 1) are indigenous to tropical West Africa. The *Dioscoreophyllum cumminsii* Diels plant grows from Guinea to the Cameroons

FIG. 1. Serendipity berries (*Dioscoreophyllum cumminsii* Diels).

and is also found in Gabon, the Congo, the Sudan, and Southern Rhodesia. The Yoruba names for this fruit are: *Ito-Igbin* and *Auyn-Ita*. It grows in the rain forest during the rainy season from approximately July to October. The serendipity berries are borne by hairy climbing vines sometimes 15 ft long and $\frac{1}{8}$–$\frac{3}{16}$ in in diameter. These vines are necessarily supported by other vines; its leaves, which are heart-shaped with ragged edges and measure about 3 in from the tip to the stem, are attached at 6 in intervals on the vine. The berries are red in color, approximately $\frac{1}{2}$ in long, and grow in grapelike clusters with approximately 50–100 berries in each bunch. The tough outer skin of the berry encloses a white, semi-solid, mucilaginous material surrounding a friable thorny seed. In spite of its intense sweetness, the fruit is not commonly cultivated or used by the natives of Nigeria. The tubers of the plant are reported to be eaten in some parts of Africa and used medicinally. The fruit has remarkable stability properties, keeping for several weeks at room temperature (Inglett and May, 1968).

B. MACROMOLECULAR PROPERTIES

Separation of the serendipity berry sweetener on Sephadex G-50 (Fig. 2) indicated the presence of a macromolecule (Inglett and Findlay, 1967;

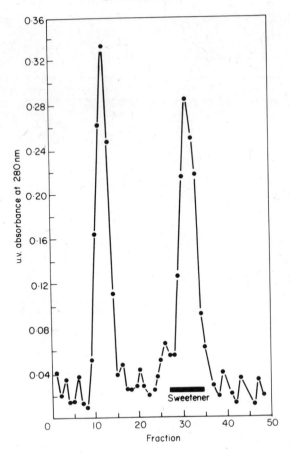

FIG. 2. Gel filtration of serendipity berry sweetener on Sephadex-50.

Inglett and May, 1969). Sedimentation rates of the sweetener gave an approximate molecular weight around 10 000 (Jansen, 1965). Studies on pectinase-treated berry extracts indicated that the sweetener was a protein or in a protein-containing fraction based on the high nitrogen values (Inglett *et al.*, 1965b).

The primary objective of the natural sweetener program was to obtain a low molecular weight sweetener. It was reasoned that if the *Dioscoreophyllum cumminsii* sweetener were a protein or at least partly protein, proteolytic enzymes should degrade it to a lower molecular weight material. On treatment of the serendipity berry sweetener with trypsin or papain, two different sweeteners were eluted from a Sephadex 50 column, the second one being of lower molecular weight. However, the higher molecular weight sweetener was present in much greater quantity, indicating that fractionation or cleavage of the original sweetener was not extensive. Bromelain, however, degraded the molecule in such a manner that only a lower molecular weight fraction was sweet tasting (Fig. 3). The large absorbance in ultraviolet light at 278 nm by this smaller molecule indicated the presence of a polypeptide (Inglett and

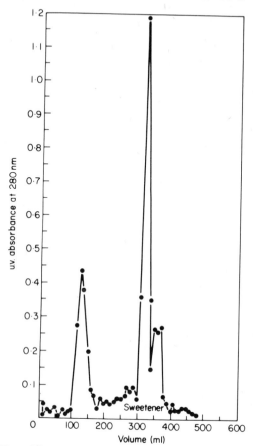

FIG. 3. Bromelain-treated serendipity berry sweetener on Sephadex-50.

Findlay, 1966). Quantitative data on proteolytic enzyme treated serendipity berry sweetener, separated by Sephadex G-100, are shown in Table I.

Disc gel electrophoresis studies on various sweetener fractions are shown in Table II (Inglett and Findlay, 1967; Inglett and May, 1969).

All sweetener fractions showed protein bands except fraction 51, which may have been a polypeptide. Parker (1973) reported that proteolytic enzyme treatment of *Dioscoreophyllum cumminsii* sweetener gave a polypeptide sweetener with a molecular weight around 6000. Inglett earlier reported his fraction to be 1500 times sweeter than sugar, and Parker gave a sweetness range between 800 and 3200.

Researches at Monell Senses Center and Unilever Research Laboratory,

TABLE I

Cleavage of the serendipity berry sweetener by proteolytic enzymes and separation of a polypeptide sweetener by Sephadex G-100 chromatography[a]

Sweetener mixture		Fractions containing sweetness	Dry weight of eluted sweetener (mg)
Enzyme added to column treatments (mg)			
Trypsin	130	11–21	39
		29–33	1
Papain	133	12–18	91
		31–33	3
Bromelain	92	31–32	6

[a] Source: Inglett and Findlay (1966).

TABLE II

Protein bands obtained by disc gel electrophoresis of sweetener fractions

Fractionated sweetener description	Number and nature of bands
From G-50 (726 µg)	3 strong, 1 weak
Fraction 51 from G-200 (215 µg)	0
From G-50 after incubated 24 h with 13·7 µg papain (200 µg)	2 weak
From G-50 after incubated 24 h with 8 µg bromelain	2 weak
From G-50 after incubated 24 h with 8 µg trypsin	1–2 diffuse, weak

working independently, confirmed the protein nature of the serendipity sweetener (Morris and Cagan, 1972; Van der Wel, 1972a). Morris and Cagan called their sweetener "Monellin" in honor of their organization. Since both groups published simultaneously on the serendipity berry principle, it seems appropriate to me to call the sweetener "serendip."

The protein properties of serendip were proved by Van der Wel (1973) by the following criteria: (1) the presence of almost 100% polypeptide material as determined by the biuret method; (2) the yield of 100% amino acid on acid hydrolysis; (3) the characteristic u.v. spectra; (4) their binding with Amido Black 10 B; and (5) the disappearance of the sweet taste after incubation with trypsin.

C. PHYSICAL PROPERTIES

The isoelectric point of serendip was 9·03 as determined by preparative isoelectric focusing (Table III). The molecular weight of serendip was deter mined by gel filtration on Sephadex 50 to be 11 500 (Van der Wel, 1973). Sodium dodecyl sulfate polyacrylamide gel electrophoresis, after cleavage of the disulfide bridges, also showed that it had a molecular weight of 11 500;

TABLE III

Some physical properties of serendip

Criteria	Serendip[a]	Serendip[b]
Isoelectric point	9·03	9·26
Molecular weight	11 500	10 500
$A_{1cm}^{1\%}$ (pH 5·6, 278 nm)	16·2	—
Sweetness intensity (times sweeter than sucrose):		
on a molar basis	$8·4 \times 10^4$	—
on a weight basis	2500	—
Temperature (°C) above which sweetness disappears at:		
pH 3·2	50	—
5·0	65	—
7·2	55	—

[a] Van der Wel (1973).
[b] Morris et al. (1973).

therefore, it must be a single polypeptide chain. Since serendip has no circular dichroism bands at 220 and 208 nm, α-helix configuration is not present in the molecule.

Sweetness is lost on heating the protein at 50 °C (pH 3·2), 65 °C (pH 5·0), and 55 °C (pH 7·2). This unstable nature of the sweetener makes it unsuitable for many commercial applications.

D. AMINO ACID COMPOSITION

Amino acid composition of serendip was determined by Van der Wel and Loeve (1973) and Morris et al. (1973), and their results are compared in Table IV. The most outstanding observation is the complete absence of histidine from serendip. The absence of mono- or dimethyl derivatives of lysine or arginine should also be noted. Serendip does not have any apparent free sulfhydryl groups, but it apparently has two half-cystine residues per mole (Van der Wel, 1973). Morris et al. (1973) have indicated only one half-cystine residue per mole, a condition which may suggest a free sulfhydryl group. The larger number of free basic groups with respect to the number of free carboxylic acid groups in the native molecule indicates the basic character of this protein.

TABLE IV

Amino acid composition of serendip

Amino acid	Serendip[a] (amino acid residues/mole)	Serendip[b] (amino acid residues/mole)
Aspartic acid	11	10
Glutamic acid	13	12
Serine	2	2
Threonine	4	4
Proline	6	6
Glycine	9	8
Alanine	3	3
Valine	4	4
Leucine	6	6
Isoleucine	7	6
Methionine	1	1
Phenylalanine	6	5
Tyrosine	5	7
Half-cystine	2	1
Histidine	0	0
Lysine	10	8
Arginine	8	7
Tryptophan	1	1
Ammonia	12	6
Total	98	91

[a] Source: Van der Wel and Loeve (1973).
[b] Source: Morris et al. (1973).

V. PROTEIN SWEETENERS FROM KATEMFE

Besides studies on miracle fruit and the serendipity berry, a large variety of plant materials were examined systematically by Inglett and May (1968) for intensity and quality of sweetness. Another African fruit containing an intense sweetener was katemfe, or the miraculous fruit of the Sudan. Botanically the plant is *Thaumatococcus daniellii* of the family Marantaceae. Inside the fruit three large black seeds are surrounded by a transparent jelly and a light yellow aril at the base of each seed (Fig. 4). The mucilaginous material around the seeds is intensely sweet and causes other foods to taste sweet. The seeds were observed to be present in trading canoes in West Africa as early as 1839, and were reported to be used by the native tribes to sweeten bread, fruits, palm wine, and tea. Preliminary studies have indicated a substance similar to the serendipity berry sweetener (Inglett and May, 1968). Katemfe yields two sweet-tasting proteins (Van der Wel, 1972b; Van der Wel and Loeve, 1972) which they called thaumatin I and II.

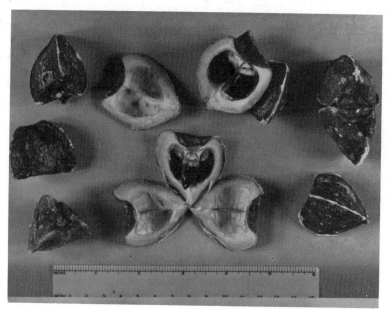

Fig. 4. Katemfe (*Thaumatococcus daniellii*).

Water-soluble components of katemfe were purified from low molecular weight material and concentrated by ultrafiltration (Van der Wel, 1973). The concentrates were submitted to gel filtration. The elution profile of the *Thaumatococcus* concentrate (Fig. 5) shows three fractions having a sweet taste (A–C). For a further purification, the sweet-tasting fractions were submitted to ion-exchange chromatography on SE-Sephadex C-25 with a linear sodium chloride gradient. In this way, two pure preparations, thaumatin I

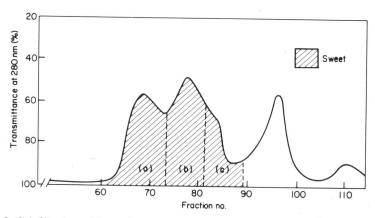

Fig. 5. Gel filtration of katemfe extract on Sephadex-50 with indicated sweet tasting fractions (A, B, and C).

276 G. E. INGLETT

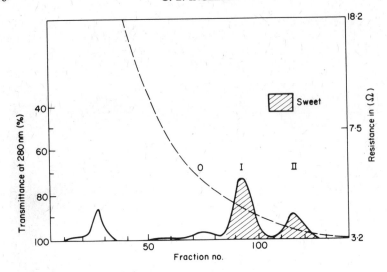

FIG. 6. Thaumatin I and II separated by ion-exchange chromatography of fraction A from
Thaumatococcus daniellii on SE-Sephadex C-25 with a linear sodium chloride gradient.
Fraction 0 was not sweet.

and thaumatin II, were obtained from the sweet-tasting gel filtrates from
Thaumatococcus (fractions A–C, see Fig. 6). The purity of the fractions was
checked by polyacrylamide gel electrophoresis, using Amido Black 10 B as
staining agent, and their protein content was assessed by the biuret method.
Physical constants for the sweet proteins of katemfe are given in Table V.
Like serendip, these protein sweeteners are heat sensitive and undergo
irreversible heat denaturation (Table III and V). Because denaturation

TABLE V

Physical properties of the sweet-tasting proteins from Katemfe[a]

Criteria	Thaumatin I	Thaumatin II
Isoelectric point	12	12
Molecular weight	21 000 ± 600	20 400 ± 600
$A_{1cm}^{1\%}$ (pH 5·6, 278 nm)	7·69	7·53
Sweetness intensity (times sweeter than sucrose):		
on a molar basis	1×10^5	1×10^5
on a weight basis	1600	1600
Temperature (°C) above which sweetness disappears at:		
pH 3·2	55	55
5·0	75	75
7·2	65	65

[a] Source: Van der Wel (1973).

coincides with the loss of sweetness, the groups underlying the conformational change must also be responsible for generating their sweet taste. At least part of the intact tertiary protein structure must be required for the sweet taste.

VI. THE PROBE THEORY OF SWEETNESS

The macromolecules responsible for sweetness either by modification of taste (miracle fruit) or direct sweetness (serendipity berry, katemfe) must have a portion of their structure that is essential for the sweet taste. I propose to call this center a "sweetness probe." The challenge of finding sweetness probes and of demonstrating their structures and utility is still before us.

The sweetness probe of the miracle fruit's active principle may lie in its bonding between the carbohydrate moiety and the protein chain. This chemical bonding could be considered analogous to aglycone–glycoside binding of such intense sweeteners as stevioside (1), osladin (5) (Jizba *et al.*, 1971), and glycyrrhizin (2). The induced sweetness of the glycoprotein in miracle fruit is attributed to the sugar portion of the molecule. The protein portion holds onto the receptor site placing the arabinose or xylose sugars close enough for inducing sweet taste. Although this assumption may seem to be reasonable, the necessity of protons for sweet taste needs additional investigation. Perhaps protons modify conformation of the glycoprotein to give the necessary shape to induce sweet taste. If the sweetness probe can be chemically defined, analogs of low molecular weight may be synthesized that would not have the residual taste deficiency of the macromolecule.

The serendipity berry sweetener may have a different type of sweetness probe. Since the *Dioscoreophyllum cumminsii* sweetener acts directly on the taste buds as a probe, a peptide linkage analogous to the aspartic acid sweeteners (Mazur *et al.*, 1969) may be an essential element. At this stage of

(5) Osladin

278   G. E. INGLETT

TABLE VI

Structural features of intensely sweet glycosides

Glycoside	Disaccharide	Hydrophobic part	Sweetness probe
Stevioside	2-*O*-β-D-Glucosyl-β-D-glucoside	Steviol (diterpene)	β-D-Glucose
Glycyrrhizin	2-*O*-β-D-Glucuronosyl-β-D-glucuronoside	Glycyrrhetinic acid (triterpene)	COOH
Osladin	2-*O*-α-L-Rhamnosyl-β-D-glucoside	Polypodosaponin	α-L-Rhamnose
Neohesperidin dihydrochalcone	2-*O*-α-L-Rhamnosyl-β-D-glucoside	Substituted phloroglucinol, —C_2H_4—	OH / OCH₃ ring
Naringin dihydrochalcone	2-*O*-α-L-Rhamnosyl-β-D-glucoside	Substituted phloroglucinol, —C_2H_4—	OH ring

development, an aspartic acid peptide probe center cannot be excluded. More information is needed on the protein structures of miracle fruit, serendipity berry, and katemfe sweeteners.

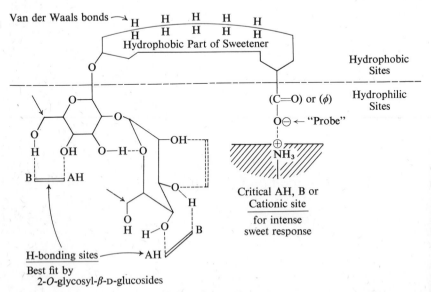

FIG. 7. Schematic diagram of the proposed types of bonding of sweet glycosides to the taste bud receptor sites.

VII. Glycosidic Sweeteners

As an extension of the sweetness probe theory, the chemical structures of the intense sweeteners—stevioside, glycyrrhizin, osladin, and the dihydrochalcones—are examined (Table VI). The obvious similarity among these intense sweeteners is the occurrence of a (1' → 2)-oxygen-linked disaccharide in the glycoside attached to an aglycone (Inglett and Hodge, 1973). Proposed binding sites of these types of sweeteners are shown in Fig. 7, which shows how the molecule could position itself between hydrophobic and hydrophilic binding sites of the taste bud. The (1' → 2)-linked carbohydrate moiety appears to be important only for positioning. An examination of nine (1' → 2)-linked disaccharides at the Northern Regional Research Laboratory revealed only low levels of sweetness (Dick et al., 1973).

VII. Summary

Sweetness is a gustatory response invoked by substances on a sweet taste bud which transmits a message to the brain indicating sweet taste. An unusual source of sweet taste is present in a West African berry known as miracle fruit (Synsepalum dulcificum). This fruit possesses a taste-modifying substance that causes sour foods—e.g. lemons, limes, or grapefruit—to taste sweet. Inglett and his associates found the active principle to be macromolecular, assumed to be glycoprotein and later confirmed by others. Until this time, only small molecules were considered sweet-invoking substances. Now macromolecules are considered capable of participating in taste perception.

The intense sweetness of the fruit of Dioscoreophyllum cumminsii, called the serendipity berry, was also revealed to be a macromolecule. In 1972, this sweetener was found to be a protein. Also in 1972, the intensely sweet principle of Thaumatococcus daniellii was reported to contain two proteins having intense sweetness.

Since intensely sweet protein sweeteners act directly on taste buds as a probe, a peptide linkage analogous to the aspartic acid sweeteners may be partly responsible for their sweetness. As an extension of the probe theory, chemical structures of the intense glycosidic sweeteners—stevioside, glycyrrhizin, osladin, and the dihydrochalcones—were examined. The obvious similarity among these intense sweeteners is the occurrence of a (1' → 2)-oxygen-linked disaccharide attached to an aglycone.

REFERENCES

Brouwer, J. N., Van der Wel, H., Francke, A. and Henning, G. J. (1968). Nature, Lond. 220, 373.
Daniell, W. F. (1852). Pharm. J. 11, 445.

Dastoli, F. R. and Harvey, R. J. (1973). Personal communication. Mirlin Corp., Hudson, Mass.

Dick, W. E., Jr., Hodge, J. E. and Inglett, G. E. (1973). Abstr. Pap. 117A, *Div. Agr. Food Chem., 166th Amer. Chem. Soc. Meeting*, Chicago, Ill.

Horowitz, R. M. and Gentili, B. (1963). U.S. Patent 3,087,821, April 30.

Inglett, G. E. (1971). "Recent Sweetener Research," 2nd ed. Botanicals, P.O. Box 3034, Peoria, Ill. 61614.

Inglett, G. E., Dowling, B., Albrecht, J. J. and Hoglan, F. A. (1964). Abstr. Pap. 1A, *Div. Agr. Food Chem., 148th Amer. Chem. Soc. Meeting*, Chicago, Ill.

Inglett, G. E., Dowling, B., Albrecht, J. J. and Hoglan, F. A. (1965A). *J. agric. Fd Chem.* **13**, 284.

Inglett, G. E., Dowling, B. and Bisgard, N. (1965B). June Report. International Minerals and Chemical Corp., Skokie, Ill. (unpublished results).

Inglett, G. E. and Findlay, J. (1966). March Report, International Minerals and Chemical Corp., Skokie, Ill. (unpublished results).

Inglett, G. E. and Findlay, J. C. (1967). Abstr. Pap. 75A, *Div. Agr. Food Chem., 154th Amer. Chem. Soc. Meeting*, Chicago, Ill.

Inglett, G. E. and May, J. F. (1968). *Econ. Bot.* **22**, 326.

Inglett, G. E. and May, J. F. (1969). *J. Fd Res.* **34**, 408.

Inglett, G. E. and Hodge, J. E. (1973). Abstr. Pap. 35A, *Div. Agr. Food Chem., 165th Amer. Chem. Soc. Meeting*, Dallas, Texas.

Jansen, Ron. (1965). Unpublished results. California Institute of Technology, Pasadena, California.

Jizba, J., Dolejs, L., Herout, V. and Sorm, F. (1971). *Tetrahedron Letters* No. 18, 1329–1332.

Kurihara, K. and Beidler, L. M. (1968). *Science, N.Y.* **161**, 1241.

Mazur, R. H., Schlatter, J. M. and Goldkamp, A. H. (1969). *J. Am. chem. Soc.* **91**, 2684.

Morris, J. A. and Cagan, R. H. (1972). *Biochim. biophys. Acta* **261**, 114.

Morris, J. A., Martenson, R., Deibler, G. and Cagan, R. H. (1973). *J. biol. Chem.* **248**, 534.

Parker, K. J. (1973). Ger. Offen. 2,224,644, Feb 1; CA **78**, 109547a.

Van der Wel, H. (1972A). *FEBS Letters* **21**, 88.

Van der Wel, H. (1972B). *In* "Olfaction and Taste IV" (D. Schneider, ed.). Wissenschaftliche Verlaggesellschaft MBH Stuttgart.

Van der Wel, H. (1973). Abstr. Pap. 33A, *Div. Agr. Food Chem., 165th Amer. Chem. Soc. Meeting*, Dallas, Texas.

Van der Wel, H. and Loeve, K. (1972). *Eur. J. Biochem.* **31**, 221.

Van der Wel, H. and Loeve, K. (1973). *FEBS Letters* **29**, 181.

CHAPTER 10

Proteins and Taxonomy

J. G. VAUGHAN

*Biology Department, Queen Elizabeth College, Campden Hill Road,
London, England*

I. TAXONOMY

Taxonomy is probably the oldest of the biological disciplines. From earliest times it has been a human activity to give names to plants and animals. The description and naming of newly discovered organisms is an important aspect of taxonomy and there is still much to be done in this field, particularly in areas, such as tropical Africa, where there has been comparatively little organized taxonomic activity. The correct identification of plants and animals is important not only to the professional taxonomist but also to workers in other biological disciplines. It would be difficult for the ecologist to carry out his task without efficient keys for the identification of those organisms found in the particular habitats under investigation. The phytochemist requires authenticated source material if his chemical findings are to be of any significance. As regards applied work, the agricultural scientist dealing with weeds and cultivars needs to be skilled in the procedures of identification. Similarly in the food industry, where much of the raw material is of plant or animal origin, quality control relies heavily on correct identification. Finally, one might mention forensic science, which deals with identification problems centred around the human species.

In addition to the identification aspect of taxonomy, classification is also important. Systems of classification and the criteria used to establish these systems have passed through a number of phases (Alston and Turner, 1963; Davis and Heywood, 1963). One often associates the first system of biological classification with the eighteenth century scientist Linnaeus. His system would now be described as "artificial" in that a very small number of characters were utilized; for example characters of the stamens and carpels were the main ones used in angiosperm classification. In the nineteenth century, more and more characters were used, this practice producing the so-called "natural" classification. Also in the nineteenth century Darwin's work on evolution profoundly affected systems of classification in that, in the years that followed, taxonomists often produced systems claimed to show evolutionary trends. These are the so-called "phylogenetic" systems. One must be careful about the validity of the phylogenetic system in that a number of conditions need to be satisfied before any evolutionary trend is positively established. Firstly, fossil evidence should be available; secondly, one may be able to witness evolution under natural conditions, and thirdly it might be possible to recreate the process. As far as many higher plants are concerned, it is difficult to satisfy these conditions.

The criteria used to establish the various systems of classification have developed in conjunction with technological advances. In the first place exomorphic characters, as seen by the naked eye or simple lenses, were used. With the development of the light microscope, internal structures became additional characters. The great systems of higher plant classification are still based essentially on structural features. However, the taxonomist finds within these systems problems which cannot be solved by the structural approach. The difficulty lies in establishing an objective empirical basis for an understanding of relationship in view of the great variation in external form, often affected by the environment, and internal structure. In recent years (Heywood, 1971; Benke and Turner, 1971) use has been made of the scanning and transmission electron microscopes in taxonomic studies. The great resolving powers of these instruments have produced further structural information but this does not obviate the basic difficulty of morphological studies in taxonomy.

In the 1920s–30s there came into prominence the discipline of cytotaxonomy. Cytotaxonomy deals with the number and behaviour of chromosomes in relation to systematic studies and it was felt that cytotaxonomy might prove a panacea for the problems experienced in structural taxonomy. Chromosomes are intimately associated with the genome and their number would provide a good empirical character. This type of study certainly advanced taxonomy in that it emphasized the importance of hybridization in the production of new species but some difficulties still remain. For example, obviously different species can have the same number of chromosomes and the number of chromosomes may vary within the same species.

Recent years have seen the development of new approaches to the problems of taxonomy (Alston and Turner, 1963; Heywood, 1973). In particular, there has been great interest in the relationships of plant chemicals to systematic problems. Consequently, there has been the appearance of the discipline often described as chemotaxonomy. The chemotaxonomist regards chemical compounds as good empirical criteria, modern methods often providing the means of exact identification. In fact one reason for the emergence of this approach in recent years has been the development of experimental techniques, such as chromatography and electrophoresis, capable of rapidly processing very small quantities of chemical constituents.

The range of substances available for study is enormous but the situation seems to have polarized so that distinction into two groups of compounds is recognized, namely "micromolecular" and "macromolecular". The micromolecular substances are very varied, for example non-protein amino acids, flavonoids, essential oils. With modern instrumentation, it is possible to identify these substances exactly and hence homologous studies are made relatively easy. However, one drawback is that some classes of micromolecular substances are not universally distributed throughout plants.

The macromolecular compounds mainly referred to in this context are the nucleic acids and the proteins. Comparative protein studies have always appealed to the taxonomist because, in the first place, they are universally distributed. Secondly, as Gibbs (1963) wrote, "it seems probable that each kind of living organism has its own set of proteins; that the proteins of nearly related species are nearly alike; that those of more distantly related ones are unlike". However, to establish homology between proteins is not as easy as with micromolecular compounds.   unds.

II. SEROLOGY

A. THE PRECIPITIN REACTION

The earliest protein studies in taxonomy involved serological or immunological techniques. The antigen/antibody reaction is highly specific, hence the interest to the taxonomist. Early work in this field was carried out by Mez and his school at Konigsberg between the two World Wars (an excellent summary of this period is given by Chester, 1937).

Mez used the precipitin reaction in his taxonomic investigations (Fig. 1). The antiserum, prepared to a particular species, is allowed to react with protein extracts from the other species under consideration. Taxonomic relationship is based on the degree of cross-reaction in that it is expected that the homologous protein extract gives the greatest amount of precipitate. Smaller amounts of precipitate are given by antigens from nearly related species and no precipitate with distantly related taxa. If one accepts this thesis, then clearly the success of the investigation depends on the accuracy of precipitate estimation. Various methods were used such as dilution series of antigen extracts, precipitate weight and the time taken for the precipitate to appear.

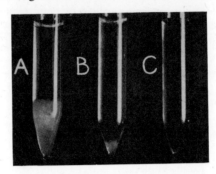

FIG. 1. Total precipitin reaction. In each tube is an antiserum to *Brassica* seed proteins. The homologous protein extract has been added in A, *Papaver* seed proteins in B and *Lactuca* seed proteins in C. The degree of cross reaction correlates with classical taxonomy (Vaughan, 1968a).

As a source of plant antigens, Mez normally used seeds. This is the procedure normally adopted today and has as its basis the necessity, in taxonomic work, of comparing organs of the same physiological stage.

Mez investigated many higher plant groups and represented his findings in the famous "Stammbaum" (Mez and Ziegenspeck, 1926). As far as one can make out, his results supported the morphological classifications of the time but he produced no innovations. Also, he was unfortunately attacked by the orthodox plant taxonomists of the period and his serological techniques were not universally accepted. Nevertheless Mez must be credited with a new approach to plant taxonomy which produced some remarkable developments.

The general technique of Mez was taken up by Boyden (Boyden and De-Falco, 1943), who used it in studies of animal systematics at Rutgers University (U.S.A.). Boyden did, however, introduce two improvements. Estimation of the amount of precipitate in the precipitin reaction had tended in the Mez period to be somewhat subjective. Boyden used the Libby photon reflectometer, a type of densitometer, to measure the precipitin reaction. He also showed that an antiserum with increasing dilutions of the homologous antigen extract gave a precipitate which rose from zero to a maximum and then fell to zero. Heterologous extracts gave the same type of curve but the peak was often at a different point to that of the homologous extract. Consequently Boyden constructed a kind of taxonomic index where the area under the homologous curve was regarded as 100% and the relative areas under the heterologous curves were taken as degrees of taxonomic similarity.

In the 1950s Boyden's methods were applied to problems of angiosperm taxonomy by M. A. Johnson (1954), also at Rutgers University. Johnson and his students investigated problems at the generic and specific levels in the Magnoliaceae, Cucurbitaceae, Ranunculaceae, Solanaceae and the Gramineae. All the papers reported some correspondence between the serological results and evidence from other taxonomic disciplines.

B. GEL DIFFUSION METHODS

A paper published in 1960 by Gell, Hawkes and Wright on the taxonomy of certain *Solanum* species used new serological methods as far as angiosperm taxonomy was concerned and the publication of this paper opened up a new era in phytoserology. Gell *et al.* (1960) pointed out that the total precipitin reaction did not distinguish between the different antigen/antibody systems and it was possible that a high percentage of a single protein would markedly affect the precipitin reaction. Consequently they used gel diffusion methods which separated the antigen/antibody systems and allowed the detection of identical and non-identical proteins, at least on immunochemical grounds.

Two gel diffusion methods were used by Gell *et al.* (1960): the double diffusion method of Ouchterlony and Elek (Crowle, 1961) and the immuno-electrophoretic method of Grabar and Williams (Grabar and Burtin, 1964). In the double diffusion method, antigens of a protein extract separate in a thin layer of agar or agarose gel and where they meet antibodies from the antiserum form a number of separate precipitin lines. If homologous and heterologous extracts are compared in this way, lateral fusion of precipitin

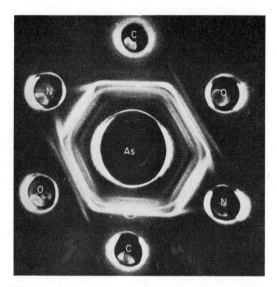

FIG. 2. Double diffusion analysis. The antiserum (As) has been made to *Brassica oleracea* seed protein and is compared with seed proteins of *B. oleracea* (O), *B. campestris* (C) and *B. nigra* (N) (Vaughan, 1968b).

lines indicates homology of proteins, at least on immunochemical grounds. Isolated lines, however, may appear—an indication of specific proteins—and "spurs" show partial identity of proteins (Fig. 2). This method is most valuable for the analysis of a small number of antigen/antibody systems but seed extracts, normally used in angiosperm taxonomy, produce a large number of systems which make taxonomic comparison difficult.

The immunoelectrophoretic method (Fig. 3) helps to obviate this difficulty in that, prior to immunodiffusion, the proteins are first separated by electrophoresis. Compared to double diffusion there is better separation of the precipitin lines but, because of the large number of systems, there is still difficulty with the taxonomic comparison. However, the technique of "adsorption" or "absorption" helps to improve the situation considerably. In comparing protein extracts A and B, if one adds extract B to antiserum A then the antibodies common to both extracts are precipitated. If one then removes the precipitate, the remaining antiserum should contain antibodies specific to A. This antiserum can then be analysed by double diffusion and immunoelectrophoretic methods. For a full taxonomic investigation the reciprocal experiment should be carried out with antiserum B and extract A.

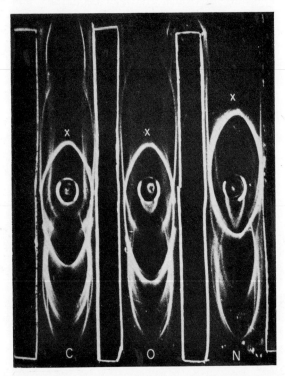

FIG. 3. Immunoelectrophoretic analysis. *B. oleracea* antiserum is compared with *B. campestris*, *B. oleracea* and *B. nigra* seed extracts (Vaughan, 1968b).

C. SEROLOGY AND CULTIVATED BRASSICA SPECIES

The gel diffusion methods have now been used by a considerable number of plant taxonomists (Fairbrothers 1969). Without a doubt these methods have mainly been used in studies of angiosperm systematics; relatively little work has been carried out with other plant groups such as the gymnosperms and algae. In the present paper, the application of these techniques is illustrated by work on cultivated *Brassica* species.

The genus *Brassica* contains a number of taxa of world-wide economic importance. Various species (*B. campestris; B. napus; B. juncea; B. carinata*) are grown as oil seed crops. Some species (*B. campestris; B. napus; B. juncea; B. pekinensis; B. chinensis; B. perviridis*) are cultivated as vegetables, either for foliage or roots. A minor, but still significant, aspect of utilization is the inclusion of *Brassica* and the related *Sinapis* species in spices and condiments (*B. nigra; B. juncea; S. alba*). Some species (e.g. *B. juncea*, Fig. 9) have been selected for all the usages described. To the plant taxonomist the genus *Brassica* presents many problems of speciation and hybridization. There is also the question of the origin of the cultivated forms. These problems have been investigated from the point of view of the morphologist (Bailey, 1930), the anatomist (Berggren, 1962) and the cytologist (Harberd, 1972). In the first instance it was decided to use gel diffusion methods in an investigation of three well accepted *Brassica* species (Vaughan *et al.*, 1966). These were *B. campestris* (turnip, turnip rape), *B. oleracea* (cabbage) and *B. nigra* (black mustard). Comparison of the seed proteins by means of the double diffusion and immunoelectrophoretic techniques (Figs 2 and 3) indicated specific differences but it was impossible to present a taxonomic relationship on the basis of these experiments.

However, reciprocal absorption of the three antisera produced considerably more information (Fig. 4 and Table I). Here, on the basis of the numbers of

TABLE I (Vaughan, 1968b)

Antiserum (As) absorbed with heterologous antigen mixture (Ag)	No of precipitin lines developing against homologous antigen mixture
AsC/AgO	5
AsC/AgN	9–10
AsO/AgC	1–2
AsO/AgN	4–6
AsN/AgC	3–5
AsN/AgO	3–5

C = *B. campestris*, O = *B. oleracea*, N = *B. nigra*.

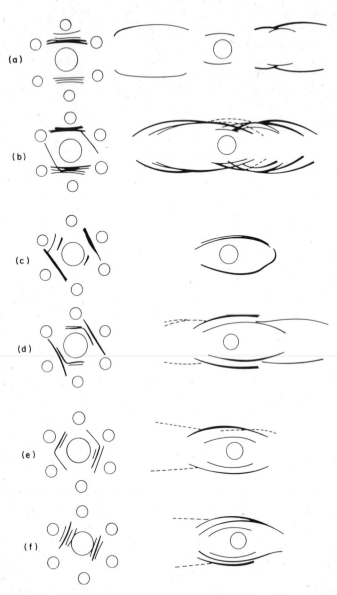

Fɪɢ. 4. Double diffusion and immunoelectrophoretic analysis of absorbed antisera: (a) and (b) *B. campestris* antiserum absorbed with *B. oleracea* and *B. nigra* respectively, (c) and (d) *B. oleracea* antiserum absorbed with *B. campestris* and *B. nigra* respectively, (e) and (f) *B. nigra* antiserum absorbed with *B. campestris* and *B. oleracea* respectively (Vaughan, 1968a).

precipitin lines, it was shown that *B. campestris* is closer to *B. oleracea* than either is to *B. nigra*. The results of the gel diffusion experiments can be assessed in an objective manner. The distinction into the three species has long been accepted by the taxonomist. This distinction is fully supported by the serological work, which also provides a scheme of relationship. At the present time, it is unlikely that a scheme of natural relationship based entirely on chemical information would be acceptable to taxonomists. However, it is interesting to note that in the classic monograph on the *Brassica* species (Schulz, 1919), *B. campestris* and *B. oleracea* were placed in the section Brassicotypus, *B. nigra* in the section Melanosinapis. The morphological relationship thus suggested by Schulz is fully supported by the serological work.

Other aspects of *Brassica* taxonomy have been investigated by the techniques described, such as generic and varietal distinction (Vaughan and Waite, 1967a) and amphidiploidy (Vaughan and Waite, 1967b).

III. GEL ELECTROPHORETIC METHODS

Although serological methods have been used widely by plant taxonomists, there are two possible disadvantages in terms of the comparison of protein profiles. Firstly, comparison is being made on the basis of the immuno-chemical properties of the proteins. This is not necessarily a full reflection of the properties of the protein. Secondly, there is the practical problem of the number of rabbits required for the production of antisera. To produce an antiserum representative of one plant taxon, one normally uses at least two rabbits. Clearly, for a survey of a range of plant taxa, even if the number is not great, then the housing and care of the experimental animals, particularly in the usual type of plant research laboratory, becomes in financial terms a serious problem.

The most successful of the protein separation methods, other than those of serology, applied to problems of plant taxonomy have been electrophoretic techniques where the proteins are separated in gels of various types. Starch gel (Vaughan and Waite, 1967a, b) has been used but, without a doubt, most plant taxonomists interested in the protein approach to chemotaxonomy have used disc polyacrylamide gel electrophoresis and these have covered a wide range of higher plants. Studies, made on a large number of leguminous species (Boulter *et al.*, 1966, 1967), indicated that the seed globulins showed patterns characteristic at the tribal level. Johnson and Hall (1965) investigated wheat and related genera; acrylamide gel results confirmed genome descriptions based on taxonomic methods and supported hypotheses of the amphidiploid origin of certain wheat cultivars. As an example of a rather more restricted taxonomic study, Ziegenfus and Clarkson (1971) investigated the relationship of some species within the genus *Acer*.

A. ELECTROPHORESIS AND CULTIVATED BRASSICA SPECIES

As with the serological techniques, reference will be made to work with cultivated *Brassica* species to illustrate the usefulness of disc polyacrylamide gel electrophoresis in studies of plant taxonomy. It has already been shown (Vaughan *et al.*, 1966) that serological gel diffusion methods provided useful information concerning the relationships of *B. campestris*, *B. oleracea* and *B. nigra*. In the same investigation, the plant material was also analysed by means of polyacrylamide gel electrophoresis. Separated proteins were recognized by means of a general protein stain. The assessment of protein homology was based on the Rp value of the protein, namely the ratio of the distance moved by the protein to that of a fast moving marker dye. The seed protein spectra of the three species were compared in terms of the Rp values. Taxonomic comparison was based on the distribution of the Rps in the same species. Both for the globulins and albumins, *B. campestris* and *B. oleracea* formed a group distinct from *B. nigra*. This was in total agreement with the serological evidence.

In the light of the success of the previous investigation, it was decided to apply polyacrylamide gel electrophoresis to a greater range of *Brassica* taxa so that problems of generic, specific and varietal distinction might be investigated, also hybridization (Vaughan and Denford, 1968). Because a reasonably wide range of taxa was investigated, it was necessary to develop methods for the handling of the Rp data in terms of taxonomic comparison, consideration of simple distribution not being satisfactory. In this investigation, albumin data were more useful than consideration of the globulins.

To compare the albumin spectra of two taxa, the following formula was used (Whitney *et al.*, 1968),

$$\% \text{ similarity} = \frac{\text{Nos. of pairs of similar bands}}{\text{Nos. of different bands and nos. of pairs of similar bands}} \times 100$$

The term "similar" applied to the same Rp. To avoid taxonomic weighting, band density was not taken into consideration. For all the material investigated, each possible pair of taxa was dealt with in this manner. To facilitate the presentation of this information, a chequer board was devised (Fig. 5). From the chequer board data, it was possible to construct a three-dimensional model, the reciprocals of the percentage similarities being used to form the arms of the model (Fig. 6). Other workers have presented their results in the form of three-dimensional models (Boyden, 1962; Sokal and Sneath, 1963).

The three-dimensional model of the *Brassica* taxa produced interesting taxonomic information. In most European systems of classification, *Brassica* and *Sinapis* are normally treated as distinct genera. This separation was supported by the protein data as was the grouping of *B. campestris* and *B.*

ALBUMINS

	C %	O %	Car %	J %	Na %	Sw %	N %	S %	A %
T	60	46	45	35	33	36	28	21	28
C		50	33	29	33	36	28	25	28
O			57	45	32	40	23	30	37
Car				53	37	33	25	30	35
J					48	35	45	24	32
Na						83	37	21	24
Sw							33	32	27
N								32	36
S									50

% percentage frequency

FIG. 5. Comparison of percentage similarities of protein spectra of various *Brassica* and *Sinapis* taxa (Vaughan and Denford, 1968). T = *B. campestris* (Turnip), C = *B. campestris* (Turnip Rape), O = *B. oleracea*, N = *B. nigra*, Car = *B. carinata*, Na = *B. napus* (Rape), Sw = *B. napus* (Swede), J = *B. juncea*, S = *S. alba*, A = *S. arvensis*.

oleracea into the section Brassicotypus and *B. nigra* into the section Melanosinapis. Some support was given for the hybrid status of *B. napus*, *B. juncea* and *B. carinata*.

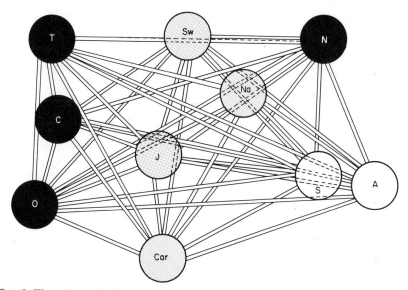

FIG. 6. Three-dimensional representation of relationships between various *Brassica* and *Sinapis* taxa, based on protein data (Vaughan and Denford, 1968). Key as in Fig. 5.

292   J. G. VAUGHAN

To complement the percentage similarity study, consideration was given to the distribution of albumin bands throughout the taxa investigated. With the technique employed, very few specific bands were found and it appears that the distinction of taxa is based mainly on the permutation of *Rps*. Figure 7 illustrates the distribution of the bands with 70% frequency and over.

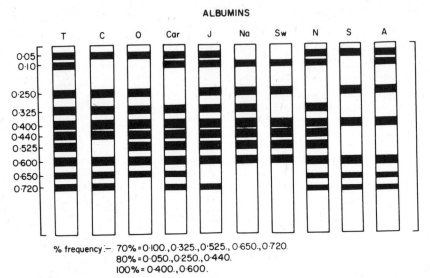

ALBUMINS

% frequency :— 70% = 0·100., 0·325., 0·525., 0·650., 0·720.
80% = 0·050., 0·250., 0·440.
100% = 0·400., 0·600.

FIG. 7. Distribution of high frequency bands in various *Brassica* and *Sinapis* taxa (Vaughan and Denford, 1968). Key as in Fig. 5.

B. campestris (turnip) is the only taxon with 100% of high-frequency bands and the hypothesis was presented that this taxon might be the closest to the archetype of the taxa examined, a suggestion which correlates with the cytological work of Sikka (1940).

B. ENZYME STAINING AND GEL ELECTROPHORESIS

In the *Brassica* investigation previously described, the complete protein complement was utilized, as far as it was revealed by the electrophoretic technique. Protein homology is based on similar *Rps* and it is this homology that is used in taxonomic assessment. In the polyacrylamide method it is sometimes difficult to establish comparative *Rp* values between different samples. The situation can be improved by attempting to identify the enzyme constituents of the profile (Wilson and Kaplan, 1962; Boulter *et al.*, 1966; Vaughan *et al.*, 1970). Enzyme identification can also be carried out in immunoelectrophoretic analysis (Grabar and Daussant, 1964; Vaughan and Gordon, 1969; see also chapter 2).

IV. Taxonomic Studies in the Genus Brassica

Most of the protein separation methods used by taxonomists have now been described. Used in isolation, each method has disadvantages. As stated, the serological method deals only with the immunochemical properties of proteins. In gel electrophoretic methods, comparison of the total protein patterns can be difficult and the serological technique of absorption is not available. When enzyme staining is used, not all the bands are identified and consideration of just one enzyme system can give a misleading taxonomic picture (Vaughan and Waite, 1967a). It is suggested, therefore, that where possible all techniques should be employed.

As stated in the introduction, chemotaxonomy as a discipline in plant

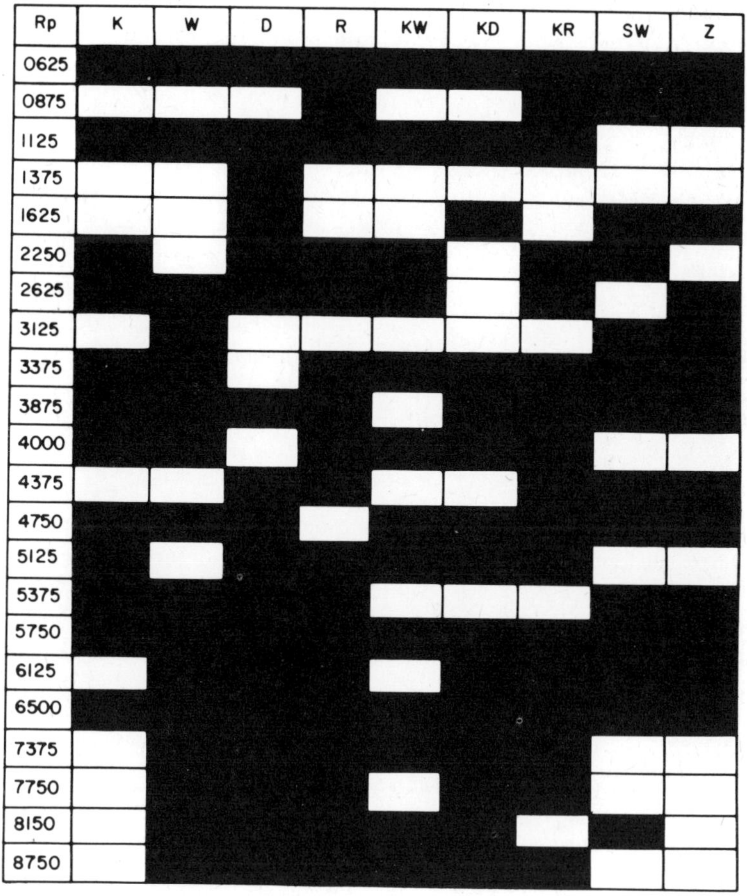

Rp	K	W	D	R	KW	KD	KR	SW	Z
0625									
0875									
1125									
1375									
1625									
2250									
2625									
3125									
3375									
3875									
4000									
4375									
4750									
5125									
5375									
5750									
6125									
6500									
7375									
7750									
8150									
8750									

Fig. 8. Protein patterns of synthesis experiments, *B. oleracea* (K) × *B. campestris* (W, D and R are different cultivars)→*B. napus* (KW, KD and KR), (Vaughan *et al.*, 1970).

systematics has developed tremendously within the last decade or so. It has confirmed taxonomic investigations based on other techniques. However, one would hope also that it would produce information on plant relationships not available from morphological and cytological studies.

A. PLANT PROTEINS AND THE ORIGIN OF HYBRIDS

One major contribution of cytotaxonomy has been to suggest from a consideration of the chromosome number that certain species evolved as hybrids. For example, within the genus *Brassica* (Morinaga, 1934; U, 1935) it has been indicated that *B. nigra* ($n = 8$), *B. oleracea* ($n = 9$) and *B. campestris* ($n = 10$) are the putative parents which have hybridized to give *B. carinata* ($n = 17$), *B. juncea* ($n = 18$) and *B. napus* ($n = 19$). There is support for this suggestion from species synthesis work (Olsson, 1960). A number of workers (for example, see Alston and Turner, 1963) have investigated this situation using protein methods and the usual result is that the protein complement of the suggested hybrid is a summation of some or all of the proteins of the putative parents.

Comparatively little work has been carried out on applying protein methods to a controlled species synthesis situation. However considerable success has been achieved at the Swedish Seed Association in Svalof in synthesizing *B. napus* from *B. campestris* and *B. oleracea*. Seed from three such experiments was analysed by means of serological and acrylamide gel (including enzymes) techniques (Vaughan *et al.*, 1970). In all cases, no new proteins were detected in the *B. napus* (Fig. 8), the protein pattern being a summation of some or all of the parental proteins. This type of result from a controlled situation can then be used to interpret protein work on suggested hybrids (Vaughan and Waite, 1967b).

B. PLANT PROTEINS AND GEOGRAPHICAL DISTRIBUTION

The plant taxonomist frequently has difficulty classifying groups that, on morphological and cytological grounds, are variously interpreted by different workers. Such a group is the 10-chromosome *Brassica campestris* complex which includes: (1) the European turnip and turnip rape; (2) the Indian oil seed forms—Toria and Sarson; and (3) the Chinese cabbages, described by Bailey (1930) as *B. pekinensis*, *B. chinensis* and *B. perviridis*. Various systems of classification have been proposed for these plants but there has been little general agreement. In a recent survey of this group (Denford, 1970) protein methods were applied to a large number of accessions. While in earlier treatments all forms were treated together as one species, the protein evidence indicated two groupings, the European and Indian forms falling into one, with the Chinese taxa forming the other.

This investigation of *B. campestris* is linked with two studies of the mustard *B. juncea* (Fig. 9; Vaughan *et al.*, 1963; Vaughan and Gordon, 1973). On the

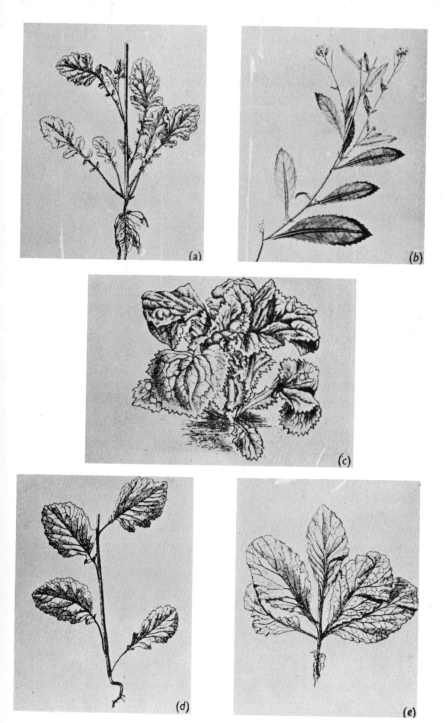

Fig. 9. Variation in *Brassica juncea* (Vaughan *et al.*, 1963, reproduced by permission of the Linnean Society).

basis of anatomy and volatile mustard oil distribution, the species appears to consist of two geographical races—Indian and Chinese. Cytological evidence suggests that *B. juncea* (*n* = 18) evolved from hybridization between *B. campestris* (*n* = 10) and *B. nigra* (*n* = 8). The presence of two geographical races could be explained by independent evolution in that an investigation of protein patterns (Denford, 1970) has revealed the existence of two groups of *B. campestris* taxa, namely European/Indian and Chinese. Since *B. campestris* is probably one of the putative parents of *B. juncea*, *B. juncea* could have evolved separately in the Indian and Chinese centres from the different stocks of *B. campestris*. In the second *B. juncea* investigation (Vaughan and Gordon, 1973) protein methods were applied to representatives of the two races but the results failed to yield any valid distinction.

V. SUMMARY

There is no doubt that chemotaxonomy now forms an important aspect of plant and animal systematics. This relatively new discipline has thrown fresh light upon many problems of biosystematics and, consequently, has attracted a large number of research workers. However, chemical techniques used in taxonomy, particularly the protein methods, can be improved. Although Turner (1969) has suggested that macromolecular data are more suitable for resolving taxonomic problems at the generic level and higher, given better methods there is no reason why proteins should not be used to equally good effect at the specific and infraspecific levels.

It is suggested that better methods of electrophoresis, adapted for taxonomic comparison, could be developed. In this connexion, isoelectric focusing might be utilized. The quantitative estimation of separated proteins has rarely been used in taxonomic studies but this could be developed for gel electrophoresis and immunoelectrophoresis (see Chapters 2 and 3). The taxonomist, from the very nature of his discipline, deals with large numbers of samples. In the future one might look forward to some automated form of scanning where reading of protein spectra, combined with some form of taxonomic formula, might be fed into a computer for analysis. Another interesting development is the use of amino acid sequences for tracing the phylogeny of higher plants (Boulter *et al.*, 1972).

In all this work the taxonomist needs guidance and assistance from the biochemist. However, the biochemist should gain equal benefit from the association in that there will be made available a great variety and range of material for chemical study.

REFERENCES

Alston, R. E. and Turner, B. L. (1963). "Biochemical Systematics". Prentice Hall, New Jersey.

Bailey, L. H. (1930). *Gentes Herb.* **2**, 211.
Benke, H. D. and Turner, B. L. (1971). *Taxon* **19**, 731.
Berggren, G. (1962). *Svensk bot. Tidskr.* **56**, 65.
Boulter, D., Ramshaw, J. A. M., Thompson, E. W., Richardson, M. and Brown, R. H. (1972). *Proc. R. Soc. B.* **181**, 441.
Boyden, A. (1962). *In* "Taxonomic Biochemistry and Serology" (C. A. Leone, ed.). Ronald Press, New York.
Boyden, A. and De Falco, R. J. (1943). *Physiol. Zool.* **16**, 229.
Boulter, D., Thurman, D. A. and Turner, B. L. (1966). *Taxon* **15**, 135.
Boulter, D., Thurman, D. A. and Derbyshire, E. (1967). *New Phytol.* **66**, 27.
Chester, K. S. (1937). *Q. Rev. Biol.* **12**, 19; 165; 294.
Crowle, A. J. (1961). "Immunodiffusion". Academic Press, New York and London.
Davis, P. H. and Heywood, V. H. (1963). "Principles of Angiosperm Taxonomy". Oliver and Boyd, Edinburgh and London.
Denford, K. E. (1970). "The study of the seed proteins of certain *Brassica* species using electrophoretic and serological techniques". Ph.D. thesis, University of London.
Fairbrothers, D. E. (1969). *Bull. serol. Mus. New Brunsw.* **41**, 1.
Gell, P. J. H., Hawkes, J. G. and Wright, S. T. C. (1960). *Proc. R. Soc. B* **151**, 364.
Gibbs, R. D. (1963). History of chemical taxonomy. *In* "Chemical Plant Taxonomy" (T. Swain, ed.), Academic Press, London and New York.
Grabar, P. and Burtin, P. (1964). "Immuno-electrophoretic Analysis". Elsevier, Amsterdam.
Grabar, P. and Daussant, J. (1964). *Cereal Chem.* **41**, 523.
Harberd, D. J. (1972). *J. Linn. Soc. (Bot.)* **65**, 1.
Heywood, V. H. (ed.) (1971). "Scannning Electron Microscopy". Academic Press, London and New York.
Heywood, V. H. (1973). *Pure appl. Chem.* **34**, 355.
Johnson, B. L. and Hall, O. (1965). *Am. J. Bot.* **52**, 506.
Johnson, M. A. (1954). *Bull. serol. Mus. New Brunsw.* **13**, 1.
Mez, C. and Ziegenspeck, H. (1926). *Bot. Arch.* **13**, 483.
Morinaga, T. (1934). *Cytologia* **6**, 62.
Olsson, G. (1960). *Hereditas* **46**, 171.
Schulz, O. E. (1919). Cruciferae-Brassiceae. *In* Engler's "Das Pflanzenreich" **1**.
Sikka, S. M. (1940). *J. Genet.* **40**, 411.
Sokal, R. R. and Sneath, P. H. A. (1963). "Principles of Numerical Taxonomy". Freeman, San Francisco.
Turner, B. L. (1969). *Taxon* **18**, 134.
U. N. (1935), *Jap. J. Bot.* **7**, 389.
Vaughan, J. G. (1968a). *Sci. Prog. Oxf.* **56**, 205.
Vaughan, J. G. (1968b). Seed protein studies of *Brassica* and *Sinapsis* species. *In* "Chemotaxonomy and Serotaxonomy" (J. G. Hawkes, ed.), Academic Press, London and New York.
Vaughan, J. G. and Denford, K. E. (1968). *J. exp. Bot.* **19**, 724.
Vaughan, J. G. and Gordon, E. I. (1969). *Phytochemistry* **8**, 883.
Vaughan, J. G. and Gordon, E. I. (1973). *Ann. Bot.* **37**, 167.
Vaughan, J. G. and Waite, A. (1967a). *J. exp. Bot.* **18**, 100.
Vaughan, J. G. and Waite, A. (1967b). *J. exp. Bot.* **18**, 269.
Vaughan, J. G., Hemingway, J. S. and Schofield, H. J. (1963). *J. Linn. Soc. (Bot.)* **58**, 435.
Vaughan, J. G., Denford, K. E. and Gordon, E. I. (1970). *J. exp. Bot.* **21**, 892.
Vaughan J. G. Waite, A., Boulter, D. and Waiters, S. (1966). *J. exp. Bot.* **17**, 332.

Whitney, P. J., Vaughan, J. G. and Heale, J. B. (1968). *J. exp. Bot.* **19**, 415.
Wilson, A. C. and Kaplan, N. O. (1962). *In* "Taxonomic Biochemistry and Serology"
    (C. A. Leone, ed.), Ronald Press, New York.
Ziegenfus, T. T. and Clarkson, R. B. (1971). *Can. J. Bot.* **49**, 1951.

Author Index

Numbers in italic are those pages on which references are listed in full

A

Aaij, C., 123, 127, *131*, 143, 145, *163*
Abbott, D. C., 36, 55, 57, *64*, *68*
Abeley, G. I., 38, 39, 40, *62*
Adams, J. M., 225, *256*
Adams, R., 222, *256*
Africa, B., 22, *26*
Agrawal, B. B. L., 72, *87*
Akabori, S., 8, *26*
Akazawa, T., 234, *356*
Albersheim, P., 72, *87*
Albrecht, J. J., 266, *280*
Alexandrescu, V., 40, 51, 52, 55, *62*
Alibert, G., 223, *256*
Aliev, K. A., 188, *205*
Allende, J. E., 113, 116, 118, 120, *131*, *133*, *134*
Aloni, Y., 125, 127, *131*, 142, 143, *162*, *163*
Alonso de Lama, J. M., *258*
Alston, R. E., 282, 283, 294, *296*
Altschul, A. M., 73, *87*
Ambler, R. P., 6, 7, 16, 24, *26*, *28*
American Society of Biological Chemists and American Physiological Society, 100, *109*
Andersen, R. A., 227, 229, *256*
Anderson, A. J., 72, *87*
Anderson, J. A., 96, 97, 98, 99, 101, *110*
Anderson, J. W., 32, *62*, 212, 216, 226, 230, 231, 233, 245, *256*, *263*
Anderson, L. E., 194, *205*
Anderson, M. B., 119, *131*, 185, *205*
Andreae, W. A., 225, *256*
Anfinsen, C. B., 15, *29*
Angeletti, P. U., 53, *67*
Angsterribbe, E., 122, *133*, 141, *162*
Antipina, A. I., 57, *66*
Apel, K., 179, 195, 196, 198, *205*, *209*
Arendzen, A. J., 149, *164*
Arglebe, C., 178, *205*

Armentrout, R. W., 6, *27*
Armstrong, D. J., 120, *134*
Armstrong, J., 195, *208*
Armstrong, J. J., 126, 130, *131*, *134*, 175, 194, 195, *205*, *209*
Armstrong, M. D., 225, *256*
Artsikhovskaya, Ye. V., 247, *262*
Asada, K., 23, *26*
Asahi, T., 161, *162*, *164*, 235, *261*
Ashwell, M., 137, 138, 143, *162*
Attardi, B., 124, *131*, 142, *163*
Attardi, G., 124, 125, 127, *131*, 142, 143, *162*, *163*
Attridge, T. H., 241, *256*
Aurand, L. W., 245, *256*
Aurich, O., 195, 196, 198, 203, *208*
Avadhani, N. G., 143, *163*, 177, *205*
Avers, C. J., 143, *165*
Avrameas, S., 45, 56, *62*
Awouters, F., 94, 96, 108, *110*

B

Badran, A. M., 32, *62*
Bafaf, K. L., 245, *256*
Bagdasarian, M., 217, *256*
Bahr, W., 252, *364*
Bailey, C. J., 8, 10, *26*
Bailey, L. H., 287, 294, *297*
Bain, J. M., 160, *163*
Baldry, C. W., 246, *256*
Balian, G., 16, *27*
Ballantyne, D. L., 22, *29*
Barath, Z., 126, 127, 128, *131*
Bard, S. A., 174, *205*
Barett, J. T., 58, *62*
Barkow, K. K., 56, *62*
Barnes, W. T., 22, *29*
Barnett, W. E., 125, 127, 128, *131*, *134*, 147, *163*, 185, 187, 188, 195, 198, *205*, *206*, *208*

Brawerman, G., 197, *209*
Brega, A., 124, *131*
Breidenbach, R. W., 161, *163*
Brewer, H. B., 22, *29*
Briantais, J. M., 56, *63*
Briarty, L. G., 121, *131*
Brieskorn, C. H., 223, *257*
Briggs, D. E., 121, *131*
Briggs, W. R., 55, *68*
Bringhurst, R. S., 247, *261*
Bronzert, T. J., 22, *29*
Brouwer, J. N., 266, *279*
Brown, B. R., 214, 250, *257*
Brown, D. H., 125, *131*, 147, *163*
Brown, H. T., 99, *109*
Brown, P. E., 250, *257*
Brown, P. J., 255, *257*
Brown, R. D., 172, 180, 203, *206*
Brown, R. H., 23, 24, 25, 27, *29*, 296, *297*
Brown, R. M., 172, *209*
Brown, S. A., *257*
Brownrigg, A., 119, 121, *134*
Brunet, P. C. J., 216, *257*
Brunt, A. A., 231, *257*
Brunton, C. J., 4, 20, *27*
Brutschek, H., 99, *110*
Bücher, Th., 126, 129, *133*, *134*
Buck, C. A., 125, 127, *133*
Bucke, C., 246, *256*
Buetow, D. E., 143, *163*, 177, *205*
Bull, A. T., 245, *257*
Burdon, R. H., 181, *206*
Burkard, G., 116, 125, 127, *131*, *132*, 148, *163*, 183, 185, 187, 188, *206*, *207*
Burke, J. P., 158, *163*
Burtin, P., 285, *297*
Burton, H., 173, *207*
Burton, W. W., 22, *264*
Bush, I. E., 4, *27*
Butler, W. L., 48, *66*
Buttrose, M. S., 108, *109*
Buzila, L., 32, 50, 57, *62*, *65*
Byrde, R. J. W., 245, *257*

C

Cadman, C. H., 231, 233, *257*
Cafiero, M., 238, *259*
Cagan, R. H., 272, 273 274, *280*
Cagnoni, G., 53, *63*
Callewaert, W., 252, *263*
Camm, E. L., 239, 241, *257*

Cammarano, P., 178, 179, *207*
Campbell, T. G., 36, 49, 52, 58, *63*
Cannon, C. G., 213, *257*
Cantagalia, P., 53, 56, *63*, *68*
Capecchi, M. R., 225, *256*
Capra, J. D., 21, *30*
Carbonara, A. O., 41, *67*
Carell, E. F., 194, *206*
Carney, W. B., 53, *67*
Carpenter, F. H., 22, *26*, *30*
Carrè, D. S., 120, *133*
Casey, J., 125, 127, *132*
Castelfranco, P., 161, *163*
Castle, J. E., 234, *261*
Cater, C. M., 222, *257*
Catsimpoolas, N., 36, 39, 41, 44, 49, 50, 52, 53, 58, *63*, 73, *87*
Cella, R., 127, 128, *133*
Chadha, K. C., 36, 51, *63*
Chaiken, I. M., 25, *28*
Chain, R. K., 28, *28*
Chakrabarti, S., 150, *163*
Chambers, D. C., 9, *29*
Chan, P., 194, *206*
Chan, S. K., 23, *27*
Chandrasekaran, A., 108, *109*
Chang, Y. H., 10, *28*
Chant, S. R., 248, *257*
Chapeville, F., 120, *133*, *134*, *135*
Chapon, L., 251, *257*
Chasson, 161
Chen, J. L., 177, 179, 197, *206*
Cherry, J. H., 119, *131*, *132*, 185, 188, *205*, *207*
Chester, K. S., 283, *297*
Chi, J. C. H., 143, *163*
Chia, L. S. Y., 119, *133*
Chiang, K., 174, 175, *205*
Chiang, K. S., 175, *206*
Chollot, B., 251, *257*
Chua, N. H., 177, 179, 196, 197, *206*
Chung, D., 14, *28*
Ciferri, O., 115, 118, 127, 128, *132*, *133*
Cividanes, I., 241, *257*
Claesson, S., 251, *257*
Clapp, S. M., 90, 99, 107, *110*
Clark, A. G., 251, 253, 254, *257*
Clark, G. M., Jr., 117, *132*
Clark, J. M., Jr., 117, *132*
Clark, M. F., 177, *206*
Clarkson, R. B., 289, *298*
Clawson, A. J., 222, *262*
Clemens, M. J., 71, *87*

Levine, R. P., 126, 130, *131, 134*, 175, 194, 195, 201, *205, 206, 208, 209*
Levy, G., 220, *261*
Lewis, D. A., 250, *257*
Lewis, J. A., 250, *260*
Li, C. H., 14, *28*
Lieberman, M., 229, *260*
Light, A., 25, *28*
Lin, C. Y., 119, *133*
Lindbloom, M., 55, *67*
Linderstrom-Lang, K., 13, *28*
Linenberg, A., 11, *28*
Linnane, A. W., 159, *164*, 172, *206*
Lipmann, F., 128, *134*, 225, *260*
Lis, H., 72, *87*
Liss, E., 238, *258*
Littauer, U. Z., 143, 145, *163*
Litvak, S., 120, *133*
Liu, T-Y., 10, *28*
Liuzzi, A., 53, *67*
Lizardi, P. M., 124, 126, *133*, 148, 158, *164*
Loening, U. E., 123, 127, *132, 133*, 176, 180, 181, 183, *207*
Loescheke, V., 73, *88*
Loeve, K., 273, 274, *280*
Loisa, M., 53, *67*
Lönnerdal, B., 72, *87*
Lontie, R., 94, 96, 98, 102, 103, 108, *110*, 252, 253, 254, *258, 263, 264*
Loomis, W. D., 32, *67*, 212, 213, 214, 226, 227, 230, 233, 245, *260*
Lothspeich, W. D., 237, *260*
Lovins, R. E., 22, *29*
Lowe, G., 25, *28*
Lowry, P. J., 10, *26*
Loyter, A., 36, *67*
Lucas-Lenard, J., 225, *260*
Luck, D. J. L., 124, 126, 127, *133, 135*, 138, 139, 141, 142, 148, 158, *164*
Ludwig, G. D., 157, *164*
Lukins, H. B., 159, *164*
Lyman, C. M., 222, *257*
Lyman, H., 178, *206*
Lyr, H., 234, *260*
Lyttleton, J. W., 114, *133*, 177, 178, *208*

M

McArthur, J. N., 242, *257*
McCarty, K. S., 123, 124, *134*
McCown, B. H., 32, *67*

McCoy, G. D., 153, *164*
McCoy, M. T., 127, 128, *132*
McCrea, M., 81, *87*
MacDonald, I. R., 154, *163*
MacDowall, M. A., 16, *28*
McFarlane, W. D., 252, 254, *261, 264*
Mache, R., 173, *208*
Macheboeuf, M., 98, *110*
Machold, O., 130, *133*, 195, 196, 198, 203, *208*
McKeenan, W. L., 177, *133*
Mackler, B., 141, *165*
Macko, V., 73, 74, 81, *87, 88*
McMartin, C., 10, *26*
Magnus, W., 31, *67*
Mahler, H. K., 138, *164*
Mahler, H. R., 141, *165*
Mahowald, T. A., 9, *29*
Maitland, P., *258*
Majak, W., 223, *260*
Majima, R., 161, *162*
Mäkinen, V., 94, 96, *109*
Mancini, G., 41, *67*
Mandels, M., 233, 245, *260*
Manigault, P., 44, 50, 51, 58, 60, *64, 68*
Mankash, E. K., 219, *260, 261*
Manning, J. E., 173, 174, 175, 176, 195, *208, 209*
Mans, R. J., 113, *133*
Manzocchi, A., 188, *208*
Marcker, K., 225, *260*
Marcker, K. A., 6, *29*
Marcus, A., 116, 117, 118, *133*, 225, *260*
Marglin, A., 16, *28*
Margoliash, E., 23, *27, 28, 29*, 226, *260*
Margulies, M. M., 126, *133*, 179, 194, 195, 198, *207, 208*
Marigo, G., 223, *256*
Maroudas, N. G., 130, *135*
Martenson, R., 273, 274, *280*
Martin, A. J. P., 10, *27*
Martin, G. J., 235, *262*
Martin, J. P., 248, *260*
Marx, H. D., 252, *264*
Mason, H. S., 214, 216, 221, *260, 261*
Mason, T., 149, 157, *164*
Massart, L., 252, *263*
Massinger, P., 129, *133*
Mastronuzzi, E., 227, *258*
Matheson, A., 59, *66*
Matheson, N. A., 217, *256*
Matile, Ph., 7, *30*
Matsubara, H., 23, *28*

Subject Index

Numbers in italic are those pages on which chemical formulae are given

A

Acetone powders, use in extracting phenols from plant proteins, 231

N-Acetylamino acids, in protein synthesis, 225

Acetyl groups, micromethod for detection, 6

Albumin, human, density gradient ultracentrifugation of, 220

Algae, gene expression in, 169

Amino acid activation, effects of phenolics on, 243–244

Amino acid analysis, 10–11

Amino acid composition, determination of, 8–11
  of alcoholic fractions of barley seed, 103, 105

Amino acid incorporation, into mitochondrial proteins, 150–156

Amino acid sequence analysis, 1–30

Amino acids,
  complete sequence in plant proteins, 24
  enzymic cleavage of, 7
  of "serendip", 273–274

Aminoacylation, of viral RNA, 120

Aminoacyl-tRNA synthetase, 119–121, 183, 186, 188–193
  in organelles, 125

δ-Aminolevulinic acid, uncoupling effects of, 202

Amino-terminal analysis, 3–6

α-Amylase activity in germinating seeds, 61

β-Amylase,
  absorption of activity, 46
  detection by specific sera, 33
  micro-complement fixation of, 48
  polymorphism of, 52

2-Anilino-5-thiazolinone derivatives, in Edman degradation procedure, 19

Animal tissues, mitochondrial DNA from, 141

Anthocyanins and anthocyanidins, inhibition of IAA oxidation by, 234

Antibiotics, as inhibitors of protein synthesis, 124

Antibody absorption, 45

Antigenic heterogeneity,
  of acid phosphatase, 51
  of proteins, 52

Antigens,
  in physiological studies, 57–61
  localization of, 56, 57

Aspergillus, cDNA in, 174

A. nidulans,
  elution pattern of total tRNA from, 187
  mitochondrial ribosomes and RNA from, 143

Astringency,
  definition, 255
  in food, 254–255

ATP, formation in *Phaseolus vulgaris*, 151

Autoradiography, in mitochondrial studies, 139

B

Bacteria,
  evolution of mitochondria from, 138
  gene expression in, 169

Barley,
  immunoelectrophoretic analysis of extracts, 36
  phenolic constituents of, 252
  proteins, 89–109

Biogenesis,
  of chloroplasts, 167–205
  of mitochondria, 137–162

Brassica,
  cytotaxonomy, 294, 296

Brassica—contd.
  distribution, 294
  economic importance, 287
  electrophoretic studies on, 290–292
  evolution, 296
  occurrence, 287
  proteins, 294
    double diffusion analysis of, 285,
      287–289
    immunoelectrophoresis, 286–289
    precipitin reaction, 284
    spectra, 291, 293
    relationships between species, 290–292
    serological studies, 287–289
    taxonomy, 293–296
B. juncea, variations in, 295
B. rapa, mitochondrial ribosomes and
  RNA from, 143
Bromelain, effect on *Dioscoreophyllum*
  sweetener, 269–70
N-Bromosuccinimide, in cleaving at
  tryptophanyl and tyrosinyl bonds,
  16

C

Cadmium–ninhydrin reagent, for loca-
  ting peptides, 17
Caffeic acid, inhibition of peroxidase by,
  234
Candida krusei, formation of hybrid
  density ribosomes in, 119
Carboxypeptidase A, determinations of,
  225
Carboxypeptidase analysis, of *C*-termi-
  nal amino acids, 6–8
Carboxypeptidases, amino acid cleavage
  by, 7
Carboxy-terminus analysis, 6–8
Casein, use in removing beer haze,
  254
Cellulases,
  extraction of water-soluble nitrogen
    fractions by, 108
  inhibition of, 245
Cereals, classification on basis of
  antigens, 49
Chelating agents, inhibiting quinoid
  formation, 226, 230
Chemotaxonomy, development of, 283
Chlamydomonas,
  chloroplast studies on, 173
  cDNA in, 174–175

Chloramphenicol, effect on amino acid
  incorporation, 154, 156, 159, 197–
  201
Chlorogenic acid, inhibition of peroxidase by, 234
Chlorogenoquinone, reactions of, 214–
  215
Chloroplast information translation,
  204
Chloroplasts,
  biogenesis, 167–205
  function, 167–186
  interrelations with cytoplasm, 203–
    205
  structure, 167–168
Chordates, quinone pigments in, 216
Chromatography,
  affinity, 36
  gas–liquid, in amino acid analysis, 11,
    22
  hydroxyapatite, 188–190
  ion-exchange,
    for amino acid analysis, 10–11
    for large scale separation of pep-
      tides, 16
    of *Thaumatococcus* proteins, 275–
      276
  polyamide sheet, for identification of
    dansyl amino acids, 4–5
Chymotrypsin, amino acid cleavage by,
  7, 13–14
Chymotrypsin inhibitor, location of, 57
Cinnamic acids, effects on enzymes,
  239–240
Classification, systems of, 282
Cleavage, chemical, of polypeptide
  chains, 14–16
Correspondence values, for indicating
  relative serological similarities,
  37
Coumarins, effects on oxidative phosphorylation in mitochondria, 236–
  238
Cyanate method, for quantitative *N*-
  group analysis, 3
Cyanogen bromide, in cleaving at
  methionyl residues, 15
Cyclamates, in food, 265
Cycloheximide
  effects on amino acid incorporation,
    154, 156, 159, 197–201
  effects on protein synthesis in eukary-
    otic cells, 115

enzymic, of proteins, 13–14
of dansyl derivatives, 3
of proteins, measurement of, 12, 13
Hydroxylamine, cleavage of Asn–Gly bonds, 16

I

IAA oxidase, activation by flavonoids, 234
Immune sera,
absorption, 45–47
preparation, 33–36
Immunization schedules for plant proteins, 34, 35
Immunochemistry,
contributions to protein research, 48–61
methods and techniques, 32–48
Immunoelectrophoresis,
discontinuous, 39
double diffusion, 38–39
of barley, 36
of *Brassica* seed proteins, 287–289
of plant proteins, 285–286
of wheat, 36
quantitative, 41–45
Information transformation, elements of 176–193
Inhibitors,
effect on protein synthesis, 197
selective, of protein synthesis, 129
Initiation phase of protein synthesis, 116, 117
Isoenzyme, definition of, 51

K

Katemfe, see *Thaumatococcus danielli*

L

Lactuca, mitochondrial DNA from, 141
Lamellae, chloroplast, immunochemical studies on, 56
Leghemoglobin, complete amino acid sequences in, 24
Leguminosae, biogenesis of mitochondria in, 160
Leucine activating enzymes, from *Euglena gracilis*, 188–189

Leucoanthocyanidin, affecting astringency, 255
"Leucosin", as defined by Osborne, 90
Leucyl-tRNA synthetase,
from *Neurospora crassa*, 128
properties of, 190–191
Lignin, hydroxycinnamic acid in biosynthesis, 223

M

Macromolecules,
as sweeteners, 267–268
as taste modifiers, 266
Magnesium requirements, of ribosomes, 124
Melanins, formation of, 248
Melanization, role of quinones in, 216
Mentha arvensis, phenolic esters of, 223
Methionine, as starting point for peptide chain formation, 116
Methionyl peptides, cleavage by cyanogen bromide, 15
Methionyl-tRNA, 225
as initiator of peptide chain synthesis, 116
from *Phaseolus vulgaris*, 148
isolation of, 116
Micro-complement fixation, 47–48
Micromolecules, as sweeteners, 267–268
Milk, effect on tea polyphenols, 255
Miracle fruit, see *Synsepalum dulcificum*
Mitochondria,
biogenesis of, 137–162
origin of, 138–139
replication of, 138–142
synthesis of specific RNA by, 142
Mitochondrial autonomy, 138–142
Mor sites, 750
Mull sites, 250
Mutants, use in studying chloroplast biogenesis, 172

N

Naringenin, *235*
as enzyme stimulator, 234
Naringenin dihydrochalcone, *268*, 278
as sweetener, 268
Neohesperidin dihydrochalcone, *268*, 278
as sweetener, 268

847631